FUNDAMENTALS OF
SEMICONDUCTOR DEVICES

FUNDAMENTALS OF SEMICONDUCTOR DEVICES

Edward S. Yang

Professor of Electrical Engineering
Columbia University

McGraw-Hill Book Company

New York St. Louis San Francisco Auckland Bogotá Düsseldorf
Johannesburg London Madrid Mexico Montreal New Delhi
Panama Paris São Paulo Singapore Sydney Tokyo Toronto

To My Family

This book was set in Times Roman.
The editors were Peter D. Nalle, Frank Cerra, and Michael Gardner;
the cover was designed by John Hite;
the production supervisor was Dennis J. Conroy.
The drawings were done by J & R Services, Inc.
Fairfield Graphics was printer and binder.

FUNDAMENTALS OF SEMICONDUCTOR DEVICES

567890 FGR FGR 8321

Library of Congress Cataloging in Publication Data

Yang, Edward S.
    Fundamentals of Semiconductor Devices

    Includes index.
    1. Semiconductors. I. Title.
TK7871.85.Y36   621.3815'2   77-11021
ISBN 0-07-072236-6

CONTENTS

PREFACE

Modern electronics is built upon the foundation of a well-established semi-conductor technology which has resulted in the silicon integrated-circuit (IC) chip. Because of the interplay between devices, networks, and systems in integrated circuits, it is important for an electrical engineer to understand the basic operation of such devices. The purpose of this text is to introduce the physical principles of semiconductor devices and their practical implementation to undergraduate and beginning graduate students in electrical engineering as well as to practicing engineers whose understanding of modern IC devices needs updating. It is intended as a follow-up to the basic electronics course offered in the sophomore or junior year. The material presented here is primarily geared to the needs of students and engineers who will be users or designers of integrated circuits and systems. The reader should have had a course in differential equations, but knowledge of quantum mechanics is not necessary.

The text is organized into four parts. The first section, Chapters 1 to 3, is an introduction to elementary semiconductor physics and to fabrication technology. Chapter 1 describes the semiconductor in equilibrium, introducing the concepts of the energy-band diagram, carrier concentration, and mobility. The recombination and transport of nonequilibrium carriers are discussed in Chapter 2. In Chapter 3, the basic fabrication processes, namely, impurity diffusion, oxidation, epitaxy, photomasking, and ion implantation, are explained.

The second part of the text is devoted to two-terminal devices. Chapter 4 presents the static and dynamic properties of the p-n junction which include the current-voltage, capacitance-voltage, and avalanche breakdown characteristics. The metal-semiconductor junctions and heterojunctions are described in Chapter 5 with emphasis placed on the Schottky-barrier diode. In

Chapter 6, the operating principles and characteristics of the solar cell and the light-emitting diode (LED) are given.

In the third section of the text, we discuss three-terminal devices, beginning with the junction field-effect transistor in Chapter 7. The metal-oxide-semiconductor structure and the MOSFET are presented in Chapter 8, in which various types of MOSFETs are described, together with the available technology. The bipolar junction transistor (BJT) and p-n-p-n devices are covered in Chapter 9. The discussion of the BJT however, is, limited to the characteristics of low-power devices.

Integrated devices are presented in the last part of the text. In Chapter 10, the integrated transistor, lateral transistor, integrated-injection logic (I^2L), and MOS inverters are covered. In addition, D-MOS, V-MOS, MNOS, and FAMOS are considered. The charge-coupled and bucket-brigade devices are described in the last chapter.

An attempt is made to link the basic device operating principles to the commercially available device structures. Thus, step-recovery and varactor diodes are included in the discussion of the p-n junction, and the Schottky-clamped transistor is covered along with the theory of metal-semiconductor junctions. Since the physical implementation of a device usually has a dominating influence on the device's characteristics, principles are generally illustrated with practical device structures or numerical examples. When it is not convenient to include a certain subject in the text, provision is made in the problem section to present the basic idea involved. For example, the photoconductor and thermistor are introduced this way. For this reason, the problems should be considered as an integral part of the text. A solution manual is available; it may be obtained from the publisher by an instructor who has adopted the text.

Considerable thought was given to the logical sequence of presentation of the material. The chapters are arranged so that one proceeds from the fundamental principles and simple devices toward the more complex and less obvious structures. For this reason, the bipolar transistor is described in Chapter 9 instead of following the section on the p-n junction described in Chapter 4. The present order has the advantage that it places equal emphasis on both the MOS and bipolar transistors. For those who prefer a different sequence, some chapters can be rearranged without loss of continuity. It is recommended, however, that the first four chapters be taught as a unit. One may then take up Chapter 9 after Chapter 4. In addition, Chapter 6 may be used as the last chapter, and Chapters 10 and 11 may be interchanged. For students who have not had a course in electronics, Appendix A serves as a link between modern physics and semiconductor theory. Although the text is written primarily for a one-semester course, it contains sufficient material for two quarters, particularly if Appendix A is covered.

This book was developed during a seven-year period of teaching seniors, first-year graduate students, and occasionally a few juniors at Columbia University. The material has been tested in the classroom and the suggestions and criticism of my colleagues, both at Columbia and elsewhere, have been very helpful. I am particularly indebted to my colleagues at Columbia—Professor Jacob Millman for his invaluable advice and encouragement and Professor Howard Card for his generous help. Special thanks are due to Professor David Hodges of the University of California at Berkeley who reviewed the entire manuscript and made many valuable suggestions. I am grateful to Professors Malvin C. Teich and Yannis Tsividis of Columbia University, Professor Sheng Li of the University of Florida, Professor John G. Skalnik of the University of California at Santa Barbara, and Dr. Roland Hung of IBM Research for their critical review and comments. My sincere appreciation is extended to George K. K. Ng and Morris C. M. Wu, doctoral students at Columbia University for their contribution in improving the accuracy and content of the text and for preparation of the problem solutions, and to Mrs. Betty Lim for her skillful typing of the various versions of the manuscript. I am indebted to Peter Nalle and Michael Gardner, my editors at McGraw-Hill, for their manifold assistance and guidance during the publication phase. Finally, I am grateful to my wife and children, whose understanding and support made possible the completion of this text.

Edward S. Yang

ELEMENTARY PHYSICAL THEORY OF SEMICONDUCTORS

In this chapter, the elementary properties of semiconductors are discussed. We begin with the crystal structure and the identification of crystal planes. The valence-bond and energy-band models are then presented. Subsequently, evaluation of carrier concentrations and the Fermi level is considered for both intrinsic and extrinsic semiconductors. The scattering and carrier transport mechanisms are also given. The concepts introduced in this chapter form the basis for later studies of all semiconductor devices.

1-1 CRYSTAL STRUCTURE

According to the structural organization, solids may be classified into *crystalline*, *polycrystalline*, and *amorphous* types. An amorphous solid does not have a well-defined structure; in fact, its distinction is its formlessness. In the past few years, amorphous solids have received increasing attention, and switching devices using them have been made. Nevertheless, their importance as materials for the active part of an electronic device is doubtful. In a polycrystalline solid, there are many small regions, each having a well-organized structure but differing from its neighboring regions. In a crystalline solid, atoms are arranged in a three-dimensional orderly array that defines a periodic structure called the *lattice*. In this solid, it is possible to specify a *unit cell* which repeats itself in three dimensions and constitutes the crystalline solid. A unit cell contains complete information regarding the arrangement of

1

atoms, and hence the unit cell can be used to describe the crystal structure. Among the many structures of crystals, we shall limit ourselves to the five cubic structures relevant to our later discussion.

1. *Simple cubic (sc) crystal.* An sc crystal is shown in Fig. 1-1*a*. Each corner of the cubic lattice is occupied by an atom which is shared by eight neighboring unit cells. The dimension *a* in the cubic unit cell is called the *lattice constant.* Few crystals exhibit this structure, and polonium is the only element crystallized in this form.
2. *Body-centered cubic (bcc) crystal.* A bcc crystal is illustrated in Fig. 1-1*b*, where, in addition to the corner atoms, an atom is located at the center of the cube. Crystals exhibiting this structure include sodium, molybdenum, and tungsten.
3. *Face-centered cubic (fcc) crystal.* Figure 1-1*c* shows an fcc crystal, which contains one atom at each of the six cubic faces in addition to the eight corner atoms. A large number of elements exhibit this crystal form, including aluminum, copper, gold, silver, nickel, and platinum.
4. *Diamond structure.* A diamond unit cell is depicted in Fig. 1-1*d*. It may be seen as two interpenetrating fcc sublattices with one displaced from the other by one-quarter of the distance along a diagonal of the cube. The top view of such two penetrating fcc lattices is shown in Fig. 1-1*e* in two-

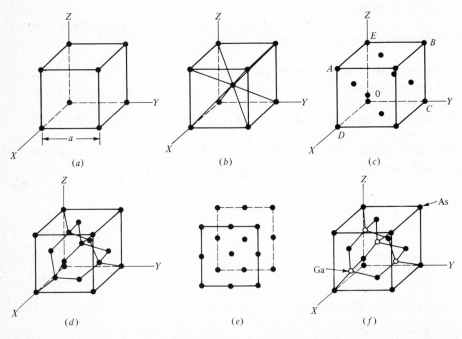

Figure 1-1 Unit cells of cubic crystals: (*a*) simple cubic, (*b*) body-centered cubic, (*c*) face-centered cubic, (*d*) diamond, (*e*) two penetrating fcc in two dimensions, (*f*) zinc blende.

dimensional form. Note that four atoms of the second fcc lattice are located within the first fcc lattice. These four atoms belong to the same unit cell of the first fcc lattice in the diamond structure. The atoms outside the original unit cell belong to the adjacent unit cells. Both silicon and germanium exhibit this structure.

5. *Zinc-blende structure.* Gallium arsenide (GaAs) exhibits the zinc-blende structure, as shown in Fig. 1-1f. It results from the diamond structure when Ga atoms are placed on one fcc lattice and As atoms on the other fcc lattice. Other materials of the zinc-blende family include gallium phosphide, zinc sulfide, and cadmium sulfide.

Let us take a further look at the fcc crystal shown in Fig. 1-1c. We notice that there are six atoms in the *ABCD* plane and five atoms in the *AEOD* plane and that the atomic spacings are different for the two planes. For these reasons, the crystal properties along different directions are different; i.e., the crystal is *anisotropic*. Therefore it is necessary to find a convenient way to specify the orientation of crystal planes and directions. In practice, a crystal plane is specified by its *Miller indices*, which are obtained by using the following procedure:

1. Determine the intercepts of the plane on the three cartesian coordinates.
2. Measure the distances of the intercepts from the origin in multiples of the lattice constant.
3. Take the reciprocals of the intercepts and then reduce to three integers having the same ratio, usually the smallest three integers.

As an example, if the intercepts of a crystal plane to the axes are 4, 2, 1, then the reciprocals are $\frac{1}{4}$, $\frac{1}{2}$, 1. Thus, the Miller indices of the crystal plane are (124). The direction of a crystal plane is normal to the plane itself and is designated by brackets of the same Miller indices. For example, the direction (111) plane is written as [111]. Some directions in a crystal lattice are equivalent because they depend only on the orientation of the axes. As an example, the directions of the crystal axes in a cubic lattice are designated

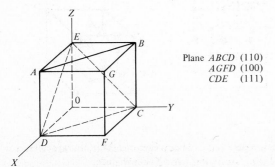

Plane *ABCD* (110)
AGFD (100)
CDE (111)

Figure 1-2 Important crystal planes represented by Miller indices.

[001], [010], and [100]. These directions are equivalent because the crystal planes they represent are the same and the only difference is the arbitrary selection of the axes. These equivalent direction indices are written as ⟨100⟩. Miller indices for three important lattice planes are shown in Fig. 1-2.

1-2 VALENCE-BOND MODEL OF SOLID

In a crystal lattice, a positively charged nucleus is surrounded by negatively charged orbiting electrons in each constituent atom. If the atoms are closely packed, the orbits of the outer-shell electrons will overlap to produce strong interatomic forces. The outer electrons, called *valence electrons*, are of primary importance in determining the electrical properties of the solid. In a metallic conductor such as aluminum or gold, the valence electrons are shared by all the atoms in the solid. These electrons are not bound to individual atoms and are free to contribute to the conduction of current upon the application of an electric field. The free-electron density of a metallic conductor is on the order of 10^{23} cm^{-3}, and the resulting resistivity is smaller than 10^{-5} Ω-cm. In an insulator such as quartz (SiO$_2$), almost all the valence electrons remain tightly bound to the constituent atoms and are not available for current conduction. As a result, the resistivity of silicon dioxide is greater than 10^{16} Ω-cm. In silicon, a crystalline semiconductor, each atom has four valence electrons to share with its four nearest-neighboring atoms, as shown schematically in two-dimensional form in Fig. 1-3*a*. The valence electrons are shared in a paired configuration called the *covalent bond*. At low temperatures, these electrons are bound, and they are not available for conduction. At high temperatures, the thermal energy enables some electrons to break the bond, and the liberated electrons are then free to contribute to current conduction. Thus, a semiconductor behaves like an insulator at low temperatures and a conductor at high temperatures. At room temperature,

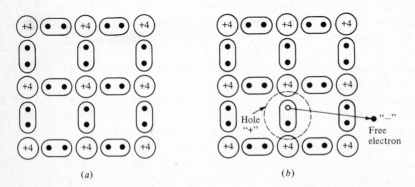

(a)   (b)

Figure 1-3 Two-dimensional schematic representations of crystal structure in silicon: (*a*) complete covalent bond and (*b*) broken covalent bond.

however, the resistivity of pure silicon is 2×10^5 Ω-cm, which is considerably higher than that of a good conductor.

Whenever a valence electron is liberated in a semiconductor, a vacancy is left behind in the covalent bond, as shown in Fig. 1-3b. This vacancy may be filled by one of its neighboring valence electrons, which would result in a shift of the vacancy location. One may then see the vacancy as moving inside the crystalline structure. As a result, this vacancy may be considered as a particle analogous to an electron. This fictitious particle is called a *hole*. It carries a positive charge and moves in the direction opposite that of an electron under an externally applied electric field.

1-3 CONCEPTS OF EFFECTIVE MASS AND ENERGY BANDS

An electron which is not captured in a covalent bond (called a *conduction electron*) or a hole is relatively free to move about in a semiconductor analogous to an electron in vacuum. However, the electron or hole is acted upon by the periodic potential of charged atomic cores. As a result, the *effective mass* of a conduction electron is different from the electronic mass of a free electron *in vacuo*. This effective mass represents the quantum-mechanical nature of motion of electrons or holes in semiconductors. It is an extremely important simplification which enables us to consider holes and conduction electrons as classical charged particles. With the simplification, the energy-momentum relationship of an electron in a crystal is given by

$$E_k = \frac{P^2}{2m_e} \tag{1-1}$$

where m_e = effective mass of electron
$\quad E_k$ = kinetic energy
$\quad P$ = crystal momentum
A similar equation can be written for holes with the hole effective mass m_h replacing m_e.

Unlike an electron in a vacuum, however, an electron in a solid can assume only restricted values of energy. Certain bands of allowed energy levels are separated by gaps of forbidden energies in which an electron cannot exist. Each allowed energy band contains a limited number of states that can accommodate a definite number of electrons. In a semiconductor the valence electrons group together to occupy a band of energy levels, called the *valence band*. The next higher band of allowed energy levels, called the *conduction band*, is separated from the former by a *forbidden gap of energy* E_g. This physical picture, called an *energy-band diagram*, is shown in Fig. 1-4 for three classes of solids. In metals, the conduction band is only partially filled by electrons, as shown in Fig. 1-4a; alternatively, the energy-band diagram for a metal may be seen as two bands which overlap so that there is no forbidden gap (Fig. 1-4b). In insulators (Fig. 1-4c) the lower band has just enough states

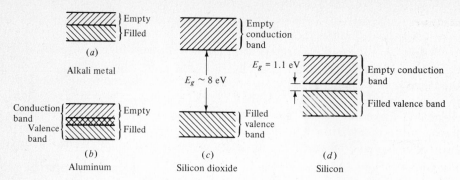

Figure 1-4 Energy-band diagrams for metals (*a* and *b*), insulator (*c*), and semiconductor (*d*) at 0 K. The vertical axis represents the energy, and the horizontal axis represents spatial position.

in it to accommodate the number of valence electrons supplied by the atoms. The valence band is completely filled, and the conduction band is empty. From quantum-mechanical considerations, it is not possible to impart any net momentum to the electrons in a completely filled band. Thus, a completely filled band does not contribute to the current flow. In terms of the band diagram, energy added thermally or by an externally applied field must contribute enough energy for the electrons in the filled valence band to jump to the conduction band in order to produce current flow. The probability of such transition is very low when E_g is large, as is true in an insulator in which $E_g > 5$ eV.

In a semiconductor (Fig. 1-4*d*) the forbidden-gap energy E_g is not too large, for example, $E_g = 1.12$ eV in silicon and 1.43 eV in GaAs. This cor-responds to the not-so-tight bonding of the valence electrons. At room temperature, some electrons in the valence band may receive enough thermal energy to overcome the forbidden gap and reach the conduction band. For each successful transition of a valence electron to the conduction band, a hole is left behind in the valence band. In other words, in the energy-band picture, a hole is an unoccupied or empty energy level in the valence band. When an external electric field is applied, the electrons in the conduction band gain kinetic energy, which results in the flow of electrons, or current conduction. At the same time the holes in the valence band also gain kinetic energy from the applied field. Thus, electrical conduction in a semiconductor may take place by two distinct and independent mechanisms: transport of negatively charged electrons with effective mass m_e in the conduction band and trans-port of positively charged holes with effective mass m_h in the valence band.

The energy-band diagram in Fig. 1-4*d* is a simplified representation of a very complex three-dimensional picture. Slightly more complicated energy-band diagrams are shown in Fig. 1-5, where the energy is plotted vs. crystal momentum based on Eq. (1-1). Normally, the top of the valence band is taken as the reference level. Note that the lowest conduction-band minimum in GaAs is located at zero momentum, directly above the valence-band maxi-

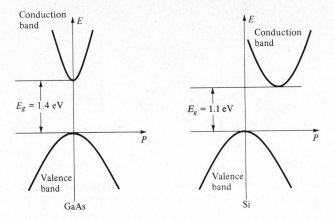

Figure 1-5 Energy-band diagram with energy vs. momentum for (*a*) GaAs (direct) and (*b*) Si (indirect).

mum. For this reason, GaAs is called a *direct* band-gap semiconductor. The lowest conduction-band minimum in Si is not at zero momentum, and silicon is called an *indirect* band-gap material. In either crystal, however, the band-gap energy E_g is the difference between the lowest conduction-band minimum and the valence-band maximum. These features are important in determining the properties of light absorption and emission, discussed in Chap. 6.

In an absolutely pure and perfect semiconductor, concentrations of electrons and holes are identical since the process of producing an electron in the conduction band creates a hole in the valence band. These concentrations are called *intrinsic* carrier concentrations n_i. The intrinsic conductivity of most semiconductors is very low and is not readily usable for device purposes. In practical devices, the crystal conductivity is controlled by adding impurities. The impurity introduced is called a *dopant*, and the doped semiconductor is said to be *extrinsic*. Let us now introduce a pentavalent element from group V in the periodic table such as phosphorus, arsenic, or antimony to substitute for some silicon atoms in an otherwise perfect silicon crystal. Each new substitutional atom has one electron more than is necessary for making the covalent bonds in the lattice, as shown in Fig. 1-6*a*. Because they are not in a valence bond, the excess electrons are rather loosely bound to their constituent atoms and can easily be set free. Since each pentavalent atom *donates* an excess electron in silicon, the pentavalent element is called a *donor*. The free electrons carry a negative charge and leave behind a fixed positive charge with the substitutional impurity atoms. Thus, the charge-neutrality condition is preserved, and there is no vacancy in the lattice site to produce a hole. When the excess electrons are set free, the atoms are said to be *ionized*. At room temperature, the thermal energy is sufficient to ionize all impurity atoms, so that the number of excess electrons equals the number of the impurity atoms. In this type of extrinsic semiconductor, called *n type*, the

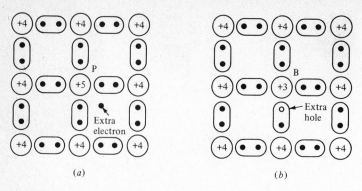

Figure 1-6 Crystal structure of silicon with a silicon atom displaced by (*a*) a pentavalent (donor) impurity atom and (*b*) a trivalent (acceptor) impurity atom.

concentration of electrons is much larger than that of holes. The conductivity of the crystal is controlled by the group V impurity concentration.

If a trivalent element, boron, aluminum, or gallium, is used as the substitutional impurity, one electron is missing from the bonding configuration, as illustrated in Fig. 1-6*b*. A hole is thus produced with a fixed negative charge remaining with the impurity atom. The number of holes is equal to the number of impurity atoms if ionization is complete. The trivalent impurity in silicon is known as an *acceptor* since it accepts an electron to produce a hole. This type of semiconductor is called *p type*.

Energy-band diagrams for both *n*- and *p*-type semiconductors are illustrated in Fig. 1-7. The energy level E_a, measured from the valence-band-edge energy E_v, is called the *ionization energy* of the acceptor impurity. The ionization energy is small because an acceptor impurity can readily accept an electron. The small ionization energy puts the impurity energy level near the valence-band edge and inside the forbidden gap. Similarly, E_d, the donor ionization energy, is measured from the band edge E_c and represents the small energy required to set the excess electron in a donor atom free. The ionization energies of some important impurities for Si, Ge, and GaAs are given in Fig. 1-8, where the levels in the lower half of the forbidden gap are

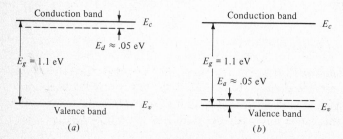

Figure 1-7 Energy-band diagram of (*a*) an *n*-type semiconductor and (*b*) a *p*-type semiconductor.

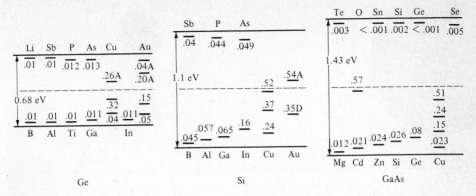

Figure 1-8 Ionization energies for various impurities in Ge, Si, and GaAs at 300 K. (*After Sze [1]*.)

measured from E_v and are acceptors unless indicated by D for a donor level. The levels in the upper half of the forbidden gap are measured from E_c and are donors unless indicated by A for an acceptor level. Although most impurity atoms introduce a single energy level, interaction between host and impurity atoms may give rise to multiple impurity levels such as gold or copper in silicon.

1-4 FREE-CARRIER CONCENTRATION IN SEMICONDUCTORS

In order to determine the electrical behavior of a semiconductor, we need to know the number of electrons and holes available for current conduction. We shall consider the electron density in the conduction band and shall apply the result to holes in the valence band by analogy. The electron density in the conduction band can be obtained if the *density-of-states function* $N(E)$ and the *distribution function* $f(E)$ are given. The first function describes the available density of energy states that may be occupied by an electron. The second function tells us the probability of occupancy of these available energy states. Thus, the density of electrons is the density of occupied states in the conduction band.

Energy and Density of States

In the energy-band diagram shown in Fig. 1-9 the lowest energy level in the conduction band, i.e., the band-edge energy E_c, is the potential energy of an electron at rest. When an electron gains energy, it moves up from band edge to energy level E. In this picture, $E - E_c$ represents the kinetic energy of the electron.

The density $N(E)$ of available states as a function of energy $E - E_c$ in the

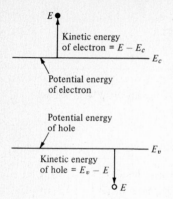

Figure 1-9 Schematic representation of kinetic and potential energy in the energy-band diagram.

conduction band is derived in Appendix A:

$$N(E) = \frac{4\pi}{h^3}(2m_e)^{3/2}(E - E_c)^{1/2} \tag{1-2}$$

where h is Planck's constant, the numerical value of which is given in Table 1-1, along with other important physical constants. The density of allowed energy states in the valence band is given by†

$$N(E) = \frac{4\pi}{h^3}(2m_h)^{3/2}(E_v - E)^{1/2} \tag{1-3}$$

Table 1-1 Physical constants

Constant	Symbol	Magnitude
Avogadro's number	N_A	6.023×10^{23} molecules/mol
Boltzmann's constant	k	1.38×10^{-23} J/K $= 8.62 \times 10^{-5}$ eV/K
Electronic charge	q	1.6×10^{-19} C
Electronvolt	eV	1.6×10^{-19} J
Free-electron mass	m	9.1×10^{-31} kg
Permittivity of free space	ϵ_o	8.854×10^{-14} F/cm
Permeability of free space	μ_o	1.257×10^{-8} H/cm
Planck's constant	h	6.625×10^{-34} J-s
Thermal voltage at 300 K	V_T	25.8 mV
Velocity of light	c	3×10^{10} cm/s

$kT = \frac{T}{11,600}$

$\cong .0259 \ (300°)$

† Strictly speaking, m_e and m_h in Eqs. (1-2) and (1-3) are known as the density-of-state effective masses.

The valence-band-edge energy E_v represents the potential energy of the hole, and $E_v - E$ represents the kinetic energy of the hole.

The Distribution Function

The probability that an energy level E is occupied by an electron is given by the Fermi-Dirac distribution function

$$f(E) = \frac{1}{e^{(E-E_f)/kT} + 1} \tag{1-4}$$

where E_f is an important parameter called the *Fermi level*, k is Boltzmann's constant, and T is the temperature in kelvins. Figure 1-10 shows the probability function at 0, 100, 300, and 400 K. There are several interesting observations we can make regarding this figure. First, at 0 K, $f(E)$ is unity for all energy smaller than E_f. This indicates that all energy levels below E_f are occupied and all energy levels greater than E_f are empty. Second, the probability of occupancy for $T > 0$ K is always $\frac{1}{2}$ at $E = E_f$, independent of temperature. Third, the function $f(E)$ is symmetrical with respect to E_f. Thus, the probability that the energy level $E_f + dE$ is occupied equals the probability that the energy level $E_f - dE$ is unoccupied. Since $f(E)$ gives the probability that a level is occupied, the probability that a level is not occupied by an electron is

$$1 - f(E) = \frac{1}{1 + e^{(E_f-E)/kT}} \tag{1-5}$$

Since a level not occupied by an electron in the valence band means that the level is occupied by a hole, Eq. (1-5) is useful in finding the density of holes in the valence band.

For all energy levels higher than $3kT$ above E_f, the function $f(E)$ can be

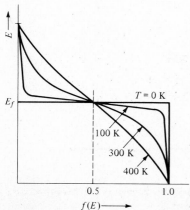

Figure 1-10 The Fermi-Dirac distribution function at different temperatures.

approximated by

$$f(E) = e^{-(E-E_f)/kT} \tag{1-6}$$

which is identical to the Maxwell-Boltzmann distribution function for classical gas particles. The electrons in the conduction band obey Eq. (1-6) if $E_c - E_f > 3kT$. For most device applications, with the exception of tunnel diodes and injection lasers, the function in Eq. (1-6) is a good approximation for $f(E)$.

Equilibrium Carrier Density

The total number of electrons in the conduction band is obtained by integrating the product of the density of states and the occupancy probability

$$n = \int_{E_c}^{\infty} f(E)N(E)\, dE \tag{1-7}$$

Substituting Eqs. (1-2) and (1-6) into Eq. (1-7) and performing the integration, we have†

$$n = \int_{E_c}^{\infty} \frac{4\pi}{h^3}(2m_e)^{3/2}(E - E_c)^{1/2}\, e^{-(E-E_f)/kT}\, dE$$

$$= \frac{4\pi}{h^3}(2m_e)^{3/2}\, e^{-(E_c-E_f)/kT} \int_{E_c}^{\infty} (E - E_c)^{1/2}\, e^{-(E-E_c)/kT}\, d(E - E_c)$$

$$= N_c\, e^{-(E_c-E_f)/kT} \tag{1-8}$$

where

$$N_c = 2\left(\frac{2\pi m_e kT}{h^2}\right)^{3/2} \tag{1-9}$$

The quantity N_c is called the *effective density of states in the conduction band*. In silicon, N_c is equal to 2.8×10^{19} cm^{-3} at 300 K (room temperature). Similarly, the density of holes in the valence band is

$$p = \int_{-\infty}^{E_v} [1 - f(E)]N(E)\, dE \tag{1-10}$$

Substituting Eqs. (1-3) and (1-6) in Eq. (1-10) and integrating yields

$$p = N_v\, e^{-(E_f-E_v)/kT} \tag{1-11}$$

where

$$N_v = 2\left(\frac{2\pi m_h kT}{h^2}\right)^{3/2} \tag{1-12}$$

† Using the definite integral

$$\int_0^{\infty} x^{1/2} e^{-ax}\, dx = \frac{1}{2a}\sqrt{\frac{\pi}{a}}$$

The quantity N_v is called the *effective density of states in the valence band*, and N_v is $10^{19} \, \text{cm}^{-3}$ for silicon at room temperature. The process of obtaining the intrinsic distribution is displayed graphically in Fig. 1-11. The intrinsic-carrier density, effective density of states, and band-gap energy of Ge, Si, and GaAs at room temperature are given in Table 1-2, along with other important properties.

The product of Eqs. (1-8) and (1-11) is

$$np = N_c N_v e^{-E_g/kT} \quad = n_i^{\,2} \tag{1-13}$$

where $E_g = E_c - E_v$, which is the energy of the forbidden gap. The band-gap energy is slightly temperature-dependent and can be described by the empirical relation

$$E_g = E_{go} - \beta T \tag{1-14}$$

where β is the temperature coefficient of the band-gap energy and E_{go} is the extrapolated value of E_g at $0 \, \text{K}$. For silicon we have $E_{go} = 1.21 \, \text{eV}$ and $\beta = 2.8 \times 10^{-4} \, \text{eV/K}$. Substituting Eqs. (1-9), (1-12), and (1-14) into Eq. (1-13) and simplifying yields

$$pn = K_1 T^3 e^{-E_{go}/kT} \quad = n_i^{\,2} \tag{1-15}$$

where K_1 is a constant. Equation (1-13) states that the np product is a constant in a semiconductor at a given temperature under *thermal equilibrium*. Thermal equilibrium is defined as the steady-state condition at a given temperature without external forces or excitation. This pn product depends only on the density of allowed energy states and the forbidden-gap energy, but it is independent of the impurity density or the position of the Fermi level.

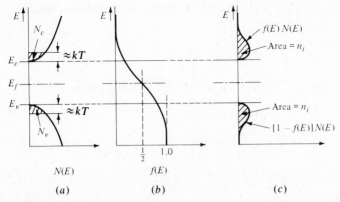

Figure 1-11 Graphical procedures for obtaining the intrinsic-carrier concentration: (*a*) the density-of-states function $N(E)$, (*b*) the Fermi-Dirac function $f(E)$, and (*c*) the density of carriers $N(E)f(E)$ at 300 K. The shaded areas in (*a*) correspond to the effective density of states; see prob. 1-11 for further explanation.

$$\frac{n_i^{\,2}(T_2)}{n_i^{\,2}(T_1)} = \left(\frac{T_2}{T_1}\right)^3 e^{\left[\frac{E_{go}}{kT_1}\left(\frac{T_2 - T_1}{T_2}\right)\right]} \qquad E_{go} \approx 1\,\text{eV}$$

Table 1-2 Properties of Ge, Si, and GaAs at 300 K (After Sze [1] and Grove [2])

Property	Ge	Si	GaAs	SiO$_2$
Atoms or molecules/cm^3	4.42×10^{22}	5.0×10^{22}	2.21×10^{22}	2.3×10^{22}
Atomic or molecular weight	72.6	28.08	144.63	60.08
Density, g/cm^3	5.32	2.33	5.32	2.27
Breakdown field, V/cm	$\sim 10^5$	$\sim 3 \times 10^5$	$\sim 3.5 \times 10^5$	$\sim 6 \times 10^6$
Crystal structure	Diamond	Diamond	Zinc blende	Amorphous
Dielectric constant	16	11.8	10.9	3.9
Effective density of states				
Conduction band N_c, cm^{-3}	1.04×10^{19}	2.8×10^{19}	4.7×10^{17}	
Valence band N_v, cm^{-3}	6.1×10^{18}	1.02×10^{19}	7.0×10^{18}	
Electron affinity χ, V	4.13	4.01	4.07	0.9

Property				
Energy gap, eV	0.68	1.12	1.43	~8
Intrinsic carrier concentration n_i, cm^{-3}	2.5×10^{13}	1.5×10^{10}	10^7	
Lattice constant, Å	5.658	5.431	5.654	
Effective mass:				
Electrons	$m_e = 0.22m,\ m_e^* = 0.12m$	$m_e = 0.33m,\ m_e^* = 0.26m$	$0.068m$	
Holes	$m_h = 0.31m,\ m_h^* = 0.23m$	$m_h = 0.56m,\ m_h^* = 0.38m$	$0.56m$	
Intrinsic mobility, cm^2/V-s:				
Electron, cm^2/V-s	3900	1350	8600	
Hole, cm^2/V-s	1900	480	250	
Temperature coefficient of expansion	5.8×10^{-6}	2.5×10^{-6}	5.8×10^{-6}	5×10^{-7}
Thermal conductivity, W/cm-°C	0.6	1.5	0.8	0.01

1-5 INTRINSIC AND EXTRINSIC SEMICONDUCTORS

In an intrinsic semiconductor, the electron density is exactly equal to the hole density. As a result,

$$n = p = n_i \qquad (1\text{-}16)$$

where n_i is the intrinsic-carrier density. Substitution of Eq. (1-16) into Eq. (1-13) yields

$$np = n_i^2 \qquad (1\text{-}17)$$

the *mass-action law*, which is valid for both intrinsic and extrinsic semiconductors under thermal equilibrium. In an extrinsic semiconductor, the increase of one type of carrier tends to reduce the number of the other type through *recombination* (described in Chap. 2) such that the product of the two remains constant at a given temperature.

By using Eqs. (1-13) and (1-17) we can write the intrinsic-carrier density as

$$n_i = \sqrt{N_c N_v}\, e^{-E_g/2kT} \qquad (1\text{-}18)$$

Near-exponential temperature dependence of n_i is obtained by using Eq. (1-18). Experimental data of the intrinsic-carrier density for Ge, Si, and GaAs as a function of temperature are shown in Fig. 1-12. From the slope of these

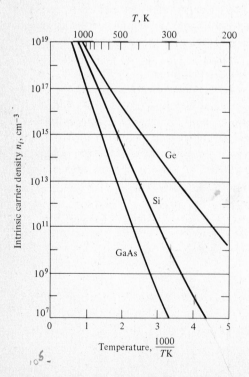

Figure 1-12 Experimental data of n_i for Ge, Si, and GaAs as a function of temperature. (*After Sze [1].*)

plots, the band-gap energy can be calculated. Note that the intrinsic-carrier density is smaller for a semiconductor having a higher band-gap energy. For the same semiconductor, a higher temperature produces a higher carrier density. We can use Fig. 1-12 to obtain $E_g = 1.12\,\text{eV}$ and $n_i = 1.5 \times 10^{10}\,\text{cm}^{-3}$ for silicon at room temperature.

The Fermi level in an intrinsic semiconductor, denoted E_i, can be obtained by equating Eqs. (1-8) and (1-11) and by setting $E_f = E_i$:

$$E_i = \tfrac{1}{2}(E_c + E_v) + \tfrac{3}{4}kT \ln \frac{m_h}{m_e} \tag{1-19}$$

If the effective masses of electron and hole are equal, the intrinsic Fermi level would be located at the middle of the forbidden gap. The small deviation of E_i from midgap due to the difference of effective masses is usually negligible. From here on, we shall use E_i to specify the midgap energy of semiconductor.

For an intrinsic semiconductor, we may set $E_f = E_i$ in Eqs. (1-8) and (1-11) and substitute them into Eq. (1-16). The result is

$$n_i = N_c e^{-(E_c - E_i)/kT} = N_v e^{-(E_i - E_v)/kT} \tag{1-20}$$

Using the foregoing relationship, we can rewrite Eqs. (1-8) and (1-11) as

$$n = n_i e^{(E_f - E_i)/kT} \tag{1-21}$$

$$p = n_i e^{(E_i - E_f)/kT} \tag{1-22}$$

In Eqs. (1-21) and (1-22), we express the electron and hole concentrations in terms of the intrinsic concentration n_i and midgap energy E_i. These equations are valid for both intrinsic and extrinsic semiconductors. For example, in an intrinsic material, we have $n = n_i = p$ since $E_f = E_i$. These two equations are sometimes more convenient to use than Eqs. (1-8) and (1-11).

With the doping of donor density N_d, the concentration of electrons in the conduction band increases from its intrinsic concentration. The result corresponds to an increase of the probability of occupancy in the conduction band. Under this condition, the Fermi level shifts upward from the midgap position. Figure 1-13 shows the graphical procedures in obtaining the carrier concentration in an n-type semiconductor. The new electron density and new position of the Fermi level can be obtained by using the following arguments. Let us assume that all donor atoms are ionized. The total number of electrons in the conduction band must equal the sum of electrons originated from the donor level and the valence band. Since each electron from the valence band leaves a hole behind, the electron density can be written as

$$n = p + N_d \tag{1-23}$$

Equation (1-23) actually describes the condition of space-charge neutrality in an n-type semiconductor. When both donors N_d and acceptors N_a are present, the space-charge neutrality condition is generalized to

$$n + N_a = p + N_d \tag{1-24}$$

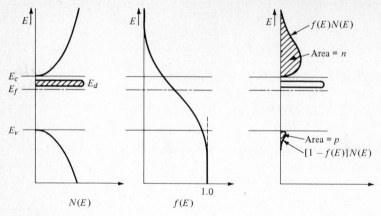

Figure 1-13 Graphical procedures for obtaining carrier concentrations in an n-type semiconductor.

By solving Eqs. (1-17) and (1-23) and by using the quadratic formula we get the equilibrium electron and hole concentrations in an n-type semiconductor:

$$n_n = \frac{\sqrt{N_d^2 + 4n_i^2} + N_d}{2} \qquad (1\text{-}25)$$

$$p_n = \frac{\sqrt{N_d^2 + 4n_i^2} - N_d}{2} \qquad (1\text{-}26)$$

The subscript n refers to the n-type semiconductor. Since the electron is the dominant carrier, it is called the *majority carrier*. The hole in the n-type semiconductor is called the *minority carrier*.

In most practical situations, the donor dopant concentration is much higher than the intrinsic concentration; that is, $n_i/N_d \ll 1$. Therefore, we can use the binomial expansion† in Eqs. (1-25) and (1-26) to obtain

$$n_n = N_d + \frac{n_i^2}{N_d} \approx N_d \qquad (1\text{-}27)$$

$$p_n = \frac{n_i^2}{N_d} \qquad (1\text{-}28)$$

The location of the Fermi level is obtained by substituting Eq. (1-27) into Eq. (1-8) and solving for E_f. Thus,

n-type

$$E_f = E_c - kT \ln \frac{N_c}{N_d} \qquad (1\text{-}29)$$

Example A silicon wafer is doped with 10^{15} phosphorus atoms/cm^{-3}. Find the carrier concentrations and Fermi level at room temperature (300 K).

SOLUTION At room temperature, we can assume complete ionization of impurity atoms. Since $N_d = 10^{15}$ cm$^{-3} \gg n_i$, we have

---

† $(1+x)^m = 1 + mx + \cdots = 1 + mx$ for $x \ll 1$.

$$n_n = N_d = 10^{15} \text{ cm}^{-3}$$

$$p_n = \frac{n_i^2}{N_d} = 2.25 \times 10^5 \text{ cm}^{-3}$$

The Fermi level is calculated by using Eq. (1-29):

$$E_c - E_f = 0.0258 \ln \frac{2.8 \times 10^{19}}{10^{15}} = 0.265 \text{ eV}$$

When the impurity concentration is very high and approaches the effective density of states, the Fermi level will coincide with E_c. The semiconductor is now *degenerate*. A more rigorous definition of a degenerate semiconductor is

$$E_c - E_f < 3kT \tag{1-30}$$

Under this condition, Eq. (1-29) is no longer valid since the assumption leading to Eq. (1-6) is no longer applicable.

By analogy, the equilibrium carrier concentrations in a p-type semiconductor are given by

$$n_p = \frac{n_i^2}{N_a} \tag{1-31}$$

$$p_p = N_a \tag{1-32}$$

The Fermi level in a p-type semiconductor is

$$E_f = E_v + kT \ln \frac{N_v}{N_a} \qquad \text{⚡ (1-33)}$$

Figure 1-14 shows the Fermi level as a function of temperature for various impurity concentrations in silicon. The temperature dependence of the majority carrier n_n is illustrated in Fig. 1-15.

At a given temperature, the impurity atoms may not be all ionized. Since each ionized donor atom gives up an electron to the conduction band, the number of ionized donor N_d^+ is equal to the number of unoccupied donor states. Using the Fermi-Dirac function of occupancy expressed in Eq. (1-4), we have

$$N_d^+ = N_d - \frac{N_d}{1 + \frac{1}{2}e^{(E_d - E_f)/kT}} \qquad = \frac{N_d}{1 + 2e^{(E_f - E_d)/kT}} \tag{1-34}$$

Similarly, the ionized-acceptor density N_a^- is equal to the occupied acceptor states†

$$N_a^- = \frac{N_a}{1 + \frac{1}{4}e^{(E_a - E_f)/kT}} \tag{1-35}$$

---

† The factor $\frac{1}{2}$ in the denominator of Eq. (1-34) is to account for the two spins of a donor level. Similarly, the factor $\frac{1}{4}$ in Eq. (1-35) is to account for the two spins of an acceptor level and the two degenerate valence bands in Ge, Si, and GaAs.

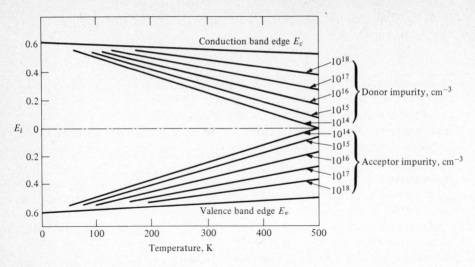

Figure 1-14 The Fermi level in silicon as a function of temperature for various impurity concentrations. (*After Grove [2].*)

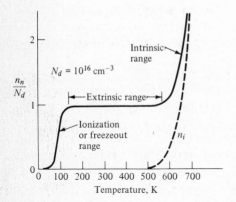

Figure 1-15 Electron density as a function of temperature in *n*-type silicon.

In an extrinsic semiconductor with partial ionization of impurity atoms, the Fermi level is calculated by replacing N_d by N_d^+ in Eq. (1-29) and N_a by N_a^- in Eq. (1-33). These changes are sometimes necessary for low-temperature design considerations.

1-6 SCATTERING AND DRIFT OF ELECTRONS AND HOLES

Under thermal equilibrium, mobile electrons and atoms of the lattice are always in random thermal motion. From the theory of statistical mechanics, a carrier at a temperature T K is estimated to have an average thermal energy of $3kT/2$ J. This thermal energy can be translated into an average velocity v_{th} by using the relationship

$$\frac{1}{2}m_e^* v_{th}^2 = \frac{3}{2}kT \tag{1-36}$$

The average velocity of thermal motion for electrons in silicon at room temperature is calculated to be approximately 10^7 cm/s. We may now visualize the electrons as moving rapidly in all directions. At the same time, the lattice atoms experience thermal vibrations that cause small periodic deviations from their mean positions. As a result, the electrons make frequent collisions with the vibrating atoms. Under thermal-equilibrium condition, the random motions of these electrons cancel out, and the average current in any direction is zero. It should be pointed out that the effect of the periodic field established by the lattice atoms has been taken into account by the effective mass. Therefore, the lattice atoms do not deflect the conduction electrons. However, the effective mass of electrons does not account for the deviation of the atoms from their periodic positions caused by the thermal energy of the system. It is these deviations from a perfect periodic lattice that deflect or scatter the free electrons. The thermal vibrations may be treated quantum-mechanically as discrete particles called *phonons*, and the collision of phonons with electrons and holes is called *lattice* or *phonon scattering*. Lattice scattering increases with increasing temperature because of the increased lattice vibration. It dominates other scattering processes at and above room temperature in lightly doped silicon. Most semiconductor devices are operated in this temperature range.

Besides lattice scattering, there are three other scattering mechanisms: (1) The ionized impurity atoms are charged centers that may deflect the free carriers. The effect of this mechanism depends on the temperature and the impurity concentration. (2) The neutral impurity atoms may introduce scattering if the concentration of these atoms is high. This scattering takes place when the free carriers are influenced by the atomic core and orbiting electrons separately. Usually, the effect of this scattering is negligible. (3) The Coulomb force between carriers (electron-electron, electron-hole) can cause scattering at high concentration of carrier densities.

Under thermal equilibrium, the random motion of electrons leads to zero current in any direction. This physical picture is shown in Fig. 1-16a. The average distance between collisions is called the *mean free path l_m*. Regardless of which scattering mechanism is dominating, the typical value of the mean free path is between 10^{-6} and 10^{-4} cm. With a velocity of 10^7 cm/s, the *mean free time τ_m* between collisions is about 1 ps (one picosecond = 10^{-12} s). If an external electric field is applied across the crystal, the electrons would be acted upon to move in the direction opposite that of the electric field. This external force superimposes on the random motion of electrons to give the picture shown in Fig. 1-16b. We notice that the electrons now *drift* in the direction opposite the field, and a current will flow.

Unlike electronic motion in a vacuum, electrons in a semiconductor under

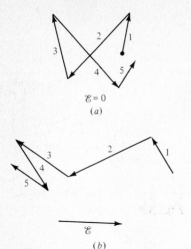

$\mathscr{E} = 0$

(a)

$\mathscr{E}$

(b)

Figure 1-16 Typical path of electron in crystal (a) without and (b) with electric field.

an external field do not achieve constant acceleration. Instead they move with a steady *drift velocity* because of the effect of scattering. The drift velocity is a function of the applied field, and at low electric fields the magnitude of the velocity can be derived by using the following simple model.

Since the acceleration of an electron is given by $\bar{a} = -q\mathscr{E}/m_e^*$ (Newton's second law) and the increase of velocity between two consecutive collisions is $\Delta v = \bar{a}\tau_m$, the average drift velocity (if zero initial velocity is assumed) is

$$v_d = \frac{\Delta v}{2} = -\frac{q\tau_m}{2m_e^*}\mathscr{E} \tag{1-37}$$

In a more accurate model including the effect of statistical distribution, the factor 2 does not appear in the denominator of Eq. (1-37). The correct expression is†

$$v_d = -\frac{q\tau_m}{m_e^*}\mathscr{E} = -\mu_n\mathscr{E} \tag{1-38}$$

where $\mu_n \equiv q\tau_m/m_e^*$ is called the *electron mobility*. A similar equation can be written for the hole:

$$v_d = \mu_p\mathscr{E} \tag{1-39}$$

where μ_p is the *hole mobility*. The negative sign is missing in Eq. (1-39) because holes drift in the same direction as the electric field.

According to Eq. (1-38), the drift velocity is linearly related to the applied electric field. This linear relationship is valid when the electric field is low. By increasing the applied field, the drift velocity eventually approaches the thermal-velocity limit obtained from Eq. (1-36). Since the electron cannot exceed the thermal velocity, further increases of the electric field would not

---

† m_e^* in Eq. (1-38) is also known as the *conductivity effective mass* [1].

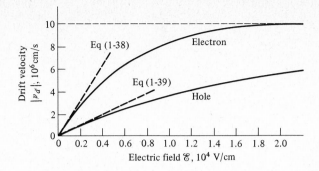

Figure 1-17 Drift velocity as a function of electric field for electron and hole. (*After Gibbons [3].*)

produce a higher drift velocity. Experimental data showing the field-dependent drift velocity in silicon are plotted in Fig. 1-17, which can be approximated by the empirical expression

$$v_d = v_{th}(1 - e^{-\mathscr{E}/\mathscr{E}_c}) \tag{1-40}$$

where $\mathscr{E}_c$ is called the *critical field*. For an *n*-type silicon wafer the critical field is approximately 1.5×10^4 V/cm.

Since the mobility takes into account the effects of scattering, the contribution of different scattering mechanisms to the mobility variation should be readily observable. It has already been mentioned that the lattice scattering increases with increasing temperature. Increased scattering results in a short-

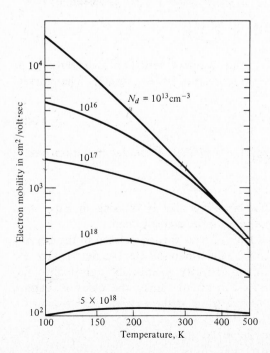

Figure 1-18 Mobility as a function of temperature in silicon.

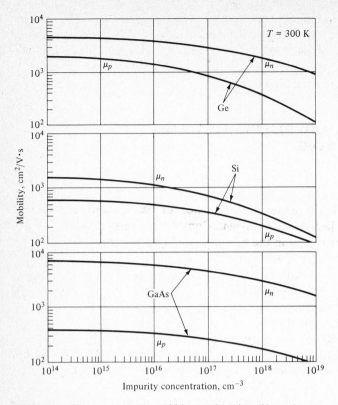

Figure 1-19 Electron and hole mobilities as a function of impurity concentration for Ge, Si, and GaAs at 300 K. (*After Sze and Irvin [4].*)

er mean free time, yielding a lower mobility. For a very pure sample, the impurity and carrier scattering mechanisms are negligible, and the temperature effect of the mobility is dominated by the lattice scattering. Experimental results show a temperature dependence ranging between $T^{-3/2}$ and $T^{-5/2}$ for the electron and hole mobilities in silicon, as demonstrated by the sample doped with 10^{13} cm^{-3} in Fig. 1-18. On the other hand, the effect of impurity scattering on the mobility is most pronounced for heavily doped samples at low temperatures, where the lattice scattering can be ignored. The low temperature reduces the thermal velocity of carriers so that electrons and holes traveling past fixed charged ions will be deflected by the Coulomb force set up by the charged ions. As the temperature is increased, the fast-moving carriers become less likely to be deflected by the charged ions and the scattering is decreased. Therefore, the mobility increases with temperature, as seen most clearly for the electron mobility with doping of 10^{18} cm^{-3} in Fig. 1-18. Note also that the mobility decreases with increasing impurity concentration at a given temperature.

Another important experimental result is shown in Fig. 1-19 for mobility as functions of donor and acceptor impurity concentrations. It is seen that the mobility is constant at low impurity concentrations. For impurity concentrations greater than $10^{16}\ \text{cm}^{-3}$ the mobility decreases as a result of impurity scattering. Note that the electron mobility is higher than that of holes, a situation found in most semiconductors.

1-7 CONDUCTION AND DIFFUSION CURRENTS IN SEMICONDUCTORS

Conduction

The transport of carriers under the influence of an applied electric field produces a current called the *drift current*. Consider the semiconductor shown in Fig. 1-20. The current flow in the bar having n electrons per unit volume is given by

$$I_n = -qAnv_d = qAn\mu_n \mathscr{E} \tag{1-41}$$

where q = magnitude of electronic charge
$\quad A$ = the cross-sectional area
$\quad L$ = length of bar
In addition Eq. (1-38) has been employed. If we now replace $\mathscr{E}$ by V/L in Eq. (1-41) and obtain the voltage-to-current ratio, we have

$$\frac{V}{I_n} = \frac{L}{qAn\mu_n} \tag{1-42}$$

The resistance of the bar is defined by

$$R \equiv \rho \frac{L}{A} \equiv \frac{V}{I_n} \tag{1-43}$$

where ρ is the *resistivity*. Substituting Eq. (1-42) into Eq. (1-43) and solving for ρ, we obtain

$$\frac{1}{\rho} = \sigma = q\mu_n n \tag{1-44}$$

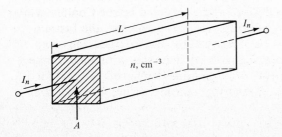

Figure 1-20 Current conduction in a semiconductor bar.

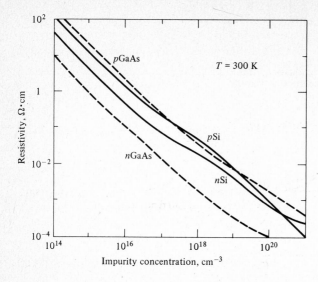

Figure 1-21 Resistivity vs. impurity concentration in Si and GaAs at 300 K. (*After Sze and Irvin [4].*)

where σ is the *conductivity*. By analogy, the hole drift current can be written as

$$I_p = qАр\mu_p\mathscr{E} \tag{1-45}$$

The overall resistivity of a semiconductor including the effects of electrons and holes becomes

$$\frac{1}{\rho} = q\mu_n n + q\mu_p p \tag{1-46}$$

The resistivity of a semiconductor is an important parameter in device design. Figure 1-21 shows the relationship between the impurity concentration and resistivity for both *n*- and *p*-type silicon and GaAs at room temperature. The deviation from linearity in these curves is caused by the nonlinear mobility effect.

Diffusion

In the preceding paragraphs we examined the transport of carriers when an applied field existed in the semiconductor. Another very important mechanism for carrier transport is *diffusion*. Diffusion of carriers occurs when there is spatial variation of concentration because carriers tend to move from regions of high concentration to low concentration. Let us assume that we introduce a number of carriers at a plane in the crystal at zero time. These carriers would experience the random collision with the lattice and move away from the center (Fig. 1-22). The initial high concentration at $x = 0$ spreads out until the carriers are uniformly distributed throughout the region. The diffusion flux obeys Fick's first law:

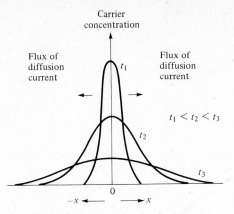

Figure 1-22 Spatial variation of carrier concentration at different times.

$$F = -D\frac{dN}{dx} \qquad (1\text{-}47)$$

where F = *flux of carriers* = number passing through 1 cm^2/s

D = diffusion constant

N = carrier density

Thus, the diffusion currents of electrons and holes are given by

$$I_n = qAD_n\frac{dn}{dx} \qquad (1\text{-}48)$$

$$I_p = -qAD_p\frac{dp}{dx} \qquad (1\text{-}49)$$

where D_n and D_p are the diffusion constants for electrons and holes, respectively. The negative sign in Eq. (1-49) indicates that the hole current flows in the direction opposite the gradient of holes. This is shown in Fig. 1-23. The diffusion constants will be shown in Chap. 2 to be related to the mobility by

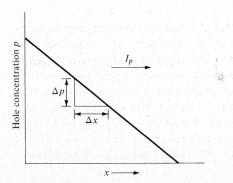

Figure 1-23 Steady-state hole concentration gradient producing diffusion current.

the Einstein relationship:

$$\frac{D_p}{\mu_p} = \frac{D_n}{\mu_n} = \frac{kT}{q} \tag{1-50}$$

For lightly doped silicon at room temperature $D_n = 38\ \text{cm}^2/\text{s}$ and $D_p = 13\ \text{cm}^2/\text{s}$.

The total electron and hole currents are obtained by adding the drift and diffusion components:

$$I_n = qA\left(\mu_n n \mathscr{E} + D_n \frac{dn}{dx}\right) \tag{1-51}$$

$$I_p = qA\left(\mu_p p \mathscr{E} - D_p \frac{dp}{dx}\right) \tag{1-52}$$

These equations will be used throughout the text to characterize semiconductor devices.

PROBLEMS

1-1 Show that the bcc lattice can be formed by two interpenetrating sc lattices.

1-2 Sketch the crystal planes (010), (011), and (001) in a cubic lattice.

1-3 Determine the Miller indices for the crystal plane shown in Fig. P1-3.

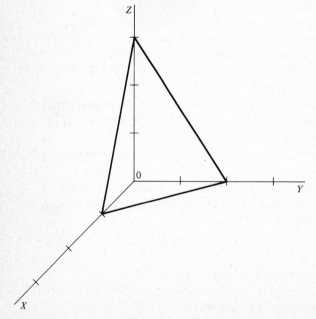

Figure P1-3.

1-4 Sketch the two-dimensional diagram of a diamond lattice if you project all the atoms onto (*a*) the (111) plane and (*b*) the (110) plane.

1-5 Sketch a unit cube to represent a cubic lattice and draw the equivalent ⟨111⟩ directions in your cube.

1-6 (*a*) Determine the number of atoms in a unit cell of an fcc crystal.

(*b*) What is the distance in units of the lattice constant *a* between two nearest-neighboring atoms?

(*c*) If each atom is assumed to be a sphere and the spherical surface of each atom comes into contact with its nearest neighbors, what percentage of the total volume of the unit cell is being occupied?

1-7 The lattice constant of silicon is 5.43 Å. Calculate (*a*) the spacing between the nearest-neighboring atoms and (*b*) the density of valence electrons.

1-8 A *mole* is defined as the quantity of a substance equal to its molecular weight in grams. In any substance, the number of molecules contained in a mole is called Avogadro's number ($N_A = 6.02 \times 10^{23}$ molecules per mole).

(*a*) Show that the number of atoms N per cubic centimeter is given by

$$N = \frac{N_A \rho}{\text{atomic weight} \times [1 \text{ g/mole}]}$$

where ρ is the density in grams per cubic centimeter.

(*b*) Calculate N for silicon and germanium.

1-9 The electron effective mass is measured in the cyclotron resonance experiment by placing a crystal sample in a dc magnetic field B. An electron moves in a circular path which is defined by equating the centrifugal force to the magnetic force qvB, where v is the electron velocity.

(*a*) Derive the expression for the angular frequency of rotation.

(*b*) The frequency is measured by applying an ac electric field (normal to the magnetic field) whose energy is absorbed if the two frequencies are in resonance. If the resonant frequency is 5 GHz and $B = 2000$ G, find m_e in terms of the free-electron mass.

1-10 (*a*) Draw the energy-band diagram for a silicon wafer doped with 10^{15} boron atoms at 77 and 600 K and room temperature. Specify the band-gap energy. Assume constant ionization energy.

(*b*) At each temperature, calculate the electron and hole concentrations and the Fermi level.

1-11 Show that the effective density of states N_c represents the density of states in a strip $1.2kT$ wide near the edge of the conduction-band edge. Explain the physical meaning of your result.

1-12 Determine the donor or acceptor concentration that produces degenerate Si and GaAs.

1-13 Carry out the integration of Eq. (1-10) to derive Eq. (1-11).

1-14 Obtain the band-gap energy of Ge, Si, and GaAs at 300 K from Fig. 1-12. Compare your results with the data in Table 1-2.

1-15 Using the data in Table 1-2, calculate the deviation of E_i from the midgap energy for Ge and Si. What conclusion do you arrive at after these calculations?

1-16 (a) Calculate the Fermi level of silicon wafers doped with 10^{15}, 10^{18}, and 10^{20} arsenic atoms/cm^3 at room temperature by assuming complete ionization.

(b) Using the Fermi level obtained in (a), show whether the assumption of complete ionization is justified in each case.

1-17 (a) Find the resistivity of intrinsic Ge and Si at 300 K.

(b) If a shallow donor impurity is added to the extent of 1 atom per 10^8 Ge or Si, find the resistivity.

1-18 (a) Use the data in Figs. 1-12 and 1-18 find the carrier concentrations, mobility, and conductivity of the sample with $N_d = 10^{13}$ at 200 and 300 K.

(b) Repeat (a) for $N_d = 10^{18}$ cm^{-3}.

(c) Obtain the temperature coefficient of conductivity for the two samples. Which one would you select as the material for a temperature-sensitive device called the *thermistor*?

1-19 Using the result of Prob. 1-8, calculate the electron mobility in aluminum, for which the density is 2.7 g/cm^3, the resistivity is 3×10^{-6} Ω-cm, and the atomic weight is 27.

1-20 (a) A silicon wafer is doped with 2×10^{16} boron and 10^{16} phosphorus atoms. Calculate the electron and hole concentrations, E_f, and resistivity at room temperature.

(b) Repeat (a) for 8×10^{15} boron atoms/cm^3.

1-21 Assuming $\mu_n = 1350$ cm^2/V-s at room temperature, calculate the electron drift velocity for an electric field of (a) 10^2 V/cm and (b) 10^5 V/cm. Comment on the validity of these results.

REFERENCES

1. S. M. Sze, "Physics of Semiconductor Devices," chap. 1, Wiley, New York, 1969.
2. A. S. Grove, "Physics and Technology of Semiconductor Devices," chap. 4, Wiley, New York, 1967.
3. J. F. Gibbons, Carrier Drift Velocities in Silicon at High Electric Field Strengths, *IEEE Trans. Electron. Devices*, **ED-14**: 37 (1967).
4. S. M. Sze and J. C. Irvin, Resistivity, Mobility, and Impurity Levels in GaAs, Ge and Si at 300°K, *Solid-State Electron.*, **11**: 599 (1968).
5. S. Li and W. R. Thurber, The Doping Density and Temperature Dependence of Electron Mobility and Resistivity in n-Type Silicon, *Solid-State Electron.*, **20**: 609 (1977).

ADDITIONAL READINGS

Adler, R. B., A. C. Smith, and R. L. Longini: "Introduction to Semiconductor Physics," SEEC Series, vol. I, Wiley, New York, 1964.
Kittel, C.: "Introduction to Solid-State Physics," 3d ed., Wiley, New York, 1966.

Millman, J., and C. C. Halkias: "Integrated Electronics," McGraw-Hill, New York, 1972.

Shockley, W.: "Electrons and Holes in Semiconductors," Van Nostrand Reinhold, New York, 1950.

Smith, R. A.: "Semiconductors," Cambridge University Press, London, 1959.

Streetman, B. G.: "Solid-State Electronic Devices," Prentice-Hall, Englewood Cliffs, N.J., 1972.

NONEQUILIBRIUM CHARACTERISTICS OF SEMICONDUCTORS

Nonequilibrium describes the condition in which free-carrier densities are different from their thermal-equilibrium values. Since most semiconductor devices are operated under nonequilibrium conditions, it is essential to know how these free carriers are introduced and transported. In addition, it is necessary to determine how thermal equilibrium is reestablished. This chapter begins with the topic of introducing excess carriers by means of generation and injection. It will be shown that, on average, the excess carriers will survive for a time before they are annihilated by recombination processes. During their lifetime, excess carriers can diffuse or drift for a useful distance to produce a current. The main object of this chapter is to establish the basic laws governing the recombination and transport of nonequilibrium electrons and holes. These basic laws will be used in later chapters to derive the current-voltage characteristics of diodes and transistors.

2-1 GENERATION, RECOMBINATION, AND INJECTION OF CARRIERS

Under thermal equilibrium, carriers in a semiconductor possess an average thermal energy corresponding to the ambient temperature. This thermal energy enables some valence electrons to reach the conduction band. The upward transition of an electron leaves a hole behind, so that an electron-hole

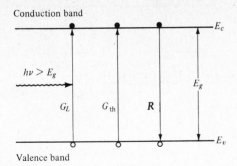

Figure 2-1 Band-to-band generation and recombination of electron-hole pairs.

pair is produced. This process is called *carrier generation* and is represented by G_{th} in Fig. 2-1. When an electron makes a transition from the conduction band to the valence band, an electron-hole pair is annihilated. This reverse process, called *recombination*, is represented by R in Fig. 2-1. Under thermal equilibrium, the generation rate and the recombination rate must be equal, so that the carrier concentrations remain constant. Thus, the condition $pn = n_i^2$ is maintained.

The equilibrium condition may be disturbed by the introduction of free carriers exceeding their thermal-equilibrium values. This process, called *carrier injection*, can be accomplished by either optical or electrical means. The optical injection involves an incident light having photon energy of monochromatic light equal to or greater than the band-gap energy E_g. The photon energy is given by the product $h\nu$, where ν is the frequency of light and h is Planck's constant. When the optical energy is absorbed by an electron in the valence band, the electron is excited to the conduction band and a hole is created in the valence band. The generation rate of electron-hole pairs by light is shown in Fig. 2-1 as G_L. The injection of carriers increases the electron and hole densities such that $pn > n_i^2$. The additional carriers are called *excess carriers*. The excess electrons and holes are always in equal number, so that space-charge neutrality is preserved. The mechanisms of injection, recombination, and transport of these excess carriers are of fundamental importance in determining the characteristics of most semiconductor devices.

2-2 INJECTION LEVELS OF CARRIERS

The number of injected carriers in a semiconductor generally controls a device's behavior. Let us consider the case of an n-type silicon wafer uniformly illuminated by a light source under steady-state conditions. Before the light source is turned on, the silicon wafer is in equilibrium and there are no excess carriers. The majority-carrier density is equal to the donor concen-

Table 2-1 n-type silicon with $N_d = 2.25 \times 10^{15}$ cm^{-3}

Carrier density, cm^{-3}	Injection condition		
	Equilibrium	Low level	High level
Excess Δn	0	10^{13}	10^{16}
Majority n_n	2.25×10^{15}	2.26×10^{15}	1.225×10^{16}
Minority p_n	10^5	10^{13}	10^{16}

tration, and the minority-carrier density can be calculated from Eq. (1-28). After the light is turned on, two different conditions may exist. If the injected-carrier density is small compared with the donor concentration, the majority-carrier density remains essentially unchanged while the minority-carrier density is equal to the injected-carrier density. This condition is called *low-level injection*. If the injected-carrier density exceeds the donor concentration, it is called *high-level injection*. These conditions are illustrated by the example shown in Table 2-1. It should be pointed out that the total carrier density always equals the sum of the equilibrium and excess-carrier densities. High-level injection usually introduces additional complexity in the mathematical analyses, but since it provides little additional physical insight into device behavior, we shall ignore high-level injection effects whenever possible.

2-3 RECOMBINATION MECHANISMS

Excess electrons in the conduction band may recombine with holes in the valence band. Thus, the optically generated electron-hole pairs may be annihilated and the thermal-equilibrium density may be reestablished. The energy released by the recombination is emitted as a photon or phonons, depending on the nature of the recombination mechanism. When a photon is emitted, the process is called *radiative recombination*. On the other hand, the lack of photon emission indicates a *nonradiative* recombination process, which emits phonons to the lattice in the form of heat dissipation.

The most important recombination processes are the *direct* and *indirect* recombinations, which may take place simultaneously in the same semiconductor. Usually, one of them dominates, and its special features can then be examined.

Direct Recombination

An electron in the conduction band may make a transition directly to the valence band to recombine with a hole. This process, called the *direct* or *band-to-band recombination*, is illustrated in Fig. 2-1. The rate of this transition is proportional to available electrons in the conduction band and the number of unoccupied states in the valence band into which the transition may take place. Since the number of unoccupied states in this case is the hole density, we can express the rate of direct recombination as

$$R = Bnp \qquad (2\text{-}1)$$

where B is a probability coefficient taking into account the probabilistic nature of recombination. In equilibrium, the recombination rate must be balanced by the thermal-generation rate. Therefore, we have $R = R_{th}$ and

$$G_{th} = R_{th} = Bn_o p_o \qquad (2\text{-}2)$$

where the subscript o represents the thermal-equilibrium condition. When a light source is applied to produce electron-hole pairs at the rate of G_L, the carrier densities are above their equilibrium values. The recombination and generation rates become

$$R = Bnp = B(n_o + \Delta n)(p_o + \Delta p) \qquad (2\text{-}3)$$

$$G = G_{th} + G_L \qquad (2\text{-}4)$$

where Δn and Δp are the excess-carrier densities, defined by

$$\Delta n \equiv n - n_o \qquad \Delta p \equiv p - p_o$$

Under steady-state illumination, the rates expressed in Eqs. (2-3) and (2-4) must be equal. Therefore, we have

$$G_L = R - G_{th} \equiv U \qquad (2\text{-}5)$$

where U is the net recombination rate. Substituting Eqs. (2-2) and (2-3) into Eq. (2-5) yields

$$G_L = B(n_o + \Delta n)(p_o + \Delta p) - Bn_o p_o \qquad (2\text{-}6)$$

Since $\Delta n = \Delta p$, so that space-charge neutrality is maintained, Eq. (2-6) can be rewritten as

$$B\,\Delta p^2 + B(n_o + p_o)\,\Delta p = G_L \qquad (2\text{-}7)$$

In low-level injection, we have $\Delta p \ll n_o + p_o$. As a result, Eq. (2-7) can be simplified to

$$\Delta p = \frac{G_L}{B(n_o + p_o)} = G_L \tau = U\tau \qquad (2\text{-}8)$$

where

$$\tau \equiv \frac{\Delta p}{U} = \frac{1}{B(n_o + p_o)} \qquad (2\text{-}9)$$

The quantity τ as defined in Eq. (2-9) is called the *carrier lifetime* of the excess carriers. Thus, if the generation rate is $10^{18}\,\text{cm}^{-3}\,\text{s}^{-1}$ and the minority-carrier lifetime is $100\,\mu\text{s}$, the excess-carrier density is $10^{14}\,\text{cm}^{-3}$, according to Eq. (2-8). Physically, τ represents the average time a hole remains free before it recombines with an electron. In a semiconductor with a direct energy-band structure such as GaAs, the band-to-band recombination is the dominant mechanism. The energy released by the recombination is emitted in the form of photons, and the process is radiative. For this reason, GaAs is used to fabricate light-emitting devices, discussed in Chap. 6.

Indirect Recombination

In silicon or germanium, the lowest energy in the conduction band does not coincide with the highest energy in the valence band in the momentum space (Fig. 1-5b). Therefore, an electron in the conduction band making a downward transition needs to change its momentum as well as its energy to satisfy the conservation principles. Consequently, the probability of direct band-to-band recombination in silicon or germanium is very low. In fact, carriers in silicon recombine much more readily through *recombination centers*, or *traps*, situated inside the forbidden gap. A recombination center may capture a hole and then an electron sequentially to complete a recombination cycle. Recombination centers may be caused by impurity atoms or defects in the crystal lattice. For example, it is known that gold in silicon and copper in GaAs act as recombination centers. It is also found that radiation-induced defects in solar cells in space applications increase the number of recombination centers and reduce the cell efficiency. Effects of recombination centers on device performance will be discussed in later chapters.

A schematic diagram showing the various steps involved in the recombination via an intermediate center is shown in Fig. 2-2, where the energy of these centers is E_t and the density of the centers is N_t. Four possible transition processes are shown: (1) an electron is captured by an empty center, (2) an electron is emitted from an occupied center, (3) an occupied center captures a hole, and (4) an empty center emits a hole. Since the

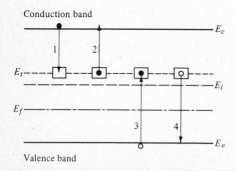

Figure 2-2 Generation and recombination through intermediate states.

probability of occupancy of a center follows the Fermi-Dirac function expressed by Eq. (1-4), we have

$$f_t^o = (e^{(E_t - E_f)/kT} + 1)^{-1} \qquad (2\text{-}10)$$

The number of occupied centers is therefore $N_t f_t^o$, and the number of empty centers is $N_t(1 - f_t^o)$. The superscript ^{o} specifies the equilibrium condition.

By following the argument used in deriving the direct-recombination rate, we find the capture rate of an electron by an empty center to be

$$R_1 = c_n n N_t(1 - f_t) \qquad (2\text{-}11)$$

where n is the electron density in the conduction band and c_n is the capture coefficient,† which has a typical value of 10^{-8} cm^3/s. The rate of emitting an electron to the conduction band from an occupied center is given by

$$R_2 = e_n N_t f_t \qquad (2\text{-}12)$$

where e_n is the emission coefficient. By analogy, the capture and emission rates of holes by the centers may be written as

$$R_3 = c_p p N_t f_t \qquad (2\text{-}13)$$

$$R_4 = e_p N_t(1 - f_t) \qquad (2\text{-}14)$$

where c_p and e_p are the hole capture and emission coefficients respectively.

Under thermal equilibrium, the number of electrons emitted from the centers must be the same as that of the captured electrons, that is, $R_1 = R_2$. Equating Eqs. (2-11) and (2-12) and using Eq. (1-21), we obtain

$$c_n n_i e^{(E_f - E_i)/kT} N_t(1 - f_t) = e_n N_t f_t \qquad (2\text{-}15)$$

Furthermore, since

$$\frac{1 - f_t}{f_t} = e^{(E_t - E_f)/kT} \qquad (2\text{-}16)$$

we have $\qquad e_n = c_n n_i e^{(E_t - E_i)/kT} \qquad (2\text{-}17)$

Using the same procedure and Eq. (1-22), we find the hole emission rate

$$e_p = c_p n_i e^{(E_i - E_t)/kT} \qquad (2\text{-}18)$$

Since the capture and emission probabilities are independent of equilibrium and nonequilibrium conditions, Eqs. (2-17) and (2-18) are valid at nonequilibrium although they have been derived under the thermal-equilibrium condition. However, the probability of occupancy expressed by Eq. (2-10) is not applicable at nonequilibrium because the Fermi level E_f is meaningful only at equilibrium.

---

† The electron capture coefficient is the product of the thermal velocity v_{th} and the electron-capture cross section σ_{cn}. σ_{cn} is a measure of the closeness of an electron to an empty center, and has a typical value of 10^{-15} cm^2. A hole-capture cross section is defined in the same manner.

Let us now consider the nonequilibrium case by applying an external source of energy, e.g., a light source, so that a generation rate G_L exists uniformly throughout the semiconductor. Under steady-state conditions, the electrons entering and leaving the conduction band in Fig. 2-3 must be equal. This is called the *principle of detailed balance*, and it yields

$$G_L = R_1 - R_2 \tag{2-19}$$

Similarly, the detailed balance of holes in the valence band leads to

$$G_L = R_3 - R_4 \tag{2-20}$$

Equating Eqs. (2-19) and (2-20), we have

$$R_1 - R_2 = R_3 - R_4 \tag{2-21}$$

We can now substitute Eqs. (2-11) to (2-14) into the foregoing expression to obtain

$$c_n n N_t (1 - f_t) - e_n N_t f_t = c_p p N_t f_t - e_p N_t (1 - f_t) \tag{2-22}$$

By assuming $c_n = c_p = c$ and using Eqs. (2-17) and (2-18), we can derive f_t from Eq. (2-22), yielding

$$f_t = \frac{n + n_i e^{(E_i - E_t)/kT}}{n + p + 2n_i \cosh[(E_t - E_i)/kT]} \tag{2-23}$$

Therefore, the net recombination rate is

$$U \equiv R_1 - R_2 = \frac{c N_t (pn - n_i^2)}{n + p + 2n_i \cosh[(E_t - E_i)/kT]} \tag{2-24}$$

$u < 0$ generation

$u > 0$ recombination

From Eq. (2-24), we find that at equilibrium, that is, $pn = n_i^2$, the net recombination rate is zero. It is also interesting to note that the maximum recombination rate occurs when $E_t = E_i$. In other words, the most efficient recombination centers are located at or near the center of the forbidden gap. Away from the level E_i, the centers would be less effective because it is more probable to capture one type of carrier but less probable to capture the other type.

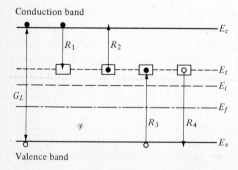

Figure 2-3 Generation and recombination processes under illumination.

When we apply the foregoing result to an n-type silicon with a center located at E_i, the net recombination rate for low-level injection ($n_n \approx n_{no}$) becomes

$$U = c_p N_t \frac{p_n n_n - n_i^2}{n_n + p_n + 2n_i} \approx c_p N_t (p_n - p_{no}) \tag{2-25}$$

The last expression in Eq. (2-25) is obtained by using $n_i^2 = n_{no} p_{no}$ and $n_n = n_{no} \gg p_n + 2n_i$ for an n-type semiconductor. From the definition of the carrier lifetime in Eq. (2-9), we have

$$\tau_p = \frac{1}{c_p N_t} \tag{2-26}$$

The lifetime τ_p is independent of the capture of the majority carriers since they are abundant and the limitation of recombination rate is the capture of minority carriers. Therefore, τ_p is also called the *minority-carrier lifetime*. Similarly, the minority-carrier lifetime in a p-type semiconductor is given by

$$\tau_n = \frac{1}{c_n N_t} \tag{2-27}$$

The carrier lifetimes expressed in Eqs. (2-26) and (2-27) are valid for low-level injection. The general expression for the carrier lifetime is rather complicated and will not be presented here. It is observed in Eqs. (2-26) and (2-27) that the minority-carrier lifetime is inversely proportional to N_t. In practical device design, a small lifetime may be desirable for high-speed switching applications. Nanosecond (10^{-9} s) switching speed is realizable by using gold doping to increase N_t in silicon diodes and transistors.

2-4 TRANSIENT RECOMBINATION RESPONSE

The physical meaning of the carrier lifetime is best illustrated by the transient response after the sudden removal of the light source. The general equation that describes the rate of change of excess carriers is

$$\frac{d\Delta p}{dt} = G_L - U \tag{2-28}$$

which can be rewritten for direct recombination under low-level injection as

$$\frac{d\Delta p}{dt} = G_L - \frac{\Delta p}{\tau_p} \tag{2-29}$$

The steady-state excess-carrier concentration is $\Delta p = G_L \tau_p$ at $t \leq 0$ because the derivative of the excess carrier is zero in Eq. (2-29). If we use this relation as the initial condition, Eq. (2-29) can be solved to yield

$$\Delta p = G_L \tau_p e^{-t/\tau_p} \tag{2-30}$$

Figure 2-4 shows the plot of Eq. (2-30). The analysis of the transient response

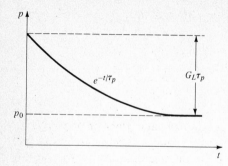

Figure 2-4 Transient decay of excess carriers after removal of light source.

for the indirect recombination in the general case is more involved. It does not necessarily lead to a solution with a simple exponential as expressed in Eq. (2-30). However, if we assume that the conditions leading to Eq. (2-26) prevail, Eq. (2-30) is a good approximation for the transient decay of carriers in indirect recombination as well.

The *photoconductivity* decay experiment is shown in Fig. 2-5. The conductivity is given by

$$\sigma = q[\mu_n n(t) + \mu_p p(t)] \tag{2-31}$$

Since $n(t) = n_o + \Delta n(t)$, $p(t) = p_o + \Delta p(t)$ and $\Delta n = \Delta p$, we have

$$\sigma(t) = \sigma_o + q(\mu_n + \mu_p)\,\Delta p(t) \tag{2-32}$$

where

$$\sigma_0 = q(\mu_n n_o + \mu_p p_o) \tag{2-33}$$

In the experimental setup, the voltage measured by the oscilloscope is given by

$$v = \frac{R_s V}{R + R_s} \approx \frac{R_s V}{R} \qquad \text{if } R \gg R_s \tag{2-34}$$

Using $R = L/\sigma A$, we rewrite Eq. (2-34) as

$$v = \frac{\sigma A R_s V}{L} \tag{2-35}$$

Since σ varies linearly with Δp in Eq. (2-32) and Δp varies exponentially with

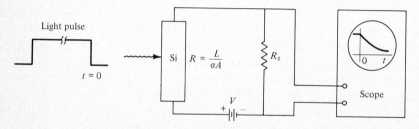

Figure 2-5 The photoconductivity decay experiment.

t/τ_p in Eq. (2-30), we can obtain the carrier lifetime from the oscilloscope waveform.

2-5 SURFACE RECOMBINATION

The recombination processes described previously take place in the bulk of the semiconductor. It is conceivable that similar carrier activity occurs at the semiconductor surface. In fact, the presence of discontinuity in the lattice structure at the surface introduces a large number of energy states in the forbidden gap. These energy states, called *surface states*, greatly enhance the recombination rate at the surface region. In addition to the surface states, other imperfections exist resulting from adsorbed ions, molecules, or mechanical damage in the layer next to the surface. The adsorbed ions, for example, may be charged so that a space-charge layer is formed near the surface. Regardless of the origin of the surface imperfections, the recombination rate at the surface per unit area can be written in analogy to Eq. (2-25) as

$$U_s = cN_{ts}[p_n(0) - p_{no}] \qquad (2\text{-}36)$$

where $p_n(0)$ is the average surface minority-carrier density and N_{ts} is the average recombination-center density *per unit area* in the surface layer. Since the product cN_{ts} has the dimension of centimeters per second, it is called the *surface recombination velocity S*. Thus Eq. (2-36) becomes

$$U_s = S[p_n(0) - p_{no}] \qquad (2\text{-}37)$$

Equation (2-37) defines a hole current density qU_s flowing into the surface from the bulk when excess carriers exist. The higher surface recombination rate leads to a lower carrier concentration at the surface. This gradient of holes yields a diffusion current density, Eq. (1-49), which is equal to the surface recombination current

$$-qD_p\frac{dp_n}{dx}\bigg|_{x=0} = qU_s = qS[p_n(0) - p_{no}] \qquad (2\text{-}37a)$$

However, an equal number of electrons is needed at the surface to accomplish recombination. Consequently, the electron and hole currents exactly cancel out, so that the net surface current is zero.

The numerical value of the surface recombination velocity may vary over a wide range depending on atmospheric conditions and the surface treatment received. In early development of transistors, surface leakage current and breakdown were serious problems in device performance. Modern planar silicon devices using silicon oxide passivation have alleviated this difficulty.

2-6 ELECTROSTATIC FIELD AND POTENTIALS

The electric field $\mathscr{E}$ is defined as the negative gradient of the potential V as expressed by

$$\mathscr{E} = -\frac{dV}{dx} \tag{2-38}$$

and the potential is related to the potential energy by

$$-qV = E \tag{2-39}$$

In a semiconductor, the potential energy for an electron in the conduction band is its lowest possible energy E_c. If an electron is located above E_c, the excess energy can only be in the form of kinetic energy. Similarly, the energy E_v denotes the potential energy of holes in the valence band, and a hole located below E_v has a kinetic-energy component. If there is no external applied field, Fig. 2-6a illustrates the relationship between energies and carrier positions in the energy-band diagram. When an external electric field is applied across the semiconductor as shown in Fig. 2-6b, the band diagram is tilted to impart kinetic energy to electrons and holes. Since E_c and E_v are always in parallel with E_i and we are interested only in the potential gradient, we can express the electric field as

$$\mathscr{E} = \frac{1}{q}\frac{dE_i}{dx} = -\frac{d\psi}{dx} \tag{2-40}$$

where

$$\psi \equiv -\frac{E_i}{q} \tag{2-41}$$

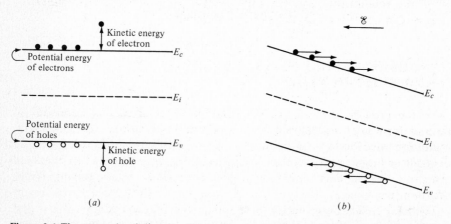

(a) (b)

Figure 2-6 The energy-band diagram of a semiconductor under (a) zero electric field and (b) an electric field.

The symbol ψ denotes the *electrostatic potential*. Similarly, we define the *Fermi potential* as

$$\varphi \equiv -\frac{E_f}{q} \qquad (2\text{-}42)$$

Substitution of Eqs. (2-41) and (2-42) into Eqs. (1-21) and (1-22) yields

$$n = n_i e^{(\psi - \varphi)/V_T} \qquad (2\text{-}43)$$

$$p = n_i e^{(\varphi - \psi)/V_T} \qquad (2\text{-}44)$$

where

$$V_T \equiv \frac{kT}{q}$$

At 300 K, V_T assumes a value of 25.8 mV. The foregoing equations are very convenient in some applications. For example, we can use them to verify the Einstein relationship as follows. Under thermal equilibrium, the Fermi potential is constant, and it can be taken as the zero reference. Therefore, Eq. (2-44) is simplified to

$$p = n_i e^{-\psi/V_T} \qquad (2\text{-}45)$$

The hole current in the semiconductor expressed in Eq. (1-52) must be zero at equilibrium; thus,

$$I_p = qA\left(\mu_p p \mathscr{E} - D_p \frac{dp}{dx}\right) = 0 \qquad (2\text{-}46)$$

Differentiating Eq. (2-45) and substituting p, dp/dx, and Eq. (2-40) into Eq. (2-46), we find

$$\frac{D_p}{\mu_p} = V_T = \frac{kT}{q} \qquad (2\text{-}47)$$

This is the relationship expressed in Eq. (1-50). A similar relationship for electrons can be verified by the same procedure.

2-7 INHOMOGENEOUS SEMICONDUCTOR AND BUILT-IN FIELD

During the growth of semiconductor crystals, special care is necessary to obtain uniform impurity distribution throughout the semiconductor. However, nonuniform distributions of impurities are sometimes introduced either accidentally or intentionally. An inhomogeneous impurity distribution results in a *built-in* electric field and is a useful technique for improving device performance.

The relationship between the built-in field and doping distribution can best be understood by means of the energy-band diagram. Let us consider an *n*-type silicon wafer with impurity distribution shown in Fig. 2-7a. The

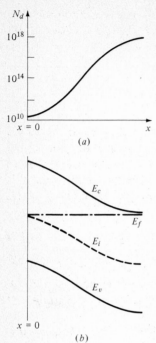

(a)

(b)

Figure 2-7 (a) Donor distribution and (b) corresponding energy-band diagram of a nonuniformly doped semiconductor.

impurity concentration is limited to below 10^{18} cm^{-3}, so that no part of the semiconductor is degenerate. The energy-band diagram is constructed by first taking the Fermi level E_f as the reference since E_f is constant at equilibrium. Assuming that all impurity atoms are ionized, the electron density n is equal to $N_d(x)$ in Fig. 2-7a. Using this relation, we can solve Eq. (1-21) for the midgap energy E_i to yield

$$E_i = E_f - kT \ln \frac{N_d(x)}{n_i} \tag{2-48}$$

where $E_i = E_f$ if $N_d = n_i$. For any value of N_d greater than n_i, the midgap energy E_i is below E_f and the difference of $E_f - E_i$ increases with increasing N_d. Thus E_i for the inhomogeneous wafer is illustrated in Fig. 2-7b. Since the band-gap energy E_g is a constant for a nondegenerate semiconductor, the energy levels E_c and E_v are plotted in parallel to E_i.

Using Eqs. (2-41) and (2-48) with E_f taken to be zero, we can write the electrostatic potential as

$$\psi = V_T \ln \frac{N_d(x)}{n_i} \tag{2-49}$$

Since the electric field is the gradient of the potential, we have

$$\mathscr{E} = -\frac{V_T}{N_d} \frac{dN_d}{dx} \tag{2-50}$$

From Eq. (2-50) we observe that a nonuniform spatial distribution of impurities produces a built-in electric field in the semiconductor. This built-in field is frequently introduced to improve device characteristics.

2-8 QUASI-FERMI LEVELS

Under thermal equilibrium, the densities of electrons and holes are specified by the position of the Fermi level through Eqs. (1-21) and (1-22), but these equations do not apply in the nonequilibrium case because injected carriers render the Fermi level meaningless. Under nonequilibrium conditions, it is possible to define two quantities E_{fn} and E_{fp} to replace the Fermi level in Eqs. (1-21) and (1-22) such that

$$n = n_i e^{(E_{fn}-E_i)/kT} = n_i e^{(\psi-\varphi_n)/V_T} \tag{2-51}$$

$$p = n_i e^{(E_i-E_{fp})/kT} = n_i e^{(\varphi_p-\psi)/V_T} \tag{2-52}$$

where E_{fn} and E_{fp} are called the *quasi-Fermi levels* for electrons and holes, respectively, and φ_n and φ_p are the corresponding *quasi-Fermi potentials*.

Example Calculate the quasi-Fermi levels at 300 K (room temperature) for a semiconductor with $N_a = 10^{16}\,\mathrm{cm}^{-3}$, $\tau_n = 10\,\mu\mathrm{s}$, $n_i = 10^{10}\,\mathrm{cm}^{-3}$, and $G_L = 10^{18}\,\mathrm{cm}^{-3}\,\mathrm{s}^{-1}$.

SOLUTION $\qquad\qquad \Delta n = \Delta p = \tau_n G_L = 10^{13}\,\mathrm{cm}^{-3}$

Therefore $\qquad p = p_o + \Delta p = N_a + \Delta p \approx 10^{16}\,\mathrm{cm}^{-3}$

$$n = n_o + \Delta n = \frac{n_i^2}{N_a} + \Delta n = 10^4 + 10^{13} \approx 10^{13}\,\mathrm{cm}^{-3}$$

Rewrite Eq. (2-51) and use $kT = 26$ meV at 300 K to obtain

$$E_{fn} - E_i = kT \ln \frac{n}{n_i} = 0.026 \ln \frac{10^{13}}{10^{10}} = 0.18\,\mathrm{eV}$$

Similarly, we obtain the quasi-Fermi level for a hole by using Eq. (2-52):

$$E_i - E_{fp} = 0.026 \ln \frac{10^{16}}{10^{10}} = 0.36\,\mathrm{eV}$$

Note that E_{fn} is above E_i and E_{fp} is below E_i.

The *pn* product under nonequilibrium is therefore

$$pn = n_i^2 \exp \frac{\varphi_p - \varphi_n}{V_T} \tag{2-53}$$

At equilibrium, $\varphi_n = \varphi_p = \varphi$ and $pn = n_i^2$. With the increase of injection, the difference of $E_{fn} - E_i$ in the first expression of Eq. (2-51) increases with n;

thus E_{fn} moves away from E_i toward E_c. Similarly, from Eq. (2-52), E_{fp} moves away from E_i toward E_v with increasing injection. Using the second expressions of Eqs. (2-51) and (2-52), we can rewrite the current transport equations in simpler forms. Differentiation of Eq. (2-51) leads to

$$\frac{dn}{dx} = \frac{n}{V_T}\left(\frac{d\psi}{dx} - \frac{d\varphi_n}{dx}\right)$$

(2-54)

When we substitute Eqs. (1-50), (2-40), and (2-54) into Eqs. (1-51) and (1-52), the current transport equations become

$$J_n = \frac{I_n}{A} = -q\mu_n n \frac{d\varphi_n}{dx} = -\sigma_n(x)\frac{d\varphi_n}{dx}$$

(2-55)

$$J_p = \frac{I_p}{A} = -q\mu_p p \frac{d\varphi_p}{dx} = -\sigma_p(x)\frac{d\varphi_p}{dx}$$

(2-56)

where J_n and J_p are the electron- and hole-current density respectively. These are the *modified Ohm's laws*, in which the combined effects of carrier drift and diffusion are incorporated. From the modified Ohm's laws, we observe that the constant quasi-Fermi levels represent the zero-current condition.

2-9 BASIC GOVERNING EQUATIONS IN SEMICONDUCTORS

The mechanisms of carrier transport, generation, and recombination in a semiconductor have been discussed in the previous sections. These mechanisms are related to one another by the condition of current continuity. Figure 2-8 depicts the one-dimensional hole-current flow within a small increment Δx in a semiconductor *having unity cross-sectional area*. The continuity of carrier flow (the fact that charge can neither be created nor destroyed) requires that the rate of change of the number of holes in Δx be equal to the holes recombining plus holes leaving the increment. The net hole flow out of the increment is

$$\frac{J_p(x + \Delta x)}{q} - \frac{J_p(x)}{q} = \frac{1}{q}\frac{\partial J_p}{\partial x}\Delta x$$

(2-57)

The net recombination in the increment Δx, according to Eqs. (2-25) and

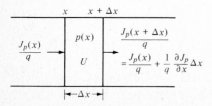

Figure 2-8 Current continuity in a semiconductor.

(2-26), is

$$U \Delta x = \frac{p - p_o}{\tau_p} \Delta x \tag{2-58}$$

and the rate of change of holes in the increment is

$$-\frac{\partial p}{\partial t} \Delta x \tag{2-59}$$

The negative rate of change indicates a decrease of carriers. By adding Eqs. (2-57) and (2-58) and setting it equal to Eq. (2-59), we obtain the equation for the continuity of hole current as

$$\frac{1}{q}\frac{\partial J_p}{\partial x} + \frac{p - p_o}{\tau_p} = -\frac{\partial p}{\partial t} \tag{2-60}$$

Similarly, the continuity of electron current is

$$-\frac{1}{q}\frac{\partial J_n}{\partial x} + \frac{n - n_o}{\tau_n} = -\frac{\partial n}{\partial t} \tag{2-61}$$

The current components I_n and I_p are related to the carrier concentrations by Eqs. (1-51) and (1-52), repeated here for convenient reference:

$$J_p = \frac{I_p}{A} = q\left(\mu_p p \mathscr{E} - D_p \frac{dp}{dx}\right) \tag{2-62}$$

$$J_n = \frac{I_n}{A} = q\left(\mu_n n \mathscr{E} + D_n \frac{dn}{dx}\right) \tag{2-63}$$

Example Find the steady-state minority-carrier distribution in a semi-infinite homogeneous slab of n-type semiconductor if a density of excess carriers $p_n(0) - p_{no}$ is generated at $x = 0$.

SOLUTION In the homogeneous semiconductor, $\mathscr{E} = 0$, and Eq. (2-62) becomes

$$J_p = -qD_p\frac{dp_n}{dx}$$

Substituting this expression into Eq. (2-60) for the time-independent case yields

$$D_p\frac{d^2p_n}{dx^2} - \frac{p_n - p_{no}}{\tau_p} = 0$$

the general solution of which is

$$p_n - p_{no} = Ae^{x/L_p} + Be^{-x/L_p}$$

where $L_p = \sqrt{D_p\tau_p}$ and is known as the *diffusion length* for holes. Since no excess carriers are generated at $x = \infty$, we have

$$p_n - p_{no} = 0 \qquad \text{at } x = \infty \text{ or } A = 0$$

Using $p_n = p_n(0)$ at $x = 0$, we obtain $B = p_n(0) - p_{no}$. Thus, the solution is

$$p_n = p_{no} + [p_n(0) - p_{no}]e^{-x/L_p}$$

This equation is plotted in Fig. 2-9.

In addition to the foregoing equations, we must consider the effect of space charge on the electric field. The semiconductor as a whole is charge-neutral. However, localized charged regions do exist. We can use Poisson's equation to describe the charged regions and their relationship to the electric field by

$$\frac{d\mathscr{E}}{dx} = \frac{\rho}{K\epsilon_o} \tag{2-64}$$

where ρ = net space-charge density
$\quad K$ = dielectric constant
$\quad \epsilon_o$ = permittivity of free space

The relationship between the charge distribution and the electric field is obtained by integrating Poisson's equation

$$\mathscr{E} = \int \frac{\rho \, dx}{K\epsilon_o} \tag{2-65}$$

Therefore, knowledge of the charge density leads to the magnitude of the electric field, as shown in Fig. 2-10, within a constant.

The net space charge in a semiconductor is the sum of the positive charges minus the sum of negative charges. Since an ionized donor atom has a fixed positive charge and an ionized acceptor atom has a fixed negative

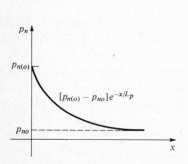

Figure 2-9 The steady-state minority-carrier distribution for a semiconductor slab with injection at $x = 0$.

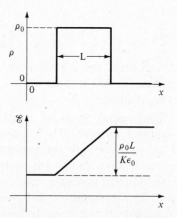

Figure 2-10 A box-type charge distribution and the corresponding electric field.

charge, the net space charge is given by

$$\rho = q[p + N_d - (n + N_a)] \tag{2-66}$$

Substituting Eqs. (2-40) and (2-66) into Eq. (2-64), and rearranging the result, we obtain

$$\frac{d^2\psi}{dx^2} = \frac{q}{K\epsilon_o} [(n - p) + (N_a - N_d)] \tag{2-67}$$

Equations (2-60) to (2-63) and Eq. (2-67) constitute a complete set of equations which describe the carrier, current, and field distributions. They can be solved when appropriate boundary and initial conditions are available. In most cases, these equations will be simplified on the basis of physical approximations before a solution is attempted.

PROBLEMS

2-1 (a) From the definition of direct recombination determine the average time an electron stays in the conduction band and the average time a hole stays in the valence band.

(b) What is the relationship between the carrier lifetime τ and the average times obtained in (a). Discuss this relationship for intrinsic and extrinsic semiconductors.

2-2 Derive the carrier lifetime in direct recombination under steady-state high-level injection conditions.

2-3 Show that the indirect recombination rate expressed by Eq. (2-24) can be rewritten as

$$U = \frac{pn - n_i^2}{(n + n_1)\tau_{po} + (p + p_1)\tau_{no}}$$

where $\qquad n_1 = N_c \exp\left(-\frac{E_c - E_t}{kT}\right) \qquad$ and $\qquad p_1 = N_v \exp\left(-\frac{E_t - E_v}{kT}\right)$

2-4 Calculate the electron and hole concentration under steady-state illumination in an n-type silicon with $G_L = 10^{16} \text{ cm}^{-3} \text{ s}^{-1}$, $N_d = 10^{15} \text{ cm}^{-3}$, and $\tau_n = \tau_p = 10 \ \mu\text{s}$.

2-5 The energy level of electron traps can be measured by the thermally stimulated current (TSC) experiment, in which the semiconductor is first cooled to a very low temperature and then exposed to a strong light filling all the traps. After the steady state is reached, the light source is removed. The semiconductor is heated at a constant rate slowly in the dark, and its conductance is measured at different temperatures. The dark current must be subtracted in the measurement. A conductance peak σ_m is measured at temperature T_m, at which point the Fermi level aligns with E_t to release all the trapped electrons.

(*a*) Show that the trap level is given by

$$E_c - E_t = kT_m \ln q \frac{\mu_n N_c(T_m)}{\sigma_m}$$

(*b*) A TSC measurement in a GaAs crystal (with cross-sectional area of 0.1 cm² and length of 1 cm) yields $T_m = -33°C$, and the thermally stimulated current peaks at 140 nA when biased with a voltage of 100 V. Calculate E_t.

2-6 In each of the following cases, obtain an expression for the indirect recombination rate to show whether there is net recombination or generation of carriers.

(*a*) The electron and hole densities are much smaller than n_i.

(*b*) The electron and hole densities are equal and are much larger than n_i.

2-7 Derive the recombination rate for the following recombination-center distributions.

(*a*) The density of recombination centers D_t (cm⁻² eV⁻¹) is uniformly distributed in energy between E_v and E_c.

(*b*) The density of recombination centers increases linearly with energy starting from $D_t = 0$ at $E = E_i$ to $2D_t$ (cm⁻² eV⁻¹) at E_c and E_v.

2-8 (*a*) Show that the change of photoconductance of a semiconductor with illumination is

$$\Delta\sigma = q(\tau_n\mu_n + \tau_p\mu_p)G_L$$

(*b*) Compare the sensitivity of devices using GaAs and Si. Assume $\tau_n = \tau_p = 1$ μs for Si and $\tau_n = \tau_p = 10$ ns for GaAs.

2-9 Following the stated procedure, derive the Einstein relationship for both electrons and holes.

2-10 Derive the electric field in an *n*-type semiconductor if (*a*) $N_d = ax$, where *a* is a constant; (*b*) $N_d = N_o e^{-ax}$.

2-11 Plot the energy-band diagrams of Prob. 2-10*a* and *b*.

2-12 (*a*) Calculate the conductivity and quasi-Fermi levels for a silicon sample with $N_d = 10^{15}$ cm⁻³, $\tau_p = 1$ μs, and $G_L = 5 \times 10^{19}$ cm⁻³ s⁻¹.

(*b*) Find the value of G_L that produces 10^{15} holes/cm³. What are the conductivity and quasi-Fermi levels?

2-13 The injection of carriers gives rise to the splitting of the electron and hole quasi-Fermi levels. Show that the nonequilibrium product (*pn*) of a semiconductor with an energy gap of E_g is the same as the equilibrium product $p_o n_o$ of a semiconductor with a band gap of $E_g - (E_{fn} - E_{fp})$.

2-14 Sketch the electric field diagram for the charge distribution shown in Fig. P2-14.

2-15 (*a*) A semi-infinite slab of *n*-type silicon is uniformly illuminated with a generation rate of G_L. Obtain the hole-continuity equation under these conditions.

(*b*) If the surface recombination velocity is S at $x = 0$, solve the new

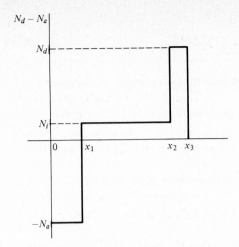

Figure P2-14.

continuity equation to show that the steady-state hole distribution is given by

$$p_n(x) = p_{no} + \tau_p G_L \left(1 - \frac{\tau_p S e^{-x/L_p}}{L_p + S\tau_p} \right)$$

2-16 (a) Write the continuity equations and current equations for electrons and holes and Poisson's equation in the form of vector equations.

(b) By assuming that $p - p_o = n - n_o = \Delta n$ and $\tau_n = \tau_p = \tau$, and by using the vector identity $\nabla \cdot a\mathbf{x} = a\nabla \cdot \mathbf{x} + \mathbf{x} \cdot \nabla a$, show that

$$\frac{\partial \Delta n}{\partial t} = -\frac{\Delta n}{\tau} + \mu^* \mathscr{E} \cdot \nabla(\Delta n) + D^* \nabla^2(\Delta n)$$

where

$$\mu^* = \frac{p - n}{p/\mu_n + n/\mu_p} \qquad D^* = \frac{p + n}{p/D_n + n/D_p}$$

This is known as the *van Roosbroech ambipolar equation.*

(c) Show that the majority-carrier effects are negligible for $p \gg n$ or $n \gg p$ by using the ambipolar equation.

ADDITIONAL READINGS

Grove, A. S.: "Physics and Technology of Semiconductor Devices," Wiley, New York. Chapter 5 on recombination theory.

Jonscher, A. K.: "Principles of Semiconductor Device Operation," Wiley, New York, 1960. Chapters 2 and 3 on recombination and transport of carriers.

Many, A., Y. Goldstein, and N. B. Grove: "Semiconductor Surfaces," Wiley, New York, 1965. Covers the topic of surface recombination in depth.

Shockley, W., and W. T. Read: Statistics of the Recombination of Holes and Electrons, *Phys. Rev.,* **87**: 835 (1952). The classic paper on recombination statistics.

CHAPTER
THREE

DEVICE-FABRICATION TECHNOLOGY

The characteristics of a semiconductor device are very much influenced by its fabrication processes. In the early development of the junction transistor, the grown-junction and alloy-junction processes were employed. These methods were limited by the lack of control of device dimensions and impurity concentrations. Planar technology, introduced in 1960, made the rapid development of modern integrated circuits possible. Planar technology comprises five independent processes: the epitaxial growth, oxidation, impurity diffusion, oxide etching and pattern definition using photolithography, and metallization. In this chapter, we describe the basic processes involved in fabricating semiconductor devices, and our emphasis is on the planar technology with silicon as the semiconductor. In addition, ion implantation is presented as an alternative doping method, and silicon nitride is shown as a masking technique. Simple evaluation of a doped layer by resistivity and junction-depth measurements is also discussed.

3-1 SUBSTRATE PREPARATION AND JUNCTION FORMATION

Elemental silicon is obtained either by chemical decomposition of compounds such as $SiCl_4$ and $SiHCl_4$ or by reduction of silicon dioxide (common sand) with carbon in a furnace. After the element is isolated, it is purified, melted, and cast into ingots. The silicon thus obtained is in polycrystalline form, and the purity is rather poor. For conversion into high-quality single-crystal silicon, the polycrystalline silicon is placed in a quartz crucible set inside a graphite susceptor and melted by radio-frequency induction heating. The complete experimental setup is shown in Fig. 3-1. A single-crystal silicon with desired orientation, known as the *seed* crystal, is dipped into the melt. A small part of the seed is allowed to melt back to eliminate possible surface imperfection. Subsequently, the seed is rotated and at the same time very slowly withdrawn from the melt. As the seed is kept at a temperature below the melting point, the molten silicon solidifies when it comes in contact with the seed. Thus, a cylindrical rod of single-crystal silicon, typically 2 to 3 in in diameter and about 10 in long, is *pulled*, as shown in Fig. 3-1. This crystal growth technique, known as the *Czochralski method*, is the approach most commonly used to obtain high-quality device-grade silicon.

The single-crystal rod is cut into thin slices, commonly known as *wafers*, with a diamond saw. Before cutting, the rod must be oriented to obtain the desired crystallographic plane. Fortunately, the orientation of the pulled crystal is the same as the seed, which is accurately oriented before the growth. After being cut, the thin wafers are mechanically lapped to remove

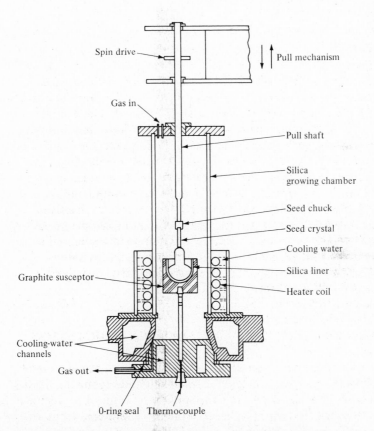

Figure 3-1 A typical crystal-pulling apparatus. (*After Runyan [1].*)

gross saw damage by means of a lapping machine. Finally, the wafers are mechanically or chemically polished to obtain a mirrorlike finish. A polished wafer is used as the *substrate* on which subsequent processing steps are performed to produce semiconductor devices or integrated circuits.

During the initial crystal growth, a silicon wafer with (111) orientation is easier to grow defect-free than the (100) crystal. For this reason, (111) wafers are less expensive and are employed as the substrate for most device and integrated-circuit fabrication. The disadvantages of the (111)-oriented crystals will be discussed in later chapters.

Impurity atoms can be added in the melt to make it a *p*-type or an *n*-type crystal. Let us assume that the melt is originally *p* type so that *p*-type silicon is grown initially. During the crystal growth, if we add donor impurities in the melt to change it into *n* type, the subsequently grown crystal will be *n* type and a *p-n* junction is formed, as shown in Fig. 3-2*a*. This *grown-junction* method was used in early investigations of *p-n* junction properties.

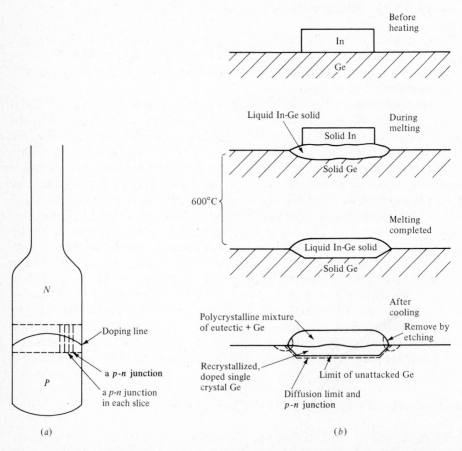

Figure 3-2 Junction-formation processes: (*a*) the grown junction and (*b*) the alloy junction.

The grown-junction method is not suitable for large-scale production, and early transistors were mostly fabricated by the *alloy-junction* method, illustrated in Fig. 3-2*b*. In this method, a metallic impurity element of the appropriate type, for example, *p*-type indium for an *n*-type germanium, is placed on the substrate, and the assembly is heated so that a liquid alloy is formed with Ge. When the semiconductor is cooled, a recrystallized *p* region is formed under the metal and a *p-n* junction is obtained. The alloy process is simple and efficient in making a single *p-n* junction, but it is not easy to control the spacing of two adjacent junctions, as required in a transistor.

At present, the overwhelming majority of junction devices are fabricated by *solid-state diffusion*. In this method, a *p*-type impurity is diffused into an *n*-type substrate in a high-temperature furnace. Alternatively, a junction can also be formed by either *ion implantation* or *epitaxial growth*. Details of these modern techniques will be described in later sections.

3-2 PLANAR TECHNOLOGY

Silicon devices are mostly fabricated by *planar technology*, involving epitaxy, oxidation, impurity diffusion, photolithography, and metallization. This section presents an overall view of the processes, using the fabrication of a *p-n* junction diode as an example.

The fabrication steps of a planar *p-n* diode are shown in Fig. 3-3. Using an n^+ silicon as the starting substrate, a thin layer of *n* silicon is grown by the epitaxial process (Fig. 3-3*a* and *b*). Subsequently, a SiO_2 layer is formed by thermal oxidation (Fig. 3-3*c*). The oxidation step is followed by the photolithographic process. A thin film of organic polymer called *photoresist* is first put on in liquid form to cover the oxide and then dried in a baking oven. The photoresist material is soluble in a special solvent unless it is polymerized. A mask with transparent and opaque regions is now placed on the photoresist. When the entire structure is exposed to ultraviolet light, the exposed photoresist is polymerized but the unexposed area is removable (Fig. 3-3*d*). Etching of the oxide follows the removal of unexposed photoresist. Then, a *p*-type impurity is diffused through the oxide opening to make a *p-n* junction (Fig. 3-3*e*). Finally a thin metal film is deposited and etched to obtain the structure shown in Fig. 3-3*f*. Using the basic steps described here, we can build a multijunction transistor or a complex integrated circuit. Since the diffusion, epitaxy, and oxidation steps control the device characteristics, we shall discuss these three processes in more detail in the following sections.

3-3 SOLID-STATE IMPURITY DIFFUSION

The most widely used technique in forming a *p-n* junction is solid-state impurity diffusion. The diffusion of impurity is, in principle, the same as

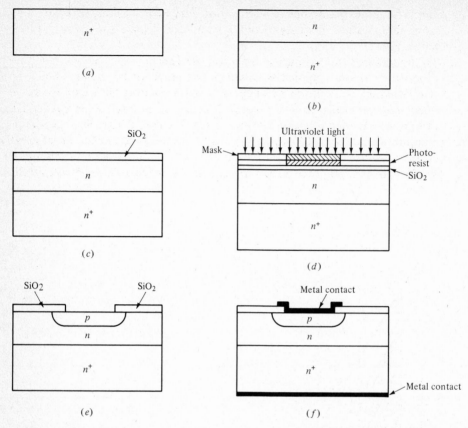

Figure 3-3 Planar technology: (*a*) the substrate, (*b*) with epitaxial layer, (*c*) after oxidation, (*d*) photolithography, (*e*) boron diffusion after oxide removal, and (*f*) the complete planar diode after metallization.

carrier diffusion, described in Sec. 1-7. Following Fick's first law, Eq. (1-47), the flux of impurity is proportional to the impurity gradient from high concentration to low concentration:

$$F = -D\frac{\partial N}{\partial x} \tag{3-1}$$

where F = flux
N = impurity density
D = impurity diffusivity
The diffusivity D follows

$$D = D_o e^{-E_a/kT} \tag{3-2}$$

where E_a is an activation energy and D_o is a constant. The diffusivity for impurity atoms is much lower than the diffusivity of carriers at room temperature. As a result, impurity atoms are essentially immobile at normal operating temperature. Diffusivities of various useful impurities in silicon are plotted in Fig. 3-4.

To understand the impurity diffusion mechanism, let us go back to examine the position occupied by an impurity atom in the crystal shown in Fig. 1-6. Most impurity atoms in silicon are situated in the lattice sites. These *substitutional impurities* can be relocated whenever empty lattice sites exist next to them. An empty lattice site in a solid is called a *vacancy*. At a high temperature, e.g., around 1000°C, many silicon atoms have moved out of their lattice sites, and a high density of vacancies exists. Consequently, impurity atoms can move through the vacancies as shown in Fig. 3-5*a*, and solid-state

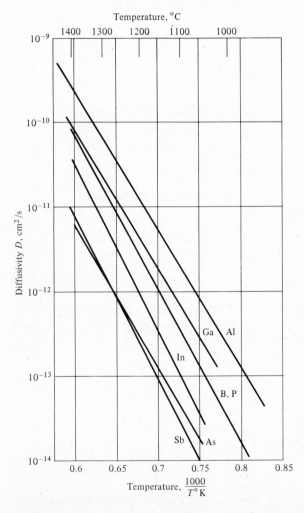

Figure 3-4 Impurity diffusivities in silicon. (*After Research Triangle Institute [2].*)

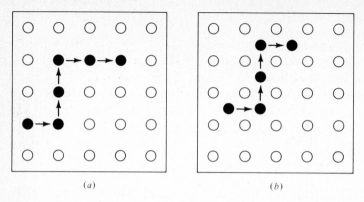

(a)   (b)

Figure 3-5 Impurity diffusion via (a) substitutional and (b) interstitial mechanisms.

diffusion becomes possible. For this reason, impurity diffusion is normally performed around 1000°C. When the crystal is cooled after the diffusion, impurity atoms which occupy lattice sites become stationary and the wafer is either *n* type or *p* type, depending on the type of impurity atoms introduced.

Impurity atoms that occupy the voids between atoms are known as *interstitial impurities*. These impurities move through the crystal lattice by jumping from one interstitial void (or site) to another. At a high temperature, the spacing between atoms is wider, so that impurities can diffuse via interstitial sites, as seen in Fig. 3-5b. When the crystal is cooled, interstitial atoms may return to substitutional sites and become electronically active. Substitution diffusion is the mechanism for boron, phosphorus, and most impurities in silicon. An important exception is gold, which diffuses primarily by the interstitial mechanism.

Using the same procedure as that for obtaining the continuity equation for carriers in Sec. 2-9, we find that the rate of change of impurity in a unit volume is equal to the difference between the incoming flux and the outgoing flux of impurity. We therefore have

$$\frac{\partial N}{\partial t} = -\frac{\partial F}{\partial x} \tag{3-3}$$

Substitution of Eq. (3-1) into the foregoing equation yields

$$\frac{\partial N}{\partial t} = D\frac{\partial^2 N}{\partial x^2} \tag{3-4}$$

This is the *diffusion equation*; solving it gives the impurity distribution of a diffused junction provided that the boundary and initial conditions are known. Frequently, a diffused junction is fabricated by using a two-step process to improve the controllability of the impurity profile. The two steps are called the *predeposition* and *drive-in diffusion*.

Predeposition

In a typical diffusion system, a wafer is placed in a high-temperature furnace in a gaseous atmosphere containing impurity atoms. The schematic diagram of a typical diffusion system is shown in Fig. 3-6. The temperature of the furnace usually ranges between 800 and 1200°C. For boron diffusion in silicon, boron tribromide may be used as the liquid source. The chemical reaction is

$$4BBr_3 + 3O_2 \longrightarrow 2B_2O_3 + 6Br_2 \tag{3-5}$$

The surface reaction between boron trioxide and silicon leads to

$$2B_2O_3 \text{ (gas)} + 3Si \text{ (solid)} \longrightarrow 4B + 3SiO_2 \text{ (solid)} \tag{3-6}$$

Thus, boron is incorporated into silicon, leaving a thin SiO_2 layer on the surface. Besides a liquid impurity source, a solid source may be employed. In boron diffusion, slices of boron nitride (BN) can be inserted between silicon wafers, and the assembly is then placed in the furnace for diffusion. Because of its simplicity, this approach has received more attention in recent years.

The number of impurity atoms taken in by the solid is linearly related to the partial pressure of the impurity species in the gaseous atmosphere. For a given impurity, there is a maximum impurity concentration that the solid can accommodate at a given temperature. This concentration is called the *solid solubility*. In most industrial processes, the partial pressure of the impurity is high enough for the impurity density inside the solid at the surface to be given by the solid solubility. The solid solubility of a few important impurities in silicon as a function of temperature is given in Fig. 3-7.

Let us assume that the number of impurities in the solid before diffusion is negligible. By using the solid-solubility limit at the surface we have fixed a boundary condition at $x = 0$ at $N = N_o$, where N_o is obtained from Fig. 3-7 for a given temperature and impurity type. An additional boundary condition is given by $N = 0$ at $x = \infty$. The second boundary condition is valid as long as impurity atoms do not diffuse through the wafer. The solution of Eq. (3-4) under these conditions is [3]

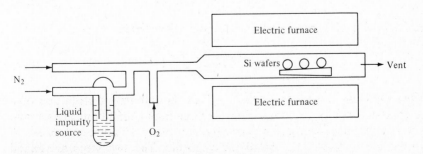

Figure 3-6 Schematic diagram of a typical diffusion apparatus.

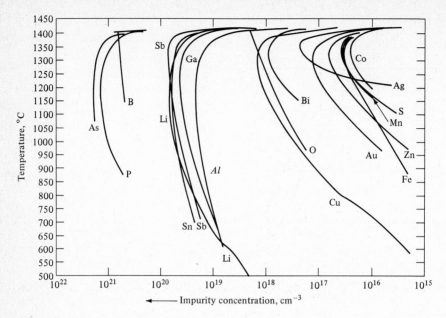

Figure 3-7 Solid solubilities of some impurity elements in silicon. (*After Hamilton and Howard [3].*)

$$N(x, t) = N_o \text{ erfc } \frac{x}{2\sqrt{Dt}} \tag{3-7}$$

The *complementary error function*, erfc, is tabulated in Appendix C, together with some of its basic properties. The term $\sqrt{Dt}$ is called the *diffusion length*. The total number of impurity atoms per unit surface is given by

$$Q(t) = \int_0^\infty N(x, t) \, dx = \frac{2}{\sqrt{\pi}} \sqrt{Dt} N_o \tag{3-8}$$

If the solid is *n* type and the diffused impurity is *p* type, a *p-n* junction is formed. The impurity distribution becomes

$$N(x, t) = N_o \text{ erfc } \frac{x}{2\sqrt{Dt}} - N_{BC} \tag{3-9}$$

where N_{BC} is the background doping density of the *n*-type solid. The junction depth from surface can be obtained by setting $N = 0$:

$$x_j = 2\sqrt{Dt} \text{ erfc}^{-1} \frac{N_{BC}}{N_o} \tag{3-10}$$

This junction depth is an important device parameter.

Example A *p-n* junction is formed by diffusing boron into *n*-type silicon having 10^{16} phosphorus atoms/cm^3 at 1200°C. The surface concentration

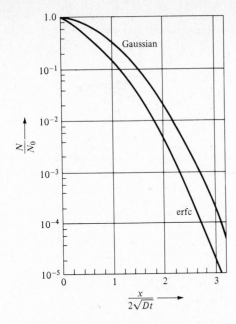

Figure 3-8 Normalized gaussian and erfc functions.

of boron is solubility-limited. Calculate the necessary diffusion time to realize a junction depth of 1 μm. What is the total number of boron atoms per square centimeter incorporated into silicon?

SOLUTION At 1200°C, the solid solubility of boron in silicon is 6×10^{20} (Fig. 3-7), and the diffusivity is $3 \times 10^{-12} \text{ cm}^2/\text{s}$ (Fig. 3-4). Setting $N(x, t) = 0$ and $x = x_j = 1 \mu$m in Eq. (3-9), we have

$$\text{erfc} \frac{x_j}{2\sqrt{Dt}} = \frac{10^{16}}{6 \times 10^{20}} = 1.66 \times 10^{-5}$$

From Fig. 3-8 or Appendix C we obtain

$$\frac{x_j}{2\sqrt{Dt}} = 3 \qquad \text{or} \qquad t = 92 \text{ s} = 1.5 \text{ min}$$

Using Eq. (3-8), we find

$$Q = \frac{2}{\sqrt{\pi}} \frac{10^{-4}}{6} (6 \times 10^{20}) = 1.13 \times 10^{16} \text{ cm}^{-2}$$

Drive-in Diffusion

After the predeposition step, the surface impurity concentration is nominally equal to the solid solubility at the diffusion temperature. For boron and phosphorus, the surface concentration is greater than 10^{20} cm^{-3}. Frequently, one would like to reduce the surface concentration and at the same time push the impurity atoms farther away from the surface into the bulk of the solid. This desirable impurity distribution can be achieved by a drive-in step after the predeposition. In a practical system, the surface of the wafer is sealed off by a thin oxide layer which prevents the escape of impurities through the surface. The boundary conditions for impurity are given by

$$\frac{\partial N}{\partial x}\bigg|_{(0,t')} = 0 \qquad N(\infty, t') = 0$$

Because the drive-in step follows the predeposition, t' is used for time here to avoid confusion. Furthermore, the total impurity Q resulting from predeposition may be assumed to be a delta function which is the initial condition. The solution of the diffusion equation in this case is

$$N(x, t') = \frac{Q}{\sqrt{\pi D't'}} e^{-x^2/4D't'} \qquad (3\text{-}11)$$

which is called the *gaussian distribution*. It is a good approximation for the carrier distribution of the two-step diffusion provided that the diffusion length of the drive-in step $\sqrt{D't'}$ is at least 3 times larger than the diffusion length $\sqrt{Dt}$ in predeposition. The normalized impurity profiles are shown in Fig. 3-8.

Gold Diffusion

In most cases, the impurity diffusion is made to achieve a specific doping profile, but sometimes it is desirable to modify the recombination lifetime without changing the doping profile. This is accomplished in silicon by introducing gold by solid-state diffusion. The introduction of gold in silicon creates energy levels deep in the forbidden gap. These deep-level energy states are effective recombination centers which reduce the minority-carrier lifetime. In switching diodes and transistors, a smaller lifetime corresponds to a faster switching speed and better performance for most digital applications. The relationship between gold diffusion and carrier lifetime is shown in Fig. 3-9. It should be pointed out that since gold diffuses very rapidly in silicon, its distribution can be assumed to be uniform. Therefore, the value of N_t in Fig. 3-9 is the solid solubility of gold in silicon at the corresponding saturation temperature.

3-4 EPITAXIAL GROWTH

The word *epitaxy* is derived from Greek roots and means "arranged upon." It describes the growth technique of arranging atoms in single-crystal fashion upon a crystalline substrate so that the lattice structure of the newly grown film duplicates that of the substrate. Probably the most important reason for

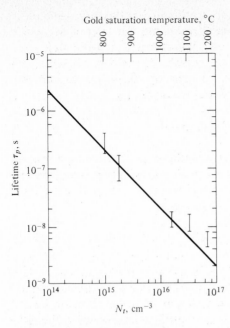

Figure 3-9 Carrier lifetime vs. gold-diffusion temperature. (*After Bakanowski and Forster [4].*)

using this growth technique is the flexibility of impurity control in the epitaxial film. The dopant in the film may be *n* type or *p* type and is independent of the substrate doping. Therefore, epitaxial growth can be used to form a lightly doped layer on a heavily doped substrate, or a *p-n* junction between the epitaxial film and the substrate. However, the *p-n* junction formed by epitaxy is over the entire substrate, and it is not suitable for localized *p-n* junction formation. The two types of epitaxial growth processes presently employed for device fabrication are known as *vapor-phase* and *liquid-phase epitaxy.*

A schematic diagram of the vapor-phase epitaxial (VPE) growth system is shown in Fig. 3-10. The hydrogen gas containing a controlled concentration of silicon tetrachloride is fed into the reactor containing silicon wafers in a graphite susceptor. The graphite is heated to a high temperature, typically above 1000°C, by a radio-frequency induction coil. The radio-frequency energy heats the graphite but not the quartz tube, so that there is no

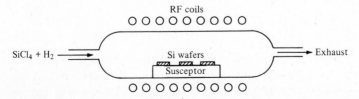

Figure 3-10 Schematic representation of an epitaxial growth system.

deposition of silicon on the tube. The high temperature is necessary for the deposited atoms to find their proper position in the lattice so as to maintain a single-crystal layer. The basic reaction in the reactor is given by

$$SiCl_4 + 2H_2 \rightleftharpoons Si\,(solid) + 4HCl \qquad (3\text{-}12)$$

Notice that the reaction is reversible. The normal reaction produces silicon solid which is grown onto the silicon substrate. The reverse reaction removes or etches the silicon substrate. This etching reaction may be used as the final cleaning step of the substrate. Typically, the substrate is etched by supplying the reactor with high-purity HCl gas before the growth is performed. Silicon tetrachloride is an excellent choice for this process because it is easy to purify and inexpensive.

The growth rate of the film as a function of the concentration of $SiCl_4$ in the gas is shown in Fig. 3-11. Note that the growth rate reaches a maximum and decreases as the $SiCl_4$ content is increased. This effect is caused by the competing chemical reaction

$$SiCl_4 + Si\,(solid) \longrightarrow 2SiCl_2 \qquad (3\text{-}13)$$

Therefore, etching of silicon will occur at high concentrations of $SiCl_4$.

In order to grow a doped epitaxial layer, impurity atoms are introduced in the gas stream. For example, phosphine (PH_3) is used for n-type doping, and diborane (B_2H_6) is used for p-type doping. The growth of a p-type silicon film is shown in Fig. 3-12, where diborane has undergone decomposition. In this figure, the boundary between the substrate and the epitaxial layer appears to be abrupt. In actual growth conditions, however, impurity atoms diffuse across the boundary so that the resulting impurity distribution is no longer a step function. Figure 3-13 shows the experimental impurity profile of an n-type epitaxial layer on an n^+ substrate. The phosphorus atoms in the substrate

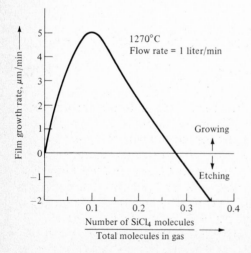

Figure 3-11 The growth rate of an epitaxial film as a function of the percentage concentration of silicon tetrachloride in gas.

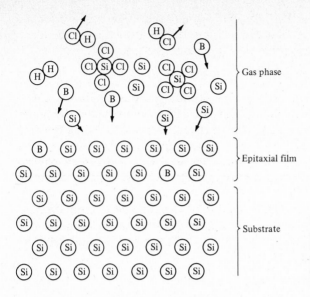

Gas phase

Epitaxial film

Substrate

Figure 3-12 Incorporation of boron atoms in the epitaxial film during the vapor-phase epitaxial growth. (*After Warner and Fordemwalt [6].*)

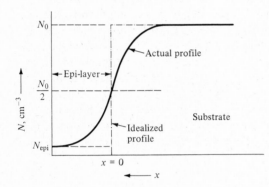

Figure 3-13 Impurity profile of the epitaxial layer and substrate interface.

have diffused into the epitaxial layer, yielding a final impurity profile of

$$N(x) = \frac{N_o}{2} \operatorname{erfc} \frac{x}{2\sqrt{Dt}} \qquad (3\text{-}14)$$

where x extends into the film with the original substrate surface chosen as $x = 0$ and N_o is the impurity concentration in the substrate.

Vapor-phase epitaxy is seldom used for the growth of more than one layer because of system complexity in the growth of multiple layers. However, the liquid-phase epitaxy (LPE) is convenient for depositing layers of different materials on the same substrate, particularly with compound semiconductors such as GaAs and GaP. An LPE apparatus is shown in Fig. 3-14 for the epitaxial growth of four different layers. In the operation, the sliding solution holder is moved to bring the substrate in contact with the

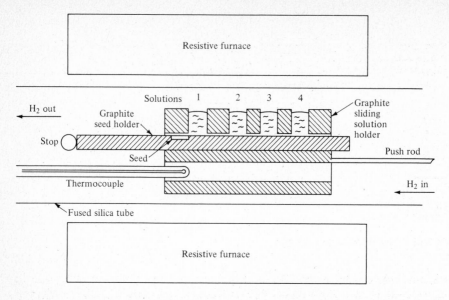

Figure 3-14 A liquid-phase epitaxial reactor. (*After Hayashi et al. [7].*)

solute when solution 1 reaches the desired growth temperature. After the specified growth period, the solution holder is moved on so that the substrate (together with its new epitaxial layer) is situated between solutions 1 and 2. Any excess solute is removed by the bottom of the sliding solution holder. The procedure is then repeated for solutions 2, 3, and 4. Multiple heterojunction† devices have been fabricated using this apparatus. The thickness of the epitaxial layers may be less than 0.5 μm, and the junctions could be very abrupt. The drawback of the LPE process is the difficulty in adapting it to mass production.

In either VPE or LPE, it is important to remove any imperfection in the substrate surface. Since any defects at the substrate surface will be duplicated, enlarged, and propagated, cleanliness of reactor and crystal perfection of the substrate are of extreme importance for successful epitaxial-layer growth.

3-5 THERMAL OXIDATION

Silicon oxide is most valuable in protecting the semiconductor surface and at the same time providing a mask for selective diffusion through the photolithographic process. Standard industrial oxidation is performed in high-temperature diffusion furnaces like the one shown in Fig. 3-15. A gas

† A junction formed by two different materials, for example, Ge–Si, is called a heterojunction. This subject will be discussed in Chap. 5.

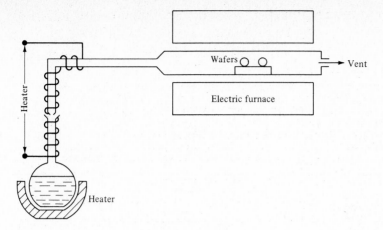

Figure 3-15 A steam-oxidation apparatus.

containing oxygen or water vapor flows through the diffusion furnace, in which chemical reaction takes place according to one of the following equations:

Dry oxidation: $\qquad$ Si (solid) $+ O_2 \longrightarrow$ SiO$_2$ (solid) $\qquad\qquad$ (3-15)

Steam oxidation: $\quad$ Si (solid) $+ 2H_2O \longrightarrow$ SiO$_2$ (solid) $+ 2H_2$ $\qquad$ (3-16)

$\qquad$ The physical theory of thermal oxidation can be explained with the help of the simple model shown in Fig. 3-16. Radioactive-tracer experiments show that the oxidizing species travels from the bulk of the gas to the oxide surface; then it diffuses through the oxide and reaches the oxide-silicon interface, where the reaction takes place. The oxidant concentration N_o at the oxide surface is determined by the temperature, the gas-flow rate, and the equivalent solid solubility in the oxide. At 1000°C, N_o is 5×10^{16} molecules/cm^3 for dry oxygen and 3×10^{19} cm^{-3} for water vapor at atmospheric pressure. We can now determine the oxide growth rate by considering the fluxes of oxidant in the oxide and at the oxide-silicon interface. From Fick's

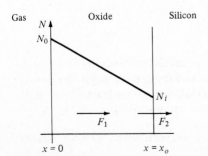

Figure 3-16 A simple model of thermal oxidation.

first law, the flux across the oxide is given by the gradient of the oxidizing species:

$$F_1 = -D\frac{dN}{dx} = \frac{D(N_o - N_i)}{x_o} \tag{3-17}$$

where N_i = oxidant concentration, molecules/cm^3, at $x = x_o$
$\quad\quad D$ = diffusion constant
$\quad\quad x_o$ = oxide thickness

The flux at the oxide-silicon interface is determined by the surface reaction-rate constant k_s and is given by

$$F_2 = k_s N_i \tag{3-18}$$

Under the steady-state condition, these fluxes must be equal, so that $F = F_1 = F_2$. Therefore, we can equate Eqs. (3-17) and (3-18) to yield

$$N_i = \frac{N_o}{1 + k_s x_o/D} \tag{3-19}$$

and

$$F = \frac{N_o}{1/k_s + x_o/D} \tag{3-20}$$

The oxide growth rate is determined by the flux and the number of oxidant molecules (N_1) needed to form a unit volume of oxide. Since there are 2.2×10^{22} SiO$_2$ molecules/cm^3 in the oxide, we need 2.2×10^{22} molecules/cm^3 of O$_2$ or 4.4×10^{22} molecules/cm^3 of H$_2$O. The equation of oxide growth is therefore

$$\frac{dx_o}{dt} = \frac{F}{N_1} = \frac{N_o/N_1}{1/k_s + x_o/D} \tag{3-21}$$

Integrating from 0 to t leads to

$$x_o - x_i + \frac{k_s}{2D}(x_o{}^2 - x_i^2) = \frac{N_o k_s}{N_1} \tag{3-22}$$

where x_i is the initial value of the oxide at $t = 0$. Solving for the oxide thickness x_o, we have

$$x_o = \frac{A}{2}\left(\sqrt{1 + \frac{t + \tau}{A^2/4B}} - 1\right) \tag{3-23}$$

where $\quad\quad A = \dfrac{2D}{k_s} \quad\quad B = \dfrac{2DN_o}{N_1} \quad\quad \tau = \dfrac{x_i^2 + Ax_i}{B}$ $\tag{3-24}$

The initial thickness x_i could be the result of a previous oxidation or a rapid initial growth. It is found that the rapid initial growth produces an x_i of 200 Å in dry oxygen, but the effect is negligible in steam oxidation.

Equation (3-23) reduces to

$$x_o = \frac{B}{A}(t + \tau) \tag{3-25}$$

for $(t + \tau) \ll A^2/4B$ and

$$x_o = (Bt)^{1/2} \tag{3-26}$$

for $t \gg A^2/4B$. Therefore, if the oxidation time is short, the oxide thickness is determined by the surface reaction-rate constant and is linearly proportional to the oxidation time. If the oxidation time is long, Eq. (3-26) is applicable and the oxide growth is determined by the diffusion constant. The oxide thickness in this case is proportional to the square root of the oxidation time. Experimental values of A and B are plotted in Fig. 3-17.

Example Find the time required to grow 2000-Å-thick oxide in dry oxygen at 1200°C.

SOLUTION Using Fig. 3-17 for dry oxidation at 1200°C, we find

$$A = 5 \times 10^{-2} \, \mu m \qquad B = 5 \times 10^{-2} \, \mu m^2/h$$

Since $x_i = 200$ Å for dry oxidation, we obtain

$$\tau = \frac{4 \times 10^4 + 500 \times 200}{5 \times 10^{-2}} \times 10^{-8} = 2.8 \times 10^{-2} \, h \qquad \text{from Eq. (3-24)}$$

Substituting these values into Eq. (3-23) yields

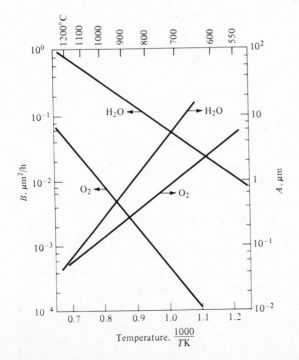

Figure 3-17 Oxidation constants A and B as functions of temperature. (*After Grove [8].*)

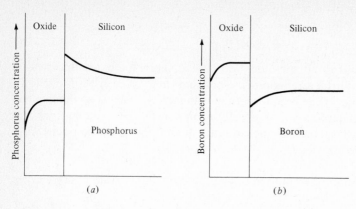

Figure 3-18 Redistribution of impurities during thermal oxidation for (a) $k > 1$ and (b) $k < 1$.

$$\frac{2000}{250} = 8 = \sqrt{1 + \frac{t + 2.8 \times 10^{-2}}{1.25 \times 10^{-2}}} - 1$$

or
$$t = (80)(1.25 \times 10^{-2}) - 2.8 \times 10^{-2} = 1.0 \text{ h}$$

During the growth of thermal oxide, silicon atoms in the surface of the wafer are incorporated into the oxide layer. Therefore, the impurity atoms which were present in the silicon surface must be redistributed between the oxide and silicon until equilibrium is reached. The equilibrium values of impurities in the oxide and silicon are related by the *segregation* or *distribution* coefficient k, which is defined by the ratio of the equilibrium concentration of an impurity in the silicon to that in the silicon oxide. For example, the segregation coefficient of phosphorus in the Si–SiO$_2$ system is roughly 10. Thus an equilibrium phosphorus concentration of 10^{16} in silicon surface leads to a concentration of 10^{15} in the SiO$_2$ side of the surface, as shown in Fig. 3-18a. On the other hand, $k < 1$ results in a higher impurity concentration in the oxide than in silicon. This is the case for boron, as illustrated in Fig. 3-18b. Up to 80 percent of predeposited boron can be taken up by the oxide during oxidation. The boron profile of a two-step diffusion yields a distribution depicted in Fig. 3-19 as a result of the boron redistribution.

3-6 OXIDE AND NITRIDE MASKS

Impurities such as boron and phosphorus may diffuse through a SiO$_2$ layer similar to that in semiconductors. However, the diffusivities of boron and phosphorus in silicon oxide are much smaller than in silicon. If a region of a silicon substrate is covered with an oxide layer and the assembly is placed in a furnace for impurity diffusion, the small diffusivity of the impurity in the

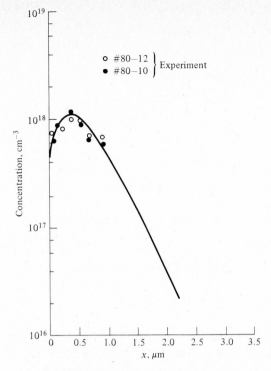

Figure 3-19 Redistribution of boron after oxidation. (*After Hamilton and Howard [3].*)

oxide prevents the impurity from penetrating the oxide layer. As shown in Fig. 3-20, the silicon oxide is used as a diffusion mask. Notice that the impurity atoms diffuse both vertically and laterally at the edge of the oxide mask. Usually we can assume that diffusion in the lateral direction is approximately 75 percent of the depth of the vertical penetration.

The masking capability of the oxide depends on the segregation coefficient, impurity concentrations, and diffusion coefficients. The minimum oxide-layer thickness for masking purposes is usually obtained from empirical experimental data. Figures 3-21 and 3-22 display the minimum thickness of the oxide required to mask against boron and phosphorus as a function of diffusion time and temperature. These curves were obtained empirically for a typical diffusion system and represent the commercial practice in device fabrication.

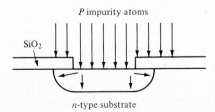

Figure 3-20 Oxide masking and two-dimensional diffusion.

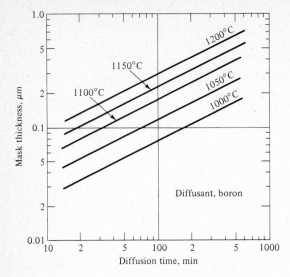

Figure 3-21 Experimentally determined curves of oxide-mask thickness for boron in silicon. (*From Ghandhi [9].*)

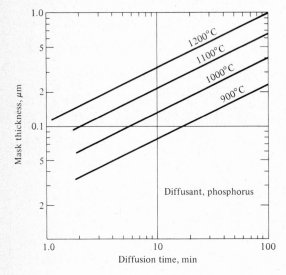

Figure 3-22 Experimentally determined curves of oxide-mask thickness for phosphorus in silicon. (*From Ghandhi [9].*)

Although a thin SiO_2 layer is adequate for masking phosphorus or boron, it is not capable of masking Ga, Al, Zn, Na, or O_2 because of their high diffusivity in SiO_2. As an example, the diffusivity of Ga is 10^{-10} cm²/s in SiO_2 and 2×10^{-12} cm²/s in silicon at 1150°C. Fortunately, silicon nitride (Si_3N_4) is found suitable for these impurities. A commonly used nitride is deposited by the chemical reaction of silane and ammonia in an excess hydrogen environment. The chemical reaction is given by

$$3SiH_4 + 4NH_3 \longrightarrow Si_3N_4 + 12H_2 \tag{3-27}$$

The substrate temperature is typically between 800 and 1100°C, and the growth rate is about 30 Å/min with the substrate at 800°C. The film thickness required for masking most common dopants in silicon is between 300 and 500 Å. The disadvantages of using Si_3N_4 include difficulty of deposition and etching. In addition, direct deposition of nitride on silicon produces high interface trapping density. A technique for circumventing this problem is to place a thin layer of SiO_2 between the silicon substrate and the nitride film.

3-7 ION IMPLANTATION

An alternate to high-temperature diffusion for introducing dopants into a semiconductor is *ion implantation.* The basic implantation apparatus is shown in Fig. 3-23. A beam of dopant ions is accelerated through a desired energy potential ranging between 10 and 500 keV. A mass-separating magnet eliminates unwanted ion species. After going through the deflection and focusing control, the ion beam is aimed at the semiconductor target so that the high-energy ions penetrate the semiconductor surface. The energetic ions will lose their energy through collisions with the target nuclei and electrons so that

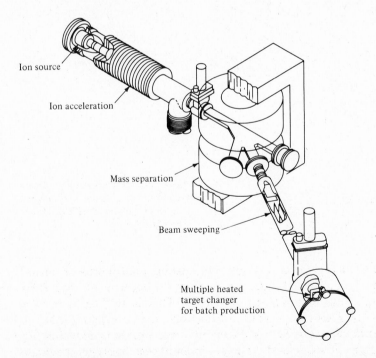

Ion source

Ion acceleration

Mass separation

Beam sweeping

Multiple heated
target changer
for batch production

Figure 3-23 Schematic drawing of an ion-implantation apparatus. A mass-separating magnet is used to select the ion species of interest, and beam-sweeping facilities are provided for large-area uniform implantation. (*From Mayer et al. [10].*)

the ions will finally come to rest. The distance traveled by the ions, i.e., the penetration depth, is called the *range*. The range is a function of the kinetic energy of the ions and the semiconductor's structural properties, e.g., lattice spacing and mass of atoms. A typical impurity-range distribution in an amorphous target is approximately gaussian, as depicted in Fig. 3-24. In a crystalline target, however, the impurity distribution shows strong orientation dependence with respect to the ion-beam direction. If an ion enters the crystal in parallel to a major crystal axis or plane, a series of sideway collisions may direct it smoothly through the lattice, as shown in Fig. 3-25a; thus the penetration depth is increased. This *channeling* effect leads to the profile

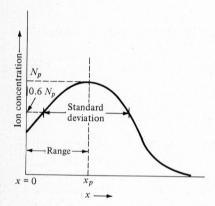

Figure 3-24 A typical ion-implanted impurity profile.

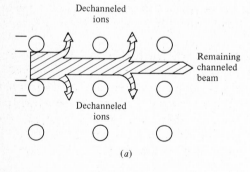

(a)

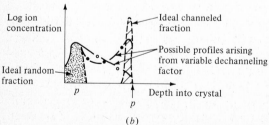

(b)

Figure 3-25 The impurity profile arising from (a) random and (b) channeled beams with a variable dechanneling factor. (*From Gibbons [11].*)

illustrated in Fig. 3-25b. The channeling effect is substantially reduced when the direction of the ion beam is misaligned with the crystal axes, and this is done in most implantations to enhance reproducibility. Thus, the ions enter the crystal and act as if they were in an amorphous solid.

The impurity distribution shown in Fig. 3-24 can be approximately described by the empirical equation

$$N(x) = N_p \exp\left[\frac{-(x - x_p)^2}{2\sigma_R^2}\right] - N_{BC} \tag{3-28}$$

where x_p = range

σ_R = standard deviation

N_p = peak impurity concentration

N_{BC} = background doping concentration

The peak impurity concentration is obtainable from the empirical relationship

$$N_p = 3 \times 10^{14} \frac{It}{A_I \sigma_R^2} \tag{3-29}$$

where I = ion-beam current

t = implantation time

A_I = beam area at target

This expression states that N_p is proportional to the product of the ion-beam current and the implantation time. Typical experimental data of the implantation range as a function of ion-beam energy are shown in Fig. 3-26 for nonchanneled boron, phosphorus, and antimony ions in a $\langle 111 \rangle$ silicon substrate. The corresponding values of the peak impurity concentration relating to the ion-beam current, implantation time, and energy are shown in Fig. 3-27. Since the standard deviation can be calculated from Eq. (3-29), the impurity concentration can be obtained by using the graphical data given in Figs. 3-26 and 3-27.

Before an incident ion loses its kinetic energy, it collides with lattice atoms and the host atoms are dislodged. As a result, a large number of vacancies are formed to turn the crystalline region into a disordered or amorphous layer. For this reason, it is necessary to anneal the semiconductor after implantation to reestablish the crystalline structure. In most cases, the semiconductor is placed in a high-temperature oven set at 200 to 800°C. The annealing temperatures are typically well below those used in solid-state diffusion.

The ion-implantation process is attractive because it can be performed at low temperatures at which impurity diffusion is negligible. In addition, the impurity concentration introduced is better controlled than with standard diffusion techniques. Another advantage is the very sharp impurity definition from an abrupt mask, as shown in Fig. 3-28, which for comparison also shows the masking in a diffused layer which cannot avoid lateral diffusion. Compared with other doping techniques, an ion-implanted layer is generally shallower, and it can be masked by a wide variety of materials. The major

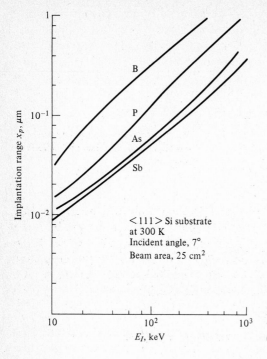

Figure 3-26 Implantation range as a function of energy for different impurity ions. (*From Wolf [12].*)

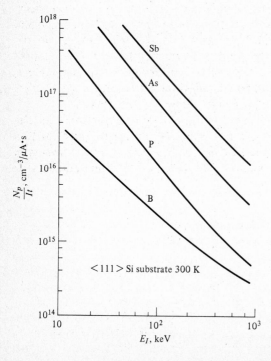

Figure 3-27 The peak concentration vs. energy for different impurity ions. (*From Wolf [12].*)

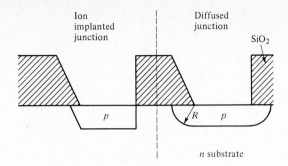

Figure 3-28 Impurity distributions for ion-implanted junction and diffused junction.

drawbacks include difficulty in controlling doping distribution exactly and high density of defects in the implanted region. These defects reduce the injected-carrier lifetime by enhancing the recombination process. For these reasons, ion-implanted regions may be undesirable for use as the critical or active regions of devices. Two of the most important applications of ion implantation are in MOS integrated circuits and microwave diodes.

3-8 CHARACTERIZATION OF A DOPED LAYER

The characteristics of a diffused or implanted layer can best be understood if the complete impurity distribution is mapped out after doping. Unfortunately, this is difficult to do. On the one hand, the theories presented are idealized, and they cannot be used to predict the exact impurity distribution. On the other hand, detailed experimental mapping of the profile is a tedious procedure. In practical situations, we frequently measure just two parameters, the junction depth and the average resistivity, to evaluate a doped layer. These two parameters are adequate for most device design purposes.

One frequently used method for junction-depth measurement is the angle-lapping technique. The silicon wafer is placed on a metal block with a small angle, and the crystal is lapped off to produce the structure shown in Fig. 3-29. For a fixed angle θ less than 1°, we have $x_j = \theta d$. The value of d is

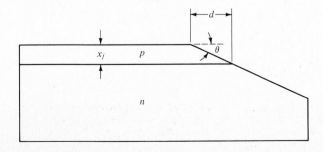

Figure 3-29 The junction-depth measurement by angle lapping and staining.

observable under a microscope after chemical staining in which the p region reacts faster with a chemical reagent to show a darker color.

Use of Eq. (1-46) gives the average resistivity $\bar{\rho}$

$$\frac{1}{\bar{\rho}} = \bar{\sigma} = \frac{q}{x_j} \int_0^{x_j} \mu_p p(x)\, dx \tag{3-30}$$

for a p-type layer. This equation is plotted in Fig. 3-30 for both gaussian and erfc functions applicable to a diffused layer. The average resistivity can be calculated if the surface concentration and N_{BC} are known. Experimentally, the average resistivity is measured with the four-point-probe method shown in Fig. 3-31. The current I is passed through the outer probes, and the potential difference between the two inner probes is measured. On the basis of electrostatic calculation, the average resistivity is obtained as

$$\bar{\rho} = 4.532 \frac{V}{I} x_j \qquad \text{for } s \gg x_j \tag{3-31}$$

where $\bar{\rho}$ is in ohm-centimeters, V is in volts, I is in amperes, and x_j is in centimeters. This equation is applicable to thin layers in which the assumption $s \gg x_j$ is valid [11].

Sheet Resistance

Frequently, a thin layer of semiconductor material is characterized by its *sheet resistance*. The sheet resistance is defined by the resistance of a square of material, shown in Fig. 3-32. Thus

$$R_s = \frac{\bar{\rho} l}{A} = \frac{\bar{\rho} l}{l x_j} = \frac{\bar{\rho}}{x_j} \qquad \Omega/\square \tag{3-32}$$

where $\Omega/\square$ means ohms per square. Notice that the size of the square l is not important as long as it is a square. In other words, the resistance is determined only by the average resistivity and thickness of the material no matter how large or small the square. In an integrated circuit, a resistor is usually fabricated on a thin diffused layer, and its resistance is given by counting the number of squares in a strip of material. For example, a resistor of 1000 Ω is obtained from material of 200 $\Omega/\square$ by putting five squares in series.

Example An integrated resistor is made by the two-step diffusion of boron into an n layer with $N_d = 10^{15}$ cm^{-3}. The junction depth is measured to be 2.5 μm, and the surface boron concentration is 5×10^{18} cm^{-3}. Obtain the sheet resistance of the diffused region and determine the size of a 1.2-kΩ resistor. Assume the minimum width of the diffused region is 0.5 mil.

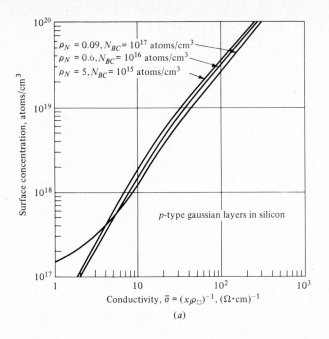

$$\rho_N = 0.09, N_{BC} = 10^{17} \text{ atoms/cm}^3$$
$$\rho_N = 0.6, N_{BC} = 10^{16} \text{ atoms/cm}^3$$
$$\rho_N = 5, N_{BC} = 10^{15} \text{ atoms/cm}^3$$

p-type gaussian layers in silicon

Surface concentration, atoms/cm^3

Conductivity, $\bar{\sigma} = (x_j \rho_{\square})^{-1}, (\Omega \cdot \text{cm})^{-1}$

(a)

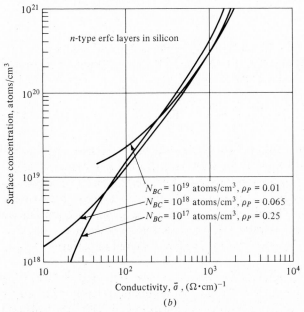

n-type erfc layers in silicon

Surface concentration, atoms/cm^3

$$N_{BC} = 10^{19} \text{ atoms/cm}^3, \rho_P = 0.01$$
$$N_{BC} = 10^{18} \text{ atoms/cm}^3, \rho_P = 0.065$$
$$N_{BC} = 10^{17} \text{ atoms/cm}^3, \rho_P = 0.25$$

Conductivity, $\bar{\sigma}, (\Omega \cdot \text{cm})^{-1}$

(b)

Figure 3-30 The Irvin conductivity curves for diffused layers. (*From Irvin [13].*)

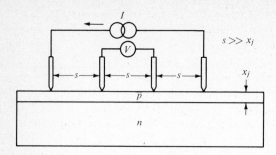

Figure 3-31 The four-point probe for resistivity measurement.

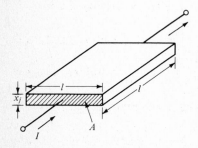

Figure 3-32 The resistance of a square semiconductor wafer.

SOLUTION The two-step diffusion yields a gaussian distribution, so that Fig. 3-30a should be used. We find that $\bar{\sigma} = 20\,(\Omega\text{-cm})^{-1}$. Since $\bar{\sigma} = 1/x_j R_s$,

$$R_s = \frac{1}{20 \times 2.5 \times 10^{-4}} = 200\ \Omega/\square$$

Thus, for 1.2 kΩ and $W = 0.5$ mil, we have

$$L = (\tfrac{1}{2})(\tfrac{1200}{200}) = 3\ \text{mils}$$

PROBLEMS

3-1 A uniformly doped n-type silicon epitaxial layer of 0.5 Ω-cm resistivity is subjected to a boron diffusion with constant surface concentration of $5 \times 10^{18}\ \text{cm}^{-3}$. It is desired to form a p-n junction at a depth of 2.7 μm. At what temperature should this diffusion be carried out if it is to be completed in 2 h?

3-2 After the predeposition step, it is found that 5×10^{15} boron atoms/cm^2 are introduced into the silicon epitaxial layer with $N_d = 10^{16}$ atoms/cm^3. Assume the diffusion constant is 3×10^{-12} cm^2/s. Calculate the junction depth at the end of 60 min. Plot the impurity profile and indicate the junction depth.

3-3 An n-type silicon substrate of 0.5 Ω-cm resistivity is subjected to a boron diffusion with constant surface concentration of 5×10^{18} cm^{-3}. The desired junction depth is 2.7 μm. Plot impurity concentrations (log scale) vs. distance from surface (linear scale) for the boron diffusion by calculation on this plot.

3-4 An *n-p-n* transistor is to be completed by diffusing phosphorus at a surface concentration of $10^{21} \, cm^{-3}$ on the structure obtained in Prob. 3-3. The new junction is to be at a depth of 2.0 μm.

(*a*) Add the phosphorus curve to your drawing from Prob. 3-3. Assume that the boron stays put during this diffusion, and scale this curve from Fig. 3-8 instead of calculating directly. (Extrapolate a bit where required.)

(*b*) Indicate emitter, base, and collector on this drawing.

(*c*) If the phosphorus diffusion is conducted at 1100°C, how long should be allowed?

(*d*) Estimate the new position of the base-collector junction by assuming that the boron surface concentration remained constant during the emitter diffusion.

3-5 (*a*) An epitaxial layer is grown with a $SiCl_4$ mole fraction of 0.02 at 1270°C for 15 min. Find the thickness of the epitaxial layer.

(*b*) Determine the time to grow the same epitaxial-layer thickness if the mole fraction is 0.05.

3-6 Find the time required to grow 2000-Å-thick oxide by steam oxidation at 900°C.

3-7 A window is open in the oxide in Prob. 3-6 for boron diffusion. After predeposition, the wafer is placed in dry oxygen atmosphere at 1200°C for 90 min. Find the oxide thickness over the window and over the area which has an initial oxide of 2000 Å.

3-8 Repeat Prob. 3-7 if steam atmosphere at 900°C is used.

3-9 A silicon wafer has an initial oxide of 1000 Å. Determine the time required to grow an additional oxide in dry oxygen at 1200°C such that the final oxide is thick enough to mask a boron diffusion of 100 min at 1100°C.

3-10 (*a*) Calculate the energy E_I of the ion beam and the current-time product It for phosphorus implantation to obtain a peak impurity concentration $N_p = 10^{18} \, cm^{-3}$ and a range of 0.3 μm at 25°C.

(*b*) If the ion beam current is 10 μA, what is the required implantation time?

3-11 By using an *n*-type substrate with a resistivity of 1 Ω-cm, boron is introduced by the two-step diffusion process: namely, predeposition at 1000°C for 15 min and driven in at 1200°C for 3.9 h. The measured junction depth is 8 μm, and the four-point probe gives a V/I ratio of 55 Ω. Calculate (*a*) the impurity profile and (*b*) the total impurity atoms incorporated per unit area.

3-12 An integrated resistor is made by diffusing phosphorus atoms into a *p*-type epitaxial layer with $N_a = 10^{17} \, cm^{-3}$. After predeposition at 1000°C the junction depth is measured to be 2.5 μm.

(*a*) Obtain the sheet resistance of the diffused region using Irvin's curves.

(*b*) Determine the length of a 2-kΩ resistor assuming that the minimum width of the diffused region is $\frac{1}{4}$ mil. Sketch the top view and cross-sectional view of your resistor. Compare your result with the example in the text.

REFERENCES

1. W. R. Runyan, "Silicon Semiconductor Technology," McGraw-Hill, New York, 1965.
2. Research Triangle Institute, "Integrated Silicon Device Technology," vol. 4, ASD-TDR-63-316, 1964.
3. D. J. Hamilton and W. G. Howard, "Basic Integrated Circuit Engineering," chap. 2, McGraw-Hill, New York, 1975.
4. A. E. Bakanowski and J. H. Forster, Electrical Properties of Gold-doped Diffused Silicon Computer Diodes, *Bell Syst. Tech. J.*, **39**: 87 (1960).
5. H. C. Theuerer, Epitaxial Silicon Films by the Hydrogen Reduction of $SiCl_4$, *J. Electrochem. Soc.*, **108**: 649 (1961).
6. R. M. Warner, Jr., and J. N. Fordemwalt, "Integrated Circuits," p. 275, McGraw-Hill, New York, 1965.
7. I. Hayashi, M. B. Panish, P. W. Foy, and S. Sumski, Junction Lasers Which Operate Continuously at Room Temperature, *Appl. Phys. Lett.*, **17**: 109 (1970).
8. A. S. Grove, "Physics and Technology of Semiconductor Devices," chaps. 1–3, Wiley, New York, 1967.
9. S. K. Ghandhi, "The Theory and Practice of Microelectronics," chap. 6, Wiley, New York, 1968.
10. J. W. Mayer, L. Eriksson, and J. A. Davies, "Ion Implantation in Semiconductors," Academic, New York, 1971.
11. J. F. Gibbons, Ion Implantation in Semiconductors, *Proc. IEEE*, **56**: 295 (1968).
12. H. F. Wolf, "Semiconductors," Wiley-Interscience, New York, 1971.
13. J. C. Irvin, Resistivity of Bulk Silicon and of Diffused Layers in Silicon, *Bell Syst. Tech. J.*, **41**: 387 (March 1962).
14. F. M. Smits, Measurement of Sheet Resistivities with the Four-Point Probe, *Bell Syst. Tech. J.*, **37**: 711 (1958).

ADDITIONAL READINGS

Burger, R. A., and R. P. Donovan: "Fundamentals of Silicon Integrated Devices," vol. 1, Prentice-Hall, Englewood Cliffs, N.J., 1967.
Grove, A. S.: "Physics and Technology of Semiconductor Devices," chaps. 1–3, Wiley, New York, 1967.
Hamilton, D. J., and W. G. Howard: "Basic Integrated Circuit Engineering," chap. 2, McGraw-Hill, New York, 1975.
Matthews, J. W. Jr. (ed.): "Epitaxial Growth," Academic Press, New York, 1975.

THE *p-n* JUNCTION

In the previous chapter, we considered techniques for the fabrication of semiconductor devices. It was shown that impurities can be introduced so that both *p*-type and *n*-type regions may exist simultaneously in a semiconductor. The metallurgical boundary between a *p* region and an *n* region is called a *p-n junction*. The most common method of forming a *p-n* junction is impurity diffusion, and the resulting impurity distribution closely conforms either to the complementary error function or the gaussian function. In practice, a diffused *p-n* junction is usually approximated by a *step* or *linearly graded* junction to avoid mathematical complexity in analyses. The impurity profile of the step junction assumes an abrupt transition between the *n*-type and *p*-type regions, as illustrated in Fig. 4-1*a*. The linearly graded junction has an impurity distribution changing gradually, as seen in Fig. 4-1*b*. Theoretical calculations using these two models are in excellent agreement with the first-order effects in practical diffused junctions. Since the step junction is easier to describe analytically, we base our analyses mainly on the step junction. Topics to be covered in this chapter include equilibrium junction properties, static and dynamic characteristics, and the junction breakdown.

4-1 THE STEP *p-n* JUNCTION AT EQUILIBRIUM

Let us assume that the *n*-type and *p*-type materials are physically separate before the junction is formed. The Fermi level is near the conduction-band

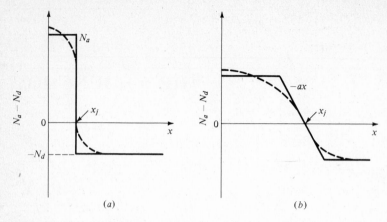

Figure 4-1 (*a*) A shallow diffused junction (*dotted*) with the step-junction approximation (*solid*) and (*b*) a deep diffused junction (*dotted*) with the linearly graded junction approximation (*solid*).

edge in the *n*-type material and near the valence-band edge in the *p*-type material, as shown in Fig. 4-2*a*. When these materials are joined, the Fermi level at equilibrium must be constant; otherwise, there would be a·current flow according to Eqs. (2-55) and (2-56). The condition of constant Fermi level is reached by transferring electrons from the *n* side to the *p* side and holes in the opposite direction. The energy-band diagram of the *p-n* junction at equilibrium is shown in Fig. 4-2*b*. The transfer of electrons and holes leaves behind uncompensated donor and acceptor ions N_d^+ and N_a^- in the *n* side and *p* side, respectively. As a result, two charged layers are set up, as depicted in Fig. 4-2*c*. Alternatively, we can arrive at this charge distribution by considering carrier diffusion and drift. When the *n*-type and *p*-type materials are brought together, holes in the *p* side will diffuse to the *n* side because there are many more holes in the *p* material. At the same time, electrons in the *n* side will diffuse in the opposite direction. An electric field is established by the uncompensated donor and acceptor ions left behind by the diffusion of electrons and holes. This electric field is in the direction to counterbalance the diffusive tendency of carriers. Under thermal equilibrium, the drift of carriers is exactly equal to the diffusion of carriers, so that the net carrier flow is zero. Thus, a unique charge distribution shown in Fig. 4-2*c* is needed to establish the necessary field.

The concepts discussed in the preceding paragraph illustrate two approaches we can adopt in analyzing semiconductor devices. The use of Fermi and quasi-Fermi levels usually leads to simple and elegant expressions as well as better physical insight, but it does not provide explicit information on the carrier and current distributions. On the other hand, the analysis of diffusion and drift of carriers gives the carrier concentrations and current components directly, but it does not supply detailed information about the internal

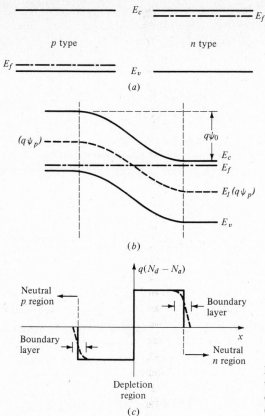

Figure 4-2 (*a*) Isolated *p*-type and *n*-type silicon before contact, (*b*) the energy-band diagram after contact, and (*c*) the space-charge distribution of (*b*).

physical mechanisms. In most engineering texts, the second approach is employed. We shall adopt this approach and describe the quasi-Fermi levels to complement our discussion.

The relationship between the charge distribution and the electrostatic potential is given by Poisson's equation (2-67), repeated here:

$$\frac{d^2\psi}{dx^2} = -\frac{\rho}{K\epsilon_o} = \frac{q}{K\epsilon_o}[(n-p)-(N_d-N_a)] \qquad (4\text{-}1)$$

The electron and hole densities are given by Eqs. (2-43) and (2-44) and are written in the following forms by taking the Fermi potential as the zero reference:

$$n = n_i e^{\psi/V_T} \qquad (4\text{-}2a)$$

$$p = n_i e^{-\psi/V_T} \qquad (4\text{-}2b)$$

Equations (4-1) and (4-2) are applicable to various regions of the *p-n* junction. These regions are (1) the *neutral* regions away from the junction, (2) the *depletion* region where there are fixed charges but no free carriers, and (3) the boundary layers between the neutral and depletion regions.

In the neutral regions, the total space-charge density is zero. Thus, Eq. (4-1) becomes

$$\frac{d^2\psi}{dx^2} = 0 \tag{4-3}$$

and

$$n - p - N_d + N_a = 0 \tag{4-4}$$

For an *n*-type neutral region, we may assume $N_a = 0$ and $p \ll n$. The potential of the *n*-type neutral region far from the junction, designated ψ_n in Fig. 4-2*b*, is derived by setting $N_a = p = 0$ in Eq. (4-4) and substituting the result into Eq. (4-2*a*):

$$\psi_n = V_T \ln \frac{N_d}{n_i} \tag{4-5}$$

By using the same procedure in the *p*-type neutral region, we obtain the potential of the *p*-type neutral region as

$$\psi_p = - V_T \ln \frac{N_a}{n_i} \tag{4-6}$$

Therefore, the potential difference between the *n*-side and the *p*-side neutral regions is

$$\psi_o = \psi_n - \psi_p = V_T \ln \frac{N_d N_a}{n_i^2} \tag{4-7}$$

where ψ_o is known as the *built-in* or *diffusion potential*. This potential difference exists in a *p-n* junction under thermal equilibrium.

For the completely depleted region, free-carrier densities are zero ($n = p = 0$), and Eq. (4-1) becomes

$$\frac{d^2\psi}{dx^2} = \frac{q}{K\epsilon_o} (N_a - N_d) \tag{4-8}$$

which will be solved in the next section.

In the boundary layers, Eqs. (4-1) and (4-2) can be solved by numerical methods since no analytic solution can be obtained. On the basis of computer calculations, the width of the boundary layer is found to be approximately 3 times a characteristic length known as the *extrinsic Debye length* L_D:†

$$L_D = \left(\frac{K\epsilon_o V_T}{q|N_d - N_a|} \right)^{1/2} \tag{4-9}$$

L_D is a measure of the sharpness of the depletion edge. For a net impurity concentration of $10^{16} \, \text{cm}^{-3}$ in silicon, for example, we have $L_D = 3 \times 10^{-6}$ cm.

---

† A lucid explanation of this concept is given in Appendix B of P. E. Gray, et al., "Physical Electronics and Circuit Models of Transistors," SEEC Series, vol. II, Wiley, New York, 1964.

Thus, the boundary layer is estimated to be less than 10^{-5} cm. This is usually smaller than the width of the depletion region, so that the boundary layer may be completely neglected. From now on, we shall assume that the *p-n* junction can be divided simply into the neutral and depletion regions. Since the depletion region will be the same as the space-charge region, the terms *depletion* and *space-charge* regions are frequently used interchangeably.

4-2 THE DEPLETION APPROXIMATION

By ignoring the boundary layers, the space-charge region of a step junction is well represented by the box-type distribution shown in Fig. 4-3a. In this region, the free carriers are negligible, so that Poisson's equation for the *n* side and *p* side can be simplified to

$$\frac{d^2\psi}{dx^2} = \begin{cases} -\dfrac{qN_d}{K\epsilon_o} & \text{for } 0 < x < x_n \qquad (4\text{-}10a) \\[2ex] \dfrac{qN_a}{K\epsilon_o} & \text{for } -x_p < x < 0 \qquad (4\text{-}10b) \end{cases}$$

$\psi_n - \psi$ always positive

$\psi - \psi_p$ always positive

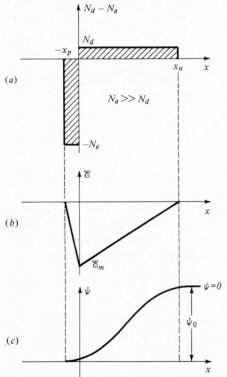

(a)

(b)

(c)

Figure 4-3 The one-sided step junction: (a) the space-charge-distribution, (b) electric field, and (c) potential diagrams.

Space-charge neutrality for the semiconductor as a whole demands that the charges on both sides of the junction be equal. Thus, *Areas are equal*

$$N_a x_p = N_d x_n \tag{4-11}$$

where x_p and x_n denote the depletion-layer width in the p side and n side, respectively. The total space-charge-layer width W is given by

$$W = x_p + x_n \quad = X_n \left(1 + \frac{N_d}{N_a} \right) \tag{4-12}$$

If the impurity concentration on one side of the junction is much higher than that of the other side, as shown in Fig. 4-3a, the junction is called a *one-sided step junction*. The one-sided step junction is an excellent approximation for a diffused junction having a shallow junction depth. As an example, if we let $N_a \gg N_d$, we have $x_n \gg x_p$ and $W \approx x_n$. Physically, this means that the space-charge layer in the heavily doped side is negligible. As a result, we can solve Poisson's equation in the lightly doped side alone to obtain the junction characteristics. Proceeding from this basis, we integrate Eq. (4-10a) once from x_n to x, yielding

$$\frac{d\psi}{dx} = -\frac{qN_d}{K\epsilon_o}(x - x_n) \tag{4-13}$$

where $d\psi/dx = 0$ at $x = x_n$ is used as the boundary condition. Since $\mathscr{E} = -d\psi/dx$, the foregoing equation can be rewritten as

$$\mathscr{E} = \mathscr{E}_m \left(1 - \frac{x}{x_n} \right) \tag{4-14}$$

where

$$\mathscr{E}_m = \frac{-qN_d x_n}{K\epsilon_o} \tag{4-15}$$

Equation (4-14) is plotted in Fig. 4-3b, and $\mathscr{E}_m$ specifies the maximum electric field in the junction.

The potential is derived by integrating Eq. (4-13) from x_n to x:

$$\psi = \frac{-qN_d x_n^2}{2K\epsilon_o} \left(1 - \frac{x}{x_n} \right)^2 \tag{4-16}$$

where we have used the boundary condition $\psi = 0$ at $x = x_n$. Then, we define

$$\psi_o = |\psi(0)| = \frac{qN_d x_n^2}{2K\epsilon_o} \tag{4-17}$$

ψ_o is the built-in potential since it is the potential difference between x_n and 0 and the small potential drop across x_p is negligible. The foregoing equation can be rearranged to express the depletion-layer width:

$$W = x_n = \left(\frac{2K\epsilon_o \psi_o}{qN_d} \right)^{1/2} \tag{4-18}$$

The spatial dependence of the potential is shown in Fig. 4-3c.

Example A silicon *p-n* step junction diode is doped with $N_d = 10^{16} \, \text{cm}^{-3}$ and $N_a = 4 \times 10^{18}$ on the *n* side and *p* side, respectively. Calculate the built-in potential, depletion-layer width, and maximum field at zero bias at room temperature.

SOLUTION $\qquad \psi_o = 0.026 \ln \dfrac{4 \times 10^{18} \times 10^{16}}{2.25 \times 10^{20}} = 0.83 \, \text{V} \qquad$ from Eq. (4-7)

$$W = x_n = \left(\frac{2 K \epsilon_o \psi_o}{q N_d} \right)^{1/2} = 3.28 \times 10^{-5} \, \text{cm} \quad \text{from Eq. (4-18)}$$

$$\mathscr{E}_m = \frac{-q N_d x_n}{K \epsilon_o} = -5 \times 10^4 \, \text{V/cm} \qquad \text{from Eq. (4-15)}$$

The foregoing calculation gives order-of-magnitude values of a typical *p-n* junction.

In a linearly graded junction, the space-charge distribution in the depletion layer is given by

$$N_d - N_a = ax \tag{4-19}$$

where *a* is the slope of the impurity concentration. Therefore, Poisson's equation can be written as

$$\frac{d^2 \psi}{dx^2} = -\frac{q}{K \epsilon_o} ax \tag{4-20}$$

By solving this equation it can be shown that the equilibrium depletion-layer width and the built-in voltage are

$$W = \left(\frac{12 K \epsilon_o \psi_o}{qa} \right)^{1/3} \tag{4-21}$$

$$\psi_o = 2 V_T \ln \frac{aW}{2 n_i} \tag{4-22}$$

Note that the exponent of $\frac{1}{3}$ in Eq. (4-21) differs from the square-root dependence of the depletion-layer width of the step junction.

4-3 BIASING OF A *p-n* JUNCTION

When an external voltage source is connected across the *p-n* junction, the thermal equilibrium is disturbed and a current will flow in the semiconductor. In general, the resistance of the space-charge region is so much higher than that of the neutral regions that the potential drops in the latter are negligible in comparison with the former. As a result, the external applied voltage is directly across the space-charge region. The magnitude of the conduction current depends strongly on the polarity of the applied voltage. If a positive voltage *V* is applied to the *p* side with respect to the *n* side, the potential-barrier height at the *p-n* junction is lowered to $\psi_o - V$, as shown in Fig. 4-4*b*. The reduced potential-barrier height allows majority carriers to diffuse through the junction so that a large current is realized. This voltage polarity is in the *forward-bias* direction, which gives a low-resistance path for the *p-n* junction. If a negative voltage $- V_R$ is applied to the *p* side with respect to the *n* side, the potential-barrier height is increased to $\psi_o + V_R$, as indicated in Fig. 4-4*c*. The increased potential barrier prevents carrier transport through the junction. Consequently, the current flow through the junction will be extremely small, and the impedance of the junction will be very high. This is known as the *reverse-bias* connection.

In Fig. 4-4*b* and *c*, we have sketched the quasi-Fermi levels for both forward- and reverse-bias conditions. The quasi-Fermi levels are related to

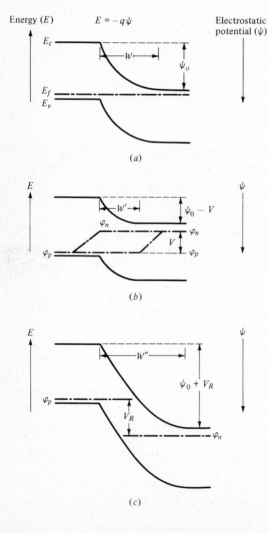

Figure 4-4 The potential distribution of a one-sided step junction under (*a*) thermal equilibrium with a depletion-layer width *W*, (*b*) forward bias *V* with a depletion-layer width $W' < W$, and (*c*) reverse bias V_R with a depletion-layer width $W'' > W$. The potential difference, e.g. ψ_0, is written without $-q$.

carrier densities by Eqs. (2-51) and (2-52), and the currents are determined by the gradient of φ_n and φ_p. The knowledge of the quasi-Fermi levels provides additional physical understanding of device behavior. For example, the constancy of φ_n and φ_p under reverse bias in Fig. 4-4c is the result of essentially zero current flow. In Fig. 4-4b, the quasi-Fermi levels for majority carriers in the neutral regions do not deviate from the Fermi level at equilibrium because the majority-carrier concentrations in these regions remain at their respective equilibrium values. However, φ_n in the *n* region is displaced from φ_p in the *p* region by the applied voltage *V*. The splitting of the quasi-Fermi levels indicates the presence of excess carriers in the neutral regions next to the depletion region. These pictures should be examined along with the carrier and current distributions to be derived later.

Under reverse bias, Eqs. (4-18) and (4-21) are valid provided that the built-in potential ψ_o is replaced by $\psi_o + V_R$. Thus the depletion-layer width becomes

$$W = \left[\frac{2K\epsilon_o(\psi_o + V_R)}{qN_d}\right]^{1/2} \qquad (N_d \ll N_a) \tag{4-23}$$

for the step junction and

$$W = \left[\frac{12K\epsilon_o(\psi_o + V_R)}{qa}\right]^{1/3} \tag{4-24}$$

for the linearly graded junction. Note that the depletion-layer width of a *p-n* junction increases with the increasing reverse-bias voltage. The depletion-layer width as a function of the reverse bias and doping is plotted in Fig. 4-5 for two diffused junctions. The results of both the step and linearly graded junctions are also plotted for comparison. Note that the one-sided step junction is a good approximation for the shallow diffused junction in Fig. 4-5a and the linearly graded junction is a good approximation for the deep diffused junction in Fig. 4-5b. Study of Probs. 4-5 and 4-6 is helpful in understanding Fig. 4-5.

Under the forward-bias condition, carriers are injected across the space-charge layer. For very small current, the injected-carrier densities do not seriously disturb the space-charge region, so that Eqs. (4-23) and (4-24) can still be used if V_R is replaced by $-V$. However, as the current is increased, the carrier concentrations in the space-charge region will become comparable with the fixed impurity-ion density. Under this condition, Eqs. (4-23) and (4-24) no longer apply. In practice Eqs. (4-23) and (4-24) cannot be used for most of the forward-current range.

4-4 INJECTION AND TRANSPORT OF MINORITY CARRIERS

Under forward bias, the applied voltage reduces the potential barrier of the *p-n* junction as depicted in Fig. 4-4b. The barrier lowering enhances electron

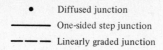

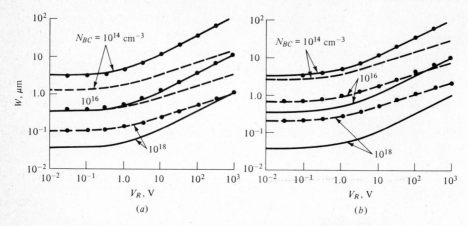

Figure 4-5 Experimental and calculated results of depletion-layer width vs. applied reverse bias for (a) $x_j = 1\,\mu\text{m}$ and (b) $x_j = 10\,\mu\text{m}$. Assume an erfc distribution and $N_o = 10^{20}/\text{cm}^3$. *(From Grove [1].)*

diffusion from the n side to p side and hole diffusion from the p side to the n side. In other words, electrons are injected into the p side, and holes are injected into the n side. Minority-carrier injection and transport are the two topics to be described in this section.

Minority-Carrier Concentrations at the Junction Boundary

Before we examine the problem of minority-carrier injection, we must determine the relationship between the built-in potential and equilibrium carrier densities. Let us again use the subscripts n and p to denote the semiconductor type and the subscript o to specify the condition of thermal equilibrium. Thus, n_{no} and n_{po} are the equilibrium electron densities in the n side and p side, respectively. The expression of the built-in potential in Eq. (4-7) can be rewritten as

$$\psi_o = V_T \ln \frac{p_{po}n_{no}}{n_i^2} = V_T \ln \frac{n_{no}}{n_{po}} \qquad (4\text{-}25)$$

where the majority-carrier densities p_{po} and n_{no} are used to replace N_a and N_d, respectively, and the mass-action law $p_{po}n_{po} = n_i^2$ has been employed. We can rearrange Eq. (4-25) to give

$$n_{no} = n_{po}e^{\psi_o/V_T} \qquad (4\text{-}26)$$

Similarly, we have

$$p_{po} = p_{no}e^{\psi_o/V_T} \tag{4-27}$$

where p_{po} is the equilibrium hole density in the p side and p_{no} is the equilibrium hole density on the n side. From Eqs. (4-26) and (4-27) it is observed that the electron densities on the two edges of the space-charge layer of a junction barrier are related through the barrier height, in this case ψ_o. It is reasonable to assume that the same rule applies when the barrier height is changed by an external voltage.

By applying a forward bias V, the junction potential is reduced to $\psi_o - V$. Thus, Eq. (4-26) is modified to

$$n_n = n_p e^{(\psi_o - V)/V_T} \tag{4-28}$$

where n_n and n_p are the electron densities at the edge of the space-charge layer in the n side and p side, respectively. For low-level injection, the injected-electron density in the n side is small compared with n_{no}, so that we can assume $n_n = n_{no}$. Substituting this condition and Eq. (4-26) into Eq. (4-28), we obtain

$$n_p = n_{po}e^{V/V_T} \tag{4-29}$$

Similarly, we have

$$p_n = p_{no}e^{V/V_T} \tag{4-30}$$

Equations (4-29) and (4-30) define the minority-carrier densities at the edges of the space-charge layer. These are the boundary conditions which specify the carrier densities uniquely for a given applied voltage.

Space-Charge Effect and the Diffusion Approximation

Let us now consider the behavior of carriers in the neutral n side with injected holes. An electric field is established momentarily by the positive charge associated with the injected excess holes. This field draws in excess electrons, which neutralize the injected holes and reestablish the space-charge neutrality. As a result, very large excess-carrier densities are possible without noticeable space-charge effect, as depicted in Fig. 4-6. However, a small field exists in the region with excess carriers. This electric field acts on both majority and minority carriers to produce the drift-current components $q\mu_n n_n \mathscr{E}$ and $q\mu_p p_n \mathscr{E}$. Since the density of majority carriers n_n is much higher than that of the minority carriers p_n, the effect of the field on the majority carriers is very strong but its effect on the minority carriers is negligible. Thus, this small local field molds the majority carriers to satisfy the space-charge neutrality while the minority carriers are hardly influenced by the same field.

With this physical picture in mind, we conclude that the injected minority carriers are the controlling force in the forward-biased p-n junction. In fact, since the majority carriers are passive and their function is just to neutralize the field introduced by the minority carriers, we can neglect the effect of the majority carriers altogether. In the region with injected carriers, we assume

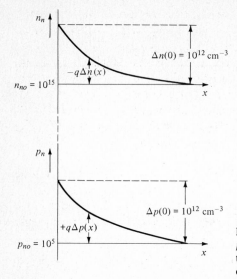

Figure 4-6 Hole injection into the n side of a p^+-n junction and the resultant electron distribution. $+q$ and $-q$ are added to illustrate charge neutrality.

that the condition of space-charge neutrality prevails and that the minority carriers are the only type of carriers present. These minority carriers, being neutralized, are *uncharged*, and they are transported by diffusion in the neutral region. This is known as the *diffusion approximation*. Mathematically, the hole current is obtained by setting $\mathscr{E} = 0$ in Eq. (2-62). Thus, we have

$$I_p = -qAD_p \frac{dp_n}{dx} \tag{4-31}$$

and the hole continuity equation in Eq. (2-60) becomes

$$\frac{\partial p_n}{\partial t} = D_p \frac{\partial^2 p_n}{\partial x^2} - \frac{p_n - p_{no}}{\tau_p} \tag{4-32}$$

which is called the *diffusion equation* for holes. The injected-hole distribution and current magnitude are obtainable by solving Eqs. (4-31) and (4-32) with appropriate boundary conditions. Similarly, the electron current and diffusion equations on the p side of the junction are

$$I_n = qAD_n \frac{dn_p}{dx} \tag{4-33}$$

$$\frac{\partial n_p}{\partial t} = D_n \frac{\partial^2 n_p}{\partial x^2} - \frac{n_p - n_{po}}{\tau_n} \tag{4-34}$$

4-5 DC CURRENT-VOLTAGE CHARACTERISTICS

A p-n junction diode refers to a packaged two-terminal rectifier in which current can flow in one direction only. In most cases a diode contains only one *p-n* junction. For this reason, the terms *junction* and *diode* are frequently used interchangeably. The current-voltage relationship in a *p-n* junction can be obtained by solving the continuity and current equations (4-31) to (4-34). Let us consider the minority carriers in the *n* side of the junction for a *long* diode, where the external contact in the lightly doped side is very far from the junction space-charge region. In other words, all the injected carriers in a long diode recombine before they reach the contact. The impurity concentration in the *n* side is assumed to be uniform. The hole diffusion equation for the time-independent case is rewritten using $p_n - p_{no} = \Delta p_n$:

$$D_p \frac{d^2 \Delta p_n}{dx^2} - \frac{\Delta p_n}{\tau_p} = \frac{\partial p_n}{\partial t} = 0 \tag{4-35}$$

where we have added $-d^2 p_{no}/dx^2$ to form the first term on the left-hand side since this added term is zero because of uniform doping. The general solution of Eq. (4-35) is

$$\Delta p = K_1 \exp\left(-\frac{x}{\sqrt{D_p \tau_p}}\right) + K_2 \exp\frac{x}{\sqrt{D_p \tau_p}} \tag{4-36}$$

where K_1 and K_2 are constants to be determined. The boundary conditions are

$$p_n = \begin{cases} p_n(0) = p_{no} e^{V/V_T} & \text{at } x = 0 \tag{4-37a} \\ p_{no} & \text{at } x = \infty \tag{4-37b} \end{cases}$$

The first boundary condition is taken from Eq. (4-30), and the second boundary condition is obtained by assuming that the injected holes recombine before reaching the external contact. Substituting the boundary conditions into Eq. (4-36) leads to

$$\Delta p_n = p_{no}(e^{V/V_T} - 1)e^{-x/L_p} \tag{4-38a}$$

where
$$L_p = \sqrt{D_p \tau_p} \tag{4-38b}$$

The symbol L_p is called the *hole diffusion length*. Physically, L_p is the mean distance traveled by an injected hole before recombination. The hole current is derived by substituting Eq. (4-38a) into Eq. (4-31):

$$I_p = qA \frac{D_p p_{no}}{L_p} (e^{V/V_T} - 1)e^{-x/L_p} \tag{4-39}$$

The hole current at the edge of the space-charge layer is evaluated at $x = 0$:

$$I_p(0) = qA \frac{D_p p_{no}}{L_p} (e^{V/V_T} - 1) \qquad e^{-(x+x_p)/L_n} \tag{4-40}$$

Thus, the hole current distribution in Eq. (4-39) can be rewritten as

$$I_p = I_p(0)e^{-x/L_p} \tag{4-41}$$

From this equation, we notice that the hole current in the *n* side decreases with distance away from the junction. Since the total current must be invariant with respect to x to satisfy current continuity, the electron current must be increasing with x to compensate for the drop of the hole current. In other words, the minority-carrier current is continuously transferred into the majority-carrier current via electron-hole pair recombination. Similarly, the electron distribution and electron current in the p side of the junction are

$$\Delta n_p = n_{po}(e^{V/V_T} - 1)e^{-x'/L_n} \qquad \text{(4-42)}$$

$$I_n = I_n(0)e^{-x'/L_n} \qquad \text{(4-43)}$$

where

$$I_n(0) = qA\frac{D_n n_{po}}{L_n}(e^{V/V_T} - 1) \qquad \text{(4-44)}$$

and

$$L_n = \sqrt{D_n \tau_n} \qquad \text{(4-45)}$$

L_n is the electron diffusion length, and x' is the spatial dimension in the p side. Again, the transfer of electron current into hole current takes place on the p side of the junction. The current and minority-carrier distributions are illustrated in Fig. 4-7, where the exponential nature of current and carrier distribution are clearly observable. Equations (4-40) and (4-44) are applicable for the reverse bias provided a negative sign is assigned to the voltage V.

In Fig. 4-7, the total current of the junction can be obtained by adding the minority-carrier-current components at the respective space-charge-layer edge. Therefore

$$I = I_n(0) + I_p(0) = I_o(e^{V/V_T} - 1) \qquad \text{(4-46)}$$

where

$$I_o = qA\left(\frac{D_p p_{no}}{L_p} + \frac{D_n n_{po}}{L_n}\right) \qquad \text{(4-47)}$$

Equation (4-46) describes the ideal current-voltage characteristic of the p-n junction depicted in Fig. 4-8. I_o is called the *saturation current*, which is the ideal reverse-bias current of the junction. It should be pointed out that the ratios of the diffusion constant to diffusion length for electrons and holes are of the same order of magnitude. Therefore, for a one-sided step junction, the equilibrium minority-carrier density of the lightly doped side is much greater than that of the heavily doped side. Consequently, the I-V characteristic is determined primarily by the lightly doped side.

Example A silicon p-n junction diode has the following parameters: $N_d = 10^{16}$, $N_a = 5 \times 10^{18}$ cm^{-3}, $\tau_n = \tau_p = 1$ μs, $A = 0.01$ cm^2. Assume that the width of the two sides of the junction are much greater than the respective minority-carrier diffusion length. Obtain the applied voltage at a forward current of 1 mA at 300 K.

SOLUTION From Fig. 1-19, we find the electron mobility in the p side ($N_a = 5 \times 10^{18}$ cm^{-3}) to be 120 cm^2/V-s and the hole mobility in the n side ($N_d = 10^{16}$ cm^{-3}) to be 1100 cm^2/V-s. Since $V_T = 26$ mV at 300 K, we use

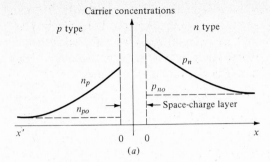

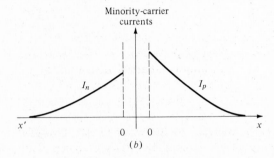

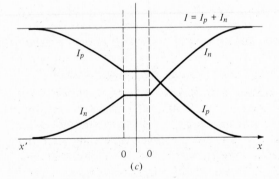

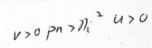

$v > 0$ $pn > n_i^2$ $u > 0$

Figure 4-7 The forward-biased *p-n* junction: (*a*) minority-carrier distributions, (*b*) minority-carrier currents, and (*c*) electron and hole currents.

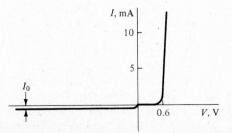

Figure 4-8 A typical current-voltage characteristic of a *p-n* junction.

Einstein's relationship to obtain

$$D_p = 0.026 \times 120 = 3.1 \text{ cm}^2/\text{s} \qquad \text{from Eq. (1-50)}$$
$$D_n = 0.026 \times 1100 = 28.6 \text{ cm}^2/\text{s}$$

Therefore $\quad L_p = \sqrt{D_p \tau_p} = 1.8 \times 10^{-3} \text{ cm} \qquad \text{from Eq. (4-38}b)$

$$L_n = \sqrt{D_n \tau_n} = 5.3 \times 10^{-3} \text{ cm} \qquad \text{from Eq. (4.45)}$$

In addition, $n_i = 1.5 \times 10^{10} \text{ cm}^{-3}$ and

$$p_{no} = \frac{n_i^2}{N_d} = 2.25 \times 10^4 \text{ cm}^{-3} \qquad n_{po} = \frac{n_i^2}{N_a} = 45 \text{ cm}^{-3}$$

Thus

$$I_o = (1.6 \times 10^{-19})(0.01)\left[\frac{(3.1)(2.25 \times 10^4)}{1.8 \times 10^{-3}} + \frac{28.6 \times 45}{5.3 \times 10^{-3}}\right] = 6.3 \times 10^{-14} \text{ A}$$

$$\text{from Eq. (4-47)}$$

Rewriting Eq. (4-46), we obtain

$$V = V_T \ln\left(\frac{I}{I_o} + 1\right) = 26 \ln \frac{10^{-3}}{6.3 \times 10^{-14}} = 610 \text{ mV}$$

In a practical junction, the current-voltage characteristic deviates significantly from Eq. (4-46) as a result of carrier recombination and generation inside the space-charge layer. This effect will be considered in the next section. In some diodes, the length of the lightly doped side W_n may be on the order of the diffusion length. This is considered a *short* diode, in which the boundary condition away from the space-charge layer is determined by the surface recombination velocity at W_n. Thus, the minority-carrier current at the far end of the n side of a p^+-n junction diode is given by

$$I_p|_{x=W_n} = qS\Delta p_n A \tag{4-48}$$

where S is the surface recombination velocity expressed in Eq. (2-37).

4-6 RECOMBINATION AND GENERATION CURRENT IN THE SPACE-CHARGE REGION

In the preceding section on the forward characteristics of a p-n junction diode, we have implicitly assumed that there is no carrier recombination within the space-charge region. However, the forward bias increases the carrier concentration at the edges of the space-charge layer, so that $pn > n_i^2$. These excess carriers are crossing the space-charge layer, so that the carrier concentrations may exceed the equilibrium values. Therefore, recombination is expected in the junction space-charge layer. The recombination current is defined by

$$I_{rec} = qA \int_0^W U \, dx \tag{4-49}$$

where U is given in Eq. (2-24). It can be shown that the maximum recombination rate occurs when $n = p$ (Prob. 4-12), and we have

$$n = p = n_i e^{V/2V_T} \tag{4-50}$$

and

$$U_{max} = \frac{n_i(e^{V/V_T} - 1)}{2\tau_o(e^{V/2V_T} + 1)} \quad \text{if } E_t = E_i \tag{4-51}$$

where τ_o is the effective minority carrier lifetime in the depletion region. For $V \gg V_T$, the maximum recombination rate becomes

$$U_{max} = \frac{n_i}{2\tau_o} e^{V/2V_T} \tag{4-52}$$

Substitution of the foregoing equation into Eq. (4-49) yields

$$I_{rec} = I_R e^{V/2V_T} \tag{4-53}$$

where

$$I_R = \frac{qAn_iW}{2\tau_o} \tag{4-54}$$

I_{rec} is derived on the basis of the worst-case condition. In both silicon and gallium arsenide diodes, the recombination current component dominates at low-current level, as shown in Fig. 4-9 for a typical silicon diode. The semilog plot shows the change of slope from $2V_T$ to V_T for increasing current, indicating the increase of the diffusion current. At high-current levels, the

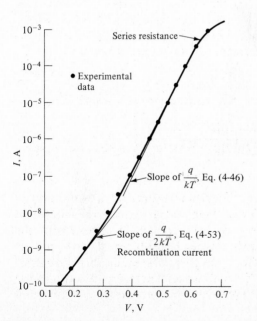

Figure 4-9 The current-voltage characteristic of a silicon-diffused junction with a substrate doping of $10^{16} \, cm^{-3}$.

series resistance provides a large ohmic drop to dominate the V-I characteristics.

In a forward-biased junction, we have the condition $pn > n_i^2$. It is therefore reasonable to expect $pn < n_i^2$ in the space-charge layer of a reverse-biased junction. Substitution of this condition into Eq. (2-24) leads to a negative recombination rate. From the physical picture shown in Chap. 2, a negative recombination rate means a positive generation rate, and the resulting current is a generation current instead of a recombination current. By using the approximation $|V/V_T| \gg 1$, the generation current is derived from the total recombination within the space-charge layer. Thus,

$$I_G = qA|U|W = \frac{qn_iA}{2\tau_o} W \tag{4-55}$$

where we have used the relation

$$U = -\frac{n_i}{2\tau_o} \tag{4-56}$$

Since the space-charge-layer width increases with reverse bias, the generation current increases with the reverse-bias voltage. Introduction of gold will reduce the carrier lifetime and thus increase the generation rate. In a practical silicon junction diode, the generation current is usually greater than the saturation current expressed in Eq. (4-47), so that the reverse current is not saturated.

4-7 TUNNELING CURRENT

In our previous discussion of the p-n junction, we considered the diffusion current resulting from injected carriers that overcame the junction potential barrier. When both the p side and n side are heavily doped with impurities, some carriers may penetrate (instead of going over) the potential barrier to produce an additional current. This mechanism, called *quantum-mechanical tunneling*, can take place when (1) the Fermi level is located within the conduction or valence band; (2) the width of the space-charge layer is so narrow that a high tunneling probability exists; and (3) at the same energy level, electrons are available in the n-type conduction band and empty states are available in the p-type valence band. These conditions are satisfied when both sides of the junction are heavily doped so that they become degenerate semiconductors. Figure 4-10a shows such a junction at 0 K without external bias. The absolute zero temperature is chosen so that all states below the Fermi level will be occupied and all states above the Fermi level will be empty. This assumption simplifies the graphical description without losing the essence of the physical reality at room temperature. When a forward bias is applied, the band diagram is changed to that of Fig. 4-10b. Note that the energy of some electrons in the conduction band of the n side is now raised

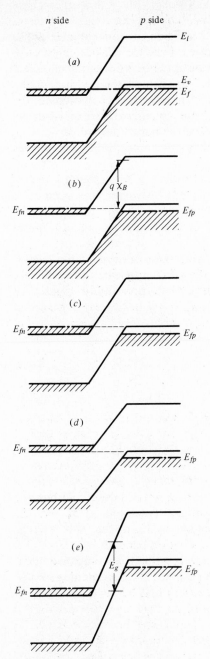

Figure 4-10 The energy-band diagram of a tunneling junction under various biasing conditions.

to a level corresponding to the level of empty states in the p side of the junction. As a result, electrons may tunnel through the junction potential barrier to produce a current. The magnitude of this tunneling current is limited either by the available electrons in the n side for tunneling or by the available empty states at the same energy level to which electrons can tunnel. Consequently, the maximum current is reached at the bias condition shown in Fig. 4-10c. Further increases of the forward bias would reduce the current since there are fewer empty states on the p side to accommodate the tunneling electrons. The condition of zero tunneling current is realized in the band diagram in Fig. 4-10d. Under reverse bias, the band diagram depicted in Fig. 4-10e leads to increasing reverse tunneling current with the reverse-bias voltage.

The analysis of the tunneling mechanism requires a background in quantum mechanics and is beyond the scope of this text. However, it is instructive to examine a simple model so as to gain physical insight into this important effect. Let us consider the potential barrier of the tunneling junction under forward bias, illustrated in Fig. 4-11, where χ_B is the potential-barrier height and is roughly equal to the band-gap energy E_g. The thickness of the barrier W is the space-charge-layer width, and n is the number of electrons available for tunneling. In this situation, the simplified tunneling probability (given here without proof) is [2]

$$T_t = \exp\left(-\frac{8\pi\sqrt{2qm_e}\chi_B^{3/2}}{3h\mathscr{E}}\right) \tag{4-57}$$

In Fig. 4-11 we have

$$\mathscr{E} = \frac{\chi_B}{W} \tag{4-58}$$

Substitution of Eq. (4-58) into (4-57) leads to

$$T_t = \exp\left[-\frac{8\pi W}{3h}(2qm_e\chi_B)^{1/2}\right] \tag{4-59}$$

Assuming that the number of empty states on the other side of the barrier, that is, $x > W$, is very large, the tunneling current is given by

$$I_t = qAv_{th}nT_t \tag{4-60}$$

where v_{th} is the velocity of the tunneling electrons. As an example, let a silicon diode be doped heavily on both sides so that the space-charge-layer

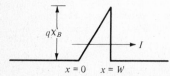

$q\chi_B$

$x = 0$ $x = W$

I

Figure 4-11 The potential barrier of the forward-biased tunneling junction corresponding to Fig. 4-10b.

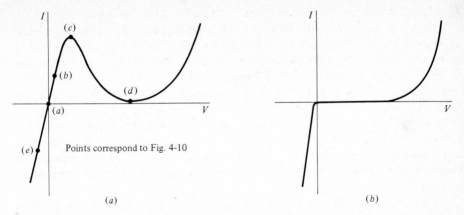

Figure 4-12 Current-voltage characteristics of (*a*) the Esaki diode and (*b*) the backward diode.

width is 50 Å. The barrier height is 1.1 eV, the diode area is 10^{-4} cm^2, $m_e = 0.2$ m, and $n = 2 \times 10^{20}$ cm^{-3}. Then the tunneling probability is 10^{-7}, and the tunneling current is approximately 2 mA. Note that the tunneling probability is quite small at a significant current level.

In addition to the tunneling-current component, the diffusion current becomes important at high forward-bias voltages. The current-voltage characteristics of a junction including the tunneling and diffusion components are summarized in Fig. 4-12*a*. A diode having this *I-V* curve is called the *tunnel diode* or *Esaki diode*. The impurity densities are typically 5×10^{19} cm^{-3}, and the depletion-layer thickness is in the order of 50 to 100 Å. If the doping densities are slightly reduced so that the forward tunneling current is negligible, the current-voltage curve will be modified to that shown in Fig. 4.12*b*. This is called a *backward diode*.

The tunneling current is relatively insensitive to temperature variation, and the negative resistance region of a silicon Esaki diode remains the same even at a temperature of 150°C. Because the time required for tunneling is very small, the switching speed of an Esaki diode is very high. Applications of this unique device include oscillators, bistable and monostable multivibrators, high-speed logic circuits, and low-noise microwave amplifiers. However, problems in using a two-terminal active device and difficulties in fabricating them in integrated-circuit form have limited the use of tunnel diodes.

4-8 TEMPERATURE DEPENDENCE OF THE *V-I* CHARACTERISTIC

The space-charge-layer recombination current and the tunneling current are comparatively less temperature-sensitive than the diffusion-current component. For this reason, the diode equation expressed in Eq. (4-46) is

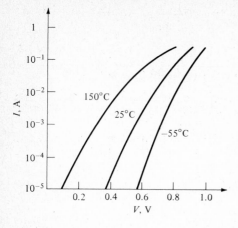

Figure 4-13 Temperature effect on the current-voltage characteristic of a silicon planar diode.

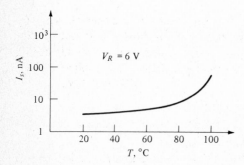

Figure 4-14 The reverse saturation current as a function of temperature in a silicon p-n diode.

adequate to describe the thermal effect on the V-I characteristics of most diodes. The temperature effect is contained implicitly in I_o and V_T. We can derive the temperature dependence of both the current and voltage of the diode using Eq. (4-46). Neglecting the unity term in comparison with the exponential term in the diode equation, we derive the relations (Prob. 4-13)

$$\left.\frac{dV}{dT}\right|_{I=\text{const}} = \frac{V}{T} - V_T\left(\frac{1}{I_o}\frac{dI_o}{dT}\right) \tag{4-61}$$

and

$$\left.\frac{dI}{dT}\right|_{V=\text{const}} = I\left(\frac{1}{I_o}\frac{dI_o}{dT} - \frac{V}{TV_T}\right) \tag{4-62}$$

The temperature dependence of the current and voltage can be obtained if the saturation current as a function of temperature is known. If we use the mass-action law, the saturation current can be written as

$$I_o = qAn_i^2\left(\frac{D_p}{L_pN_d} + \frac{D_n}{L_nN_a}\right) \tag{4-63}$$

The parameters within the parentheses are relatively insensitive to temperature variation, so that the temperature effect of the saturation current is proportional to the square of the intrinsic-carrier concentration. Using Eq. (1-15) for the temperature dependence of n_i^2, we obtain

$$I_o \propto n_i^2 \propto T^3 e^{-E_{go}/kT} \tag{4-64}$$

Differentiating this equation and dividing the result by I_o leads to

$$\frac{I}{I_o}\frac{dI_o}{dT} = \frac{3}{T} + \frac{E_{go}}{kT^2} \approx \frac{E_{go}}{kT^2} \tag{4-65}$$

since in most cases, the term $3/T$ is negligible. Substitution of this expression into Eqs. (4-61) and (4-62) yields

$$\frac{dV}{dT} = \frac{qV - E_{go}}{qT} \tag{4-66}$$

$$\frac{1}{I}\frac{dI}{dT} = \frac{E_{go} - qV}{kT^2} \tag{4-67}$$

In a silicon diode where $E_{go} = 1.21\,\text{eV}$ and the operating voltage is typically $0.6\,\text{V}$ at room temperature (300 K), the current approximately doubles every 10°C, and the voltage decreases linearly with temperature with a coefficient of approximately $-2\,\text{mV/°C}$. The temperature dependence of a silicon diode under both forward and reverse bias is plotted in Figs. 4-13 and 4-14.

4-9 SMALL-SIGNAL AC ANALYSIS

The small-signal characteristics of a p-n junction are important for circuit applications. In this section, we present the derivation of the diode impedance by first solving for the small-signal carrier distribution. With a small-signal voltage superimposed on the dc voltage V, the total applied voltage is expressed as

$$v = V + v_a e^{j\omega t} \tag{4-68}$$

The small-signal condition is satisfied if the applied ac signal v_a is small compared with V_T. By substituting Eq. (4-68) into Eq. (4-30) and using the approximation

$$\exp\frac{v_a e^{j\omega t}}{V_T} = 1 + \frac{v_a}{V_T} e^{j\omega t} \qquad \text{for } \frac{v_a}{V_T} \ll 1 \tag{4-69}$$

we have

$$p_n(0) = P_n(0) + p_{a_1} e^{j\omega t} \tag{4-70}$$

where $P_n(0)$ is given by Eq. (4-37a) and

$$p_{a_1} = \frac{p_{no} v_a}{V_T} e^{V/V_T} \tag{4-71}$$

The boundary condition can be separated into an ac component and a dc component. In addition, we can separate the hole diffusion equation into dc and ac equations by substituting

$$p_n = P_n + p_a e^{j\omega t} \tag{4-72}$$

into Eq. (4-32). The dc equation is identical with Eq. (4-35), and the ac equation is

$$D_p \frac{d^2 p_a}{dx^2} - \frac{p_a}{\tau_p} = j\omega p_a \tag{4-73}$$

for which the ac boundary conditions are

$$p_n = \begin{cases} p_{a_1} & \text{at } x = 0 \\ 0 & \text{at } x = \infty \end{cases}$$

Using these boundary conditions, we derive the expression for the ac hole density:

$$p_a = p_{a_1} \exp\left(\frac{-x}{L_p}\sqrt{1 + j\omega\tau_p}\right) \tag{4-74}$$

Differentiating this hole density and substituting the result into Eq. (4-31), we obtain the hole current at $x = 0$:

$$i_p(0) = \frac{v_a}{V_T} I_p(0)\sqrt{1 + j\omega\tau_p} \tag{4-75}$$

where $I_p(0)$ is given in Eq. (4-40). Similarly, we find the electron current in the p side as

$$i_n(0) = \frac{v_a}{V_T} I_n(0)\sqrt{1 + j\omega\tau_n} \tag{4-76}$$

The total ac current of the p-n junction is therefore

$$i = i_p(0) + i_n(0) \tag{4-77}$$

For a p^+-n junction, $i_p(0) = i$ and $I_p(0) = I$. Thus, the *diode admittance* is reduced to

$$y = \frac{i_p(0)}{v_a} = \frac{I}{V_T}\sqrt{1 + j\omega\tau_p} \approx \frac{I}{V_T} + j\frac{\omega\tau_p I}{2V_T} \tag{4-78}$$

The last equation is obtained by using $1 \gg \omega\tau_p$ for a low-frequency approximation. Note that the admittance has a conductive and a capacitive component. The capacitance in Eq. (4-78) is called the *diffusion capacitance*:

$$C_D = \frac{\tau_p I}{2V_T} \tag{4-79}$$

For the diode of the example in Sec. 4-5, we obtain the small-signal conductance $g_D = I/V_T = 0.04 \, \mho$ and $C_D = 2 \times 10^{-8}$ F. The small-signal equivalent circuit of a p-n junction diode is shown in Fig. 4-15. In addition to the diode resistance and diffusion capacitance, we have included a series resistance and a transition capacitance. The series resistance is caused by the ohmic drop across the neutral

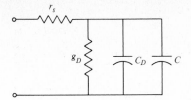

Figure 4-15 A diode equivalent circuit.

semiconductor regions and the contacts. The transition capacitance arises from the junction space-charge layer and is the subject of the next section.

4-10 TRANSITION CAPACITANCE, IMPURITY PROFILING, AND VARACTOR

In Sec. 4-3, we have shown that the depletion-layer width is a function of the bias voltage. Since the charge stored in either half of the junction is directly proportional to the depletion-layer width, we have

$$Q = qAN_dW = A\sqrt{2qK\epsilon_o(\psi_o + V_R)N_d} \qquad (4\text{-}80)$$

where Eq. (4-23) has been employed. The small-signal capacitance of the space-charge layer is defined by

$$C \equiv \frac{dQ}{dV_R} \qquad (4\text{-}81)$$

Substituting Eq. (4-80) into Eq. (4-81) leads to

$$C = A\left[\frac{qK\epsilon_oN_d}{2(V_R + \psi_o)}\right]^{1/2} \qquad (4\text{-}82)$$

C is called the *transition* or *depletion-layer* capacitance. This equation can be rewritten

$$\frac{1}{C^2} = \frac{2}{qK\epsilon_oN_dA^2}(V_R + \psi_o) \qquad (4\text{-}83)$$

An experimental plot of $1/C^2$ vs. V_R is shown in Fig. 4-16. From this figure we can calculate the donor density from the slope of the straight line. In addition, the built-in voltage can be obtained by extrapolating the line to the voltage axis. At the intercept, we have $1/C^2 = 0$ and $\psi_o = -V$ from Eq. (4-83). For the linearly graded junction, the slope of impurity profile and the built-in voltage can be derived from a $1/C^3$-vs.-V plot.

In a *p-n* junction with unknown impurity profile, the capacitance-voltage curve can be used to map out the impurity distribution of the lightly doped side. This is called *impurity profiling*. Let us consider an arbitrary impurity profile shown in Fig. 4-17. The incremental variation of charge is related to

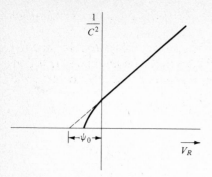

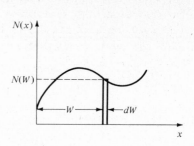

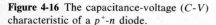

Figure 4-16 The capacitance-voltage $(C\text{-}V)$ characteristic of a $p^+\text{-}n$ diode.

Figure 4-17 An arbitrary impurity profile in the n side of a $p^+\text{-}n$ junction.

the change of space-charge-layer width by

$$dQ = qAN(W)\,dW \tag{4-84}$$

where $N(W)$ is the impurity concentration at the edge of the space-charge layer W. The incremental electric field is related to the incremental charge through Poisson's equation. Thus,

$$d\mathscr{E} = \frac{dQ}{K\epsilon_o A} \tag{4-85}$$

Furthermore, the incremental field can be expressed as a function of the incremental bias voltage by

$$d\mathscr{E} = \frac{dV}{W} \tag{4-86}$$

Solving Eqs. (4-85) and (4-86), we obtain

$$\frac{dQ}{dV} = \frac{AK\epsilon_o}{W} \equiv C \tag{4-87}$$

Substituting of Eqs. (4-85) to (4-87) into Eq. (4-84) and rearranging the result yields

$$N(W) = \frac{2}{qK\epsilon_o A^2}\frac{1}{d(1/C^2)/dV} \tag{4-88}$$

For any $p\text{-}n$ junction with arbitrary impurity concentration in the lightly doped side, we can measure the capacitance vs. reverse-bias voltage and plot $1/C^2$ vs. V. From this plot, we can take $\Delta(1/C^2)/\Delta V$ and substitute the result into Eq. (4-88) to obtain $N(W)$. At the same time, we can use Eq. (4-87) to find W. From a series of such calculations, a complete impurity profile can be mapped out. This method can be used to measure a diffused junction or an ion-implanted region to obtain the impurity profile. A word of caution should be added. If a large density of trapping centers or interface states is present, as in a gold-doped silicon, the foregoing analysis must be modified to accommodate these charged states [3].

The capacitance of a diffused junction can be calculated by solving Poisson's equation to obtain the depletion-layer width and then making use of Eq. (4-87). A computer calculation made by Lawrence and Warner gave the results shown in Fig. 4-18.

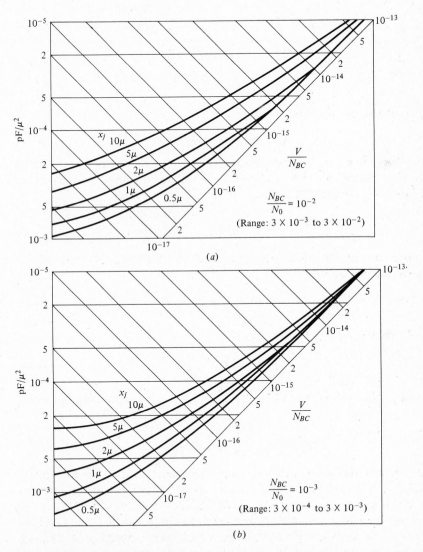

Figure 4-18 The junction capacitance of a diffused-silicon *p-n* diode C per unit area vs. the total junction voltage V divided by the background concentration N_{BC} for various junction depths x_j. *(After Lawrence and Warner [4].)*

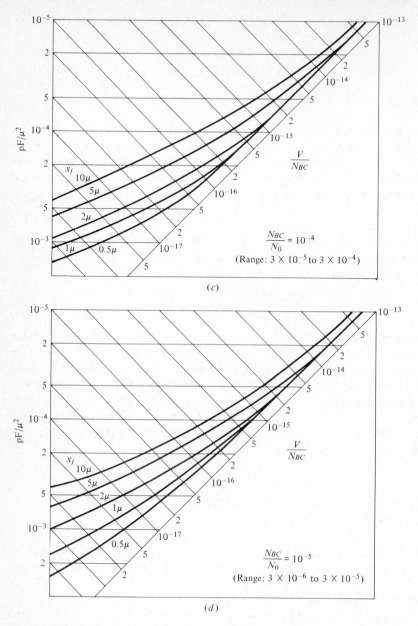

$$\frac{N_{BC}}{N_0} = 10^{-4}$$
(Range: 3×10^{-5} to 3×10^{-4})

(c)

$$\frac{N_{BC}}{N_0} = 10^{-5}$$
(Range: 3×10^{-6} to 3×10^{-5})

(d)

Figure 4-18 (*continued*).

Example Consider a p-n junction fabricated by the two-step diffusion of boron into a substrate with $N_d = 2 \times 10^{15}$ cm^{-3}. The surface concentration of boron is 10^{18} cm^{-3}, and the junction depth is 5 μm. Assume that the built-in potential is 0.8 V; obtain the junction capacitance at a reverse bias of 5 V.

SOLUTION Since $N_{BC} = 2 \times 10^{15}\,cm^{-3}$ and $N_o = 10^{18}\,cm^{-3}$, we have $N_{BC}/N_o = 2 \times 10^{-3}$. In addition

$$\frac{\psi_o + V_R}{N_{BC}} = \frac{5.8}{2 \times 10^{15}} = 2.9 \times 10^{-15}$$

Now, we use Fig. 4-18b to find $C = 4 \times 10^3\,pF/cm^2$.

The capacitance of a p-n junction biased in the reverse direction is useful in LC tuning circuits in which the resonant frequency is controlled by an external voltage. A diode made specially for this purpose is known as the *varactor* (from *vari*able re*actor*). The capacitance-voltage equation of a junction diode can be written

$$C = C_o(V_R + \psi_o)^{-n} \tag{4-89}$$

where V_R is the reverse voltage and $n = \frac{1}{2}$ for a one-sided step junction, as expressed in Eq. (4-82). The resonant frequency of an LC circuit including a p-n junction capacitor is therefore given by

$$\omega_r = \frac{1}{\sqrt{LC}} = \frac{(V_R + \psi_o)^{n/2}}{\sqrt{LC_o}} \tag{4-90}$$

In circuit applications, it is frequently desirable to have a linear relationship between the resonant frequency and the control voltage, that is, $n = 2$. This special feature is obtained in the *hyperabrupt* varactor, and the detailed impurity profile necessary to achieve this function is derived by computer calculations [5]. In designing varactors, we should minimize the series resistance and leakage current so that the selectivity Q is high. In addition, the range of capacitance variation should be large. A Q of 20 and a factor of 6 variation in capacitance are achievable design goals.

4-11 CHARGE STORAGE AND REVERSE TRANSIENT: THE STEP-RECOVERY DIODE

With a constant forward bias, carriers are injected and maintained in a p-n junction diode. The carriers so maintained in the steady-state condition cannot be removed instantaneously when the forward bias is abruptly switched to a reverse bias. Figure 4-19 shows the injected-carrier distributions at different times after the application of a step-function reverse current. The temporary charge-storage effect of minority carriers is called the *reverse recovery transient*. The current and voltage waveforms are shown in Fig. 4-20. We consider the charge-storage effect by using a long p^+-n diode.

Let us define the total stored charge in the n side as

$$Q_s = qA \int_0^{W_n} \Delta p_n\, dx \tag{4-91}$$

where W_n is the thickness of the n-side from the junction to the ohmic contact.

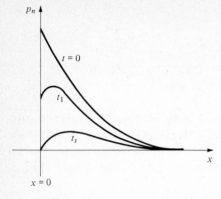

Figure 4-19 Change of minority-carrier distribution during reverse transient in a p^+-n diode. t_s is defined in Fig. 4-20, and $0 < t_1 < t_s$.

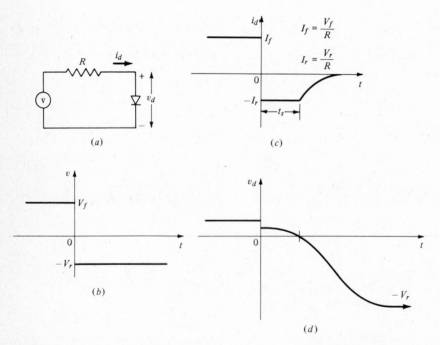

Figure 4-20 The reverse transient of a long p^+-n diode: (a) the biasing circuit, (b) the driving voltage waveform, and (c) the current and (d) voltage waveforms across the diode.

By integrating the continuity equation Eq. (2-60) once from 0 to W_n and using Eq. (4-91) we obtain

$$I_p(0) - I_p(W_n) = \frac{dQ_s}{dt} + \frac{Q_s}{\tau_p} \qquad (4\text{-}92)$$

This is called the *charge-control* equation. In a long diode, we can assume $I_p(W_n)$ is zero. Therefore, the steady-state forward current of the diode is obtained by setting $dQ_s/dt = 0$:

$$I_f = I_p(0) = \frac{Q_{sf}}{\tau_p} \tag{4-93}$$

where Q_{sf} is the stored charge under steady-state, forward-bias condition. Thus,

$$Q_{sf} = I_f\tau_p \tag{4-94}$$

Let us now assume that a negative current I_r is applied at $t = 0$ by reversing the bias voltage as shown in Fig. 4-20. The charge-control equation becomes

$$-I_r = \frac{dQ_s}{dt} + \frac{Q_s}{\tau_p} \tag{4-95}$$

Solving the foregoing equation with Eq. (4-94) as the initial condition, we derive

$$Q_s(t) = \tau_p[-I_r + (I_f + I_r)e^{-t/\tau_p}] \tag{4-96}$$

Let us now define the storage time t_s at which point when all the stored charge has been removed, that is, $Q_s \cong 0$. Consequently

$$t_s = \tau_p \ln \left(1 + \frac{I_f}{I_r}\right) \tag{4-97}$$

An exact analysis by solving the time-dependent diffusion equation yields

$$\text{erf} \sqrt{\frac{t_s}{\tau_p}} = \frac{I_f}{I_r + I_f} \tag{4-98}$$

Both results are depicted in Fig. 4-21, which shows that the charge-control

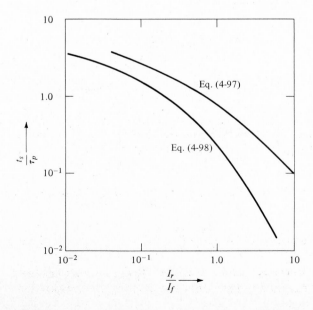

Figure 4-21 Normalized storage time vs. diode-current ratio.

analysis is a reasonable approximation. However, the tail end of the reverse recovery transient shown in Fig. 4-20 can be obtained only through the exact solution, which is not given here [6].

In general, we can assume that the time t_s defines the end of the reverse transient. In other words, the low-impedance state of a forward-biased junction is returned to the high-impedance state of a reverse-biased junction after t_s. From Eq. (4-97), we find that the minority-carrier lifetime should be small to achieve high switching speed in a p-n diode. In practice, gold doping is an effective method for lifetime reduction in computer-diode fabrication. Typical storage time of a switching diode ranges from $\frac{1}{2}$ to 10 ns. For power rectifiers the storage time ranges from 1 μs to 10 ms.

The reverse transient waveforms can be modified by introducing a built-in electric field in the diode. Such an electric field is obtained by doping the diode nonuniformly. For example, if the impurity concentration in the lightly doped side of the p^{+}-n diode is

$$N_d = N_d(0)e^{ax} \qquad (4\text{-}99)$$

where $N_d(0)$ is the impurity value at the junction and a is a constant, then, according to Eq. (2-50), the electric field is

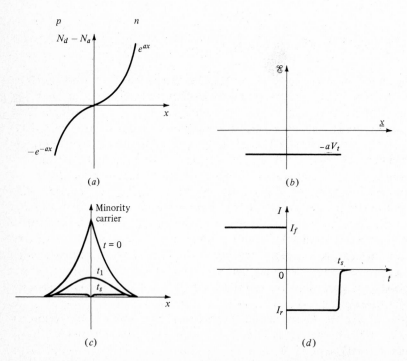

Figure 4-22 The step-recovery or snap-action diode: (a) impurity profile near the junction, (b) electric field distribution, (c) injected minority-carrier distribution, and (d) current waveform.

$$\mathscr{E} = -aV_T \tag{4-100}$$

Substituting Eq. (4-100) into (2-62), we find that both drift- and diffusion-current components have the same sign and the currents are accumulative. In other words, the direction of the field expressed in Eq. (4-100) helps hole transport and is known as the *aiding field*. If the donor-impurity distribution is described by a positive exponent in Eq. (4-99), the electric field is positive and is a *retarding* field for holes. In either case, the field-free approximation used in previous sections must be modified to incorporate the drift-current term.

An unusual application of the built-in field is demonstrated by the *step-recovery* diode shown in Fig. 4-22. The impurity distribution in the junction region provides a retarding field for both injected electrons and holes, so that carriers are confined closely to the junction region. Upon the application of a step reverse current from its forward-bias condition, the stored charge is reduced, just as in a regular diode, but the injected carriers at the junction do not reach zero until all stored charge has been removed. For this reason, the diode abruptly returns to its high-impedance state, and the current waveform shows a snap-action switching in less than 1 ns. The tail end of the current recovery in a regular diode is absent here. Step-recovery diodes are useful for fast pulse and high-frequency harmonic generation [7].

4-12 AVALANCHE BREAKDOWN IN p-n JUNCTIONS

One of the most important considerations in device design is the junction breakdown. A p-n junction reaches the breakdown condition when a slight increase of the reverse-bias voltage produces a very large current. The breakdown process is not inherently destructive, and it can be repeated as long as the maximum current is limited. In the early days of development, the junction breakdown was explained on the basis of Zener's field-emission theory. Zener proposed that the breaking of covalent bonds in the depletion region under high electric field generates electrons and holes so that some valence electrons will move from the valence band to the conduction band by quantum-mechanical tunneling. This mechanism is known as *Zener break-down*. Later, it was found that Zener's model can describe only junctions having a low breakdown voltage. For junctions that break down at a high voltage, e.g., greater than 6 V in silicon, the *avalanche* mechanism is responsible. Because most junctions reach the breakdown voltage via the avalanche process, we shall limit our discussion to this effect.

Let us consider the space-charge region in a reverse-biased junction as shown in Fig. 4-23. A strayed hole in the n region entering the space-charge layer acquires kinetic energy from the electric field as it travels toward the p region. With its high energy, the hole collides with the crystal lattice and ionizes an atom of the lattice to generate an electron-hole pair. Similar-

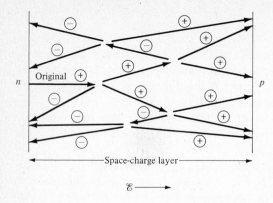

Space-charge layer

$\mathscr{E} \longrightarrow$

Figure 4-23 Avalanche multiplication in the space-charge region initiated by a hole injection from the n side.

ly, an electron traveling toward the n region may create an electron-hole pair via the collision process. After the first collision, the original and the generated carriers will continue their journey, and more collisions may take place to produce more carriers. As a result, the increase of carriers is a multiplication process called *avalanche multiplication* or *impact ionization*. The number of electron-hole pairs produced per centimeter of travel by an electron (hole) is called the *electron (hole) ionization coefficient*. The electron ionization coefficient $\alpha(x)$ and the hole ionization coefficient $\beta(x)$ are in general not equal, although they are both functions of the electric field. In the following analysis, however, we shall assume that $\alpha(x)$ and $\beta(x)$ are identical to simplify our derivation.

The avalanche multiplication mechanism can be analyzed by considering the physical picture shown in Fig. 4-24. It is noted that, without the impact ionization, carriers are introduced into the space-charge region by $I_p(0)$, $I_n(W)$, and the space-charge-generation term G. With impact ionization, holes are traveling and increasing toward the right, and electrons are traveling and increasing toward the left. The continuity of hole current per unit area within Δx demands that

$$I_p(x + \Delta x) - I_p(x) = \alpha(x)[I_n(x) + I_p(x)] \, \Delta x + qAG \, \Delta x \qquad (4\text{-}101)$$

or

$$\frac{dI_p(x)}{dx} = \alpha(x)I + qAG \qquad (4\text{-}102)$$

where

$$I = I_n(x) + I_p(x) = \text{const} \qquad (4\text{-}103)$$

Similarly, we obtain

$$-\frac{dI_n(x)}{dx} = \alpha(x)I + qAG \qquad (4\text{-}104)$$

The negative sign on the left-hand side of this equation results from the decreasing electron current from x to $x + \Delta x$.

Integration of Eq. (4-102) from 0 to x and Eq. (4-104) from x to W yields

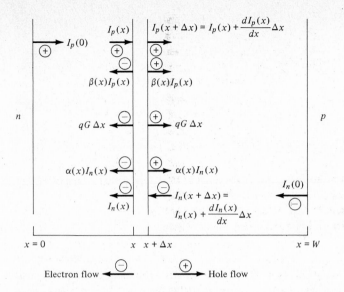

$$I_p(x+\Delta x) = I_p(x) + \frac{dI_p(x)}{dx}\Delta x$$

$$I_n(x+\Delta x) = I_n(x) + \frac{dI_n(x)}{dx}\Delta x$$

Electron flow ⊖ ⊕ Hole flow

Figure 4-24 Current components in a reverse-biased p-n junction under the avalanche condition.

$$I_p(x) - I_p(0) = I\int_0^x \alpha(x)\,dx + \int_0^x qAG\,dx \qquad (4\text{-}105)$$

$$-I_n(W) + I_n(x) = I\int_x^W \alpha(x)\,dx + \int_x^W qAG\,dx \qquad (4\text{-}106)$$

Adding Eqs. (4-105) and (4-106) and rearranging the result, we get

$$I = \frac{I_o + I_G}{1 - \displaystyle\int_0^W \alpha(x)\,dx} \qquad (4\text{-}107)$$

The symbols I_o and I_G are the same as expressed in Eqs. (4-47) and (4-55). Frequently, Eq. (4-107) is written as

$$I = M(I_o + I_G) \qquad (4\text{-}108)$$

where the *avalanche multiplication factor M* is defined by

$$M \equiv \frac{1}{1 - \displaystyle\int_0^W \alpha(x)\,dx} \qquad (4\text{-}109)$$

Theoretically, the avalanche breakdown condition is reached when M is approaching infinity. Therefore, the breakdown criterion is

$$\int_0^W \alpha(x)\,dx = 1 \qquad (4\text{-}110)$$

At this point, it is necessary to find the ionization coefficient as a function of x.

This function can be derived by first using the empirical formula

$$\alpha = A \exp \frac{-B}{|\mathscr{E}|} \tag{4-111}$$

where A and B are material constants. For silicon, $A = 9 \times 10^5 \, \text{cm}^{-1}$ and $B = 1.8 \times 10^6 \, \text{V/cm}$. The magnitude of the electric field is calculated for each junction from the solution of Poisson's equation.

Example Calculate the breakdown voltage of a one-sided step junction.

SOLUTION Let us substitute the electric field expressed in Eq. (4-14), with W replacing x_n, into Eq. (4-111), to yield

$$\alpha = A \exp \frac{-B}{|\mathscr{E}_m|(1 - x/W)} \tag{4-112}$$

Note that in Eq. (4-14) the maximum field is at $x = 0$, where most avalanche multiplication takes place. As an approximation, we may use series expansion to simplify the exponent by setting

$$\frac{1}{1 - x/W} = 1 + \frac{x}{W} + \cdots \approx 1 + \frac{x}{W} \qquad \text{for } x \sim 0 \tag{4-113}$$

Substituting Eqs. (4-112) and (4-113) into Eq. (4-110) and integrating, we obtain

$$1 = \frac{AW\mathscr{E}_m}{B} \exp \frac{-B}{|\mathscr{E}_m|} \left(1 - \exp \frac{-B}{|\mathscr{E}_m|} \right) \tag{4-114}$$

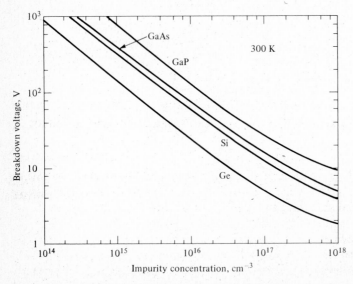

Figure 4-25 Avalanche breakdown voltage of a one-sided step junction as a function of impurity concentration in the lightly doped side. *(After Sze [8].)*

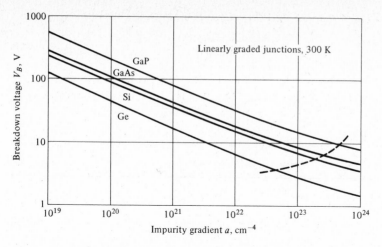

Figure 4-26 Avalanche breakdown voltage vs. impurity gradient for linearly graded junctions in Ge, Si, GaAs, and GaP. The dashed line indicates the maximum gradient beyond which the tunneling mechanism will set in. *(After Sze [8].)*

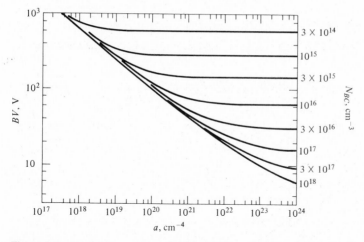

Figure 4-27 Breakdown voltage of diffused-silicon junctions with erfc distribution. *(After Grove [9].)*

When we use Eqs. (4-15) and (4-23) together with Eq. (4-114), we get the avalanche breakdown voltage as a function of impurity concentration in the lightly doped side, and the result is plotted in Fig. 4-25.

Similar calculation can be made for a diffused junction or a linearly graded junction. Figures 4-26 and 4-27 show computed results, and Fig. 4-28 displays the junction curvature effect on the breakdown voltage. Note that a small junction curvature could be the limiting factor in obtaining a high breakdown voltage.

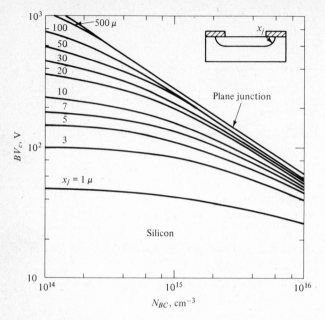

Figure 4-28 Effect of junction curvature on the breakdown voltage of a one-sided step junction. *(After Grove [9].)*

PROBLEMS

4-1 A Si step-junction diode is doped with $N_d = 10^{15}$ cm^{-3} on the n side and $N_a = 4 \times 10^{20}$ on the p side. At room temperature, calculate (*a*) the built-in potential and (*b*) the depletion-layer width and maximum field at zero bias.

4-2 An alternative method of deriving the built-in potential is to assume zero net electron or hole current at equilibrium. Derive Eq. (4-7) by this method.

4-3 If the doping levels on the two sides of a step junction are of the same order of magnitude, the built-in potential and depletion-layer width are given by

$$\psi_o = \frac{qN_aN_d(x_n + x_p)^2}{2K\epsilon_o(N_a + N_d)} \qquad x_n = \left[\frac{2K\epsilon_o\psi_oN_a}{qN_d(N_a + N_d)}\right]^{1/2} \qquad x_p = \left[\frac{2K\epsilon_o\psi_oN_d}{qN_a(N_a + N_d)}\right]^{1/2}$$

Derive these expressions.

4-4 Derive the expressions for (*a*) the electric field, (*b*) potential distribution, (*c*) depletion-layer width, and (*d*) built-in potential for a linearly graded *p-n* junction.

4-5 (*a*) Sketch the impurity distribution and depletion layer of the diffused junction with $N_{BC} = 10^{14}$ cm^{-3} in Fig. 4-5*a*. Explain why the depletion-layer-width vs. V_R curve conforms to that of the one-sided step junction.

(b) Repeat (a) for $N_{BC} = 10^{18}$ to show that the junction acts as a linear junction for small V_R and a step junction for large V_R.

4-6 Repeat Prob. 4-5 for Fig. 4-5b.

4-7 (a) Consider the one-sided step junction shown in Fig. 4-3. Obtain the boundary conditions of carriers, p_n and n_p, under the conditions that the injection into the p side is low level but injection into the n side is no longer at low level.

(b) Simplify the expressions in (a) if high-level injection is approached in the n side of the junction.

4-8 (a) Obtain the expressions of the hole density and current distribution if the boundary condition in Eq. (4-37b) is changed to $p_n = p_{no}$ at $x = W_n$, where W_n has the same order of magnitude as L_p.

(b) Plot the hole density and current distribution obtained in (a) for $W_n/L_p = 0.1$, 1, and 10.

4-9 Plot the carrier and current distributions for a reverse-biased p-n junction similar to those depicted in Fig. 4.7a and c.

4-10 (a) The hole-injection efficiency of a p-n junction is defined as I_p/I at $x = 0$. Show that this efficiency can be written

$$\gamma = \frac{I_p}{I} = \frac{1}{1 + \sigma_n L_p/\sigma_p L_n}$$

(b) What should you do to make γ approach unity in a practical diode?

4-11 In a p^+-n junction diode, the width of the n region W_n is much smaller than L_p. Using Eq. (4-48) as one of the boundary conditions, derive the carrier and current distributions. Sketch the shape of minority carriers in the n side for $S = 0$ and $S = \infty$.

4-12 Show that the maximum recombination rate occurs at $p = n$ in the junction space-charge layer and derive Eq. (4-50).

4-13 Derive Eqs. (4-61) and (4-62).

4-14 A silicon diode operates at a forward voltage of 0.5 V. Calculate the factor by which the current will be multiplied when the temperature is increased from 25 to 150°C. Assume that $I \approx I_o e^{V/2V_T}$ and that I_o doubles every 10°C.

4-15 Derive the small-signal ac hole distribution and diode admittance of a p^+-n diode if the width of the n region W_n is of the same order as the diffusion length. Assume infinite surface recombination velocity at $x = W_n$.

4-16 The capacitance of a GaAs p^+-n junction diode as a function of reverse bias is measured by using a capacitance meter at 1 MHz. The following capacitance data in picofarads are taken at $\frac{1}{2}$-V increments from 0 to 5 V: 19.9, 17.3, 15.6, 14.3, 13.3, 12.4, 11.6, 11.1, 10.5, 10.1, 9.8. Calculate ψ_o and N_d. The diode area is 4×10^{-4} cm^2.

4-17 A p^+-n silicon diode is used as a varactor. The doping concentrations on the two sides of the junction are $N_a = 10^{19}$ and $N_d = 10^{15}$, respectively. The diode area is 100 square mils.

(a) Find the diode capacitance at $V_R = 1$ and 5 V.

(b) Calculate the resonant frequencies of a tank circuit using this varactor with $L = 2$ mH.

4-18 A capacitor in an integrated circuit is fabricated by the diffusion of boron into an n-type Si substrate with $N_{BC} = 5 \times 10^{16}$. Single-step diffusion is performed at $T = 1250°C$. The surface concentration is solubility-limited.

(a) What is the time required to obtain a junction depth of 10 μm?

(b) What is the corresponding capacitance per unit area at a reverse bias of 2 V?

4-19 The hole lifetime of a p^+-n diode is measured by the diode-recovery method.

(a) For $I_f = 1$ mA and $I_r = 2$ mA, t_s is found to be 3 ns in an oscilloscope with a 0.1-ns rise time. Find τ_p.

(b) If the fast scope in (a) is not available and you have to use a slower

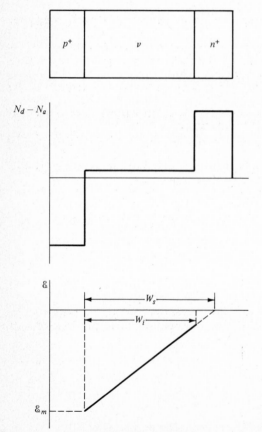

Fig. P4-21

scope with a 10-ns rise time, how can you make an accurate measurement? Describe your result.

4-20 Zener breakdown takes place when the maximum electric field approaches 10^6 V/cm in silicon. Assuming $N_a = 10^{20}$ cm^{-3} in the p side, what is the required donor concentration in the n side to obtain a Zener breakdown voltage of 2 V? Use one-sided step-junction approximation.

4-21 (*a*) For a p^+-ν-n^+ diode shown in Fig. P4.21, it is assumed that the p^+ and n^+ regions do not support any of the external applied voltage. Show that the avalanche-breakdown condition can be expressed as

$$\frac{A\mathscr{E}_m^2\epsilon}{qN_\nu B} \exp\left(-\frac{B}{|\mathscr{E}_m|}\right)\left[1 - \exp\left(-\frac{qBN_\nu W_i}{\mathscr{E}_m^2\epsilon}\right)\right] = 1$$

(*b*) Plot BV vs. $N_d - N_a$ for $W_i = 2$ μm, for Si at 300 K. Assume $\mathscr{E}_m = 6 \times 10^5$ V/cm for all impurity concentrations.

REFERENCES

1. A. S. Grove, "Physics and Technology of Semiconductor Devices," pp. 168–169, Wiley, New York, 1967.
2. S. M. Sze, "Physics of Semiconductor Devices," chap. 4, Wiley, New York, 1969.
3. E. Schibli and A. G. Milnes, Effects of Deep Impurities on n^+p Junction Reverse-biased Small-Signal Capacitance, *Solid-State Electron.*, **11**: 323 (1968).
4. H. Lawrence and R. M. Warner, Jr., Diffused Junction Depletion Layer Calculations, *Bell Syst. Tech. J.*, **39**: 389–403 (1960).
5. P. J. Kannam, S. Ponczak, and J. A. Olmstead, Design Considerations of Hyperabrupt Varactor Diodes, *IEEE Trans. Electron, Devices*, **ED-18**: 109 (1971).
6. B. Lax and S. F. Neustadter, Transient Response of a *p-n* Junction, *J. Appl. Phys.*, **25**: 1148 (1954).
7. J. L. Moll and S. A. Hamilton, Physical Modeling of the Step Recovery Diode for Pulse and Harmonic Generation Circuits, *Proc. IEEE*, **57**: 1250 (1969).
8. S. M. Sze, chap. 3 in Ref. 2.
9. A. S. Grove, pp. 194–201 in Ref. 1.

ADDITIONAL READINGS

Gray, P. E., D. DeWitt, A. R. Boothroyd, and J. F. Gibbons: "Physical Electronics and Circuit Models of Transistors," SEEC Series, vol. II, Wiley, New York, 1964.

Jonscher, A. K.: "Principles of Semiconductor Device Operation," Wiley, New York, 1960.

Shockley, W.: The Theory of *p-n* Junction in Semiconductors and *p-n* Junction Transistors, *Bell Syst. Tech. J.*, **28**: 435 (1949).

METAL-SEMICONDUCTOR JUNCTIONS

Historically, the first practical semiconductor device was the metal-semiconductor diode. In the early development of electronics, rectifiers were made by pressing a metallic whisker to a semiconductor crystal to form a contact. Since the characteristics of point-contact diodes were not reproducible, in most cases they were replaced by the *p-n* junction diode in the 1950s. Recently, modern semiconductor and vacuum technology has been employed to fabricate reproducible metal-semiconductor contacts, and the theory of these junctions is now fairly well understood. It is possible to obtain both rectifying and nonrectifying metal-semiconductor (M-S) junctions. The nonrectifying junction has a low ohmic drop regardless of the polarity of the externally applied voltage and is called the *ohmic contact*. All semiconductor devices need ohmic contacts to make connections to other devices or circuit elements. The rectifying junction is commonly known as the *Schottky-barrier* diode or *hot-carrier* diode. In this chapter, the energy-band diagram and the rectifying properties of the Schottky barrier are presented. In addition, the current-voltage characteristics are given, and the application of the M-S contact to some device structures is discussed. The method of constructing the energy-band diagram of an M-S barrier is also applied to *heterojunctions*, i.e., junctions formed by two different semiconductor materials.

5-1 ENERGY-BAND DIAGRAM OF THE SCHOTTKY BARRIER

The ideal energy-band diagram for a metal and an *n*-type semiconductor before making contact is shown in Fig. 5-1a, with $\phi_m > \phi_s$, where $q\phi_m$ and $q\phi_s$

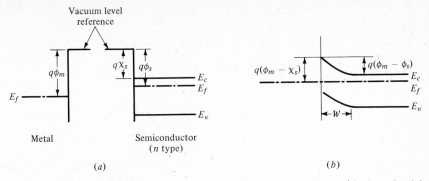

Figure 5-1 Energy-band diagram of a metal-semiconductor contact with $\phi_m > \phi_s$: (a) before contact and (b) after contact and at thermal equilibrium.

represent the *work functions* of the metal and the semiconductor, respectively. As shown in the diagram, the work function is defined as the work required to bring an electron from the Fermi level of the material to the vacuum level. The symbol χ_s, called the *electron affinity*, specifies the energy required to release an electron from the bottom of the conduction band to the vacuum level. When the metal and semiconductor are joined, electrons from the semiconductor cross over to the metal until the Fermi level of the M-S system is aligned. This condition of thermal equilibrium is shown in Fig. 5-1b. The depletion of electrons in the vicinity of the interface produces an upward band bending near the surface. As a result, a potential barrier ψ_o, called the *built-in* or *diffusion voltage*, is established at the M-S contact. The built-in voltage ψ_o is therefore given by

$$\psi_o = \phi_m - \phi_s \qquad (5\text{-}1)$$

This potential ψ_o is supported by a space-charge layer with a width W, as depicted in Fig. 5-1b. The barrier for electrons to flow from the metal to the semiconductor is given by

$$q\phi_b = q(\phi_m - \chi_s) \qquad (5\text{-}2)$$

where ϕ_b is called the *barrier height* of the M-S contact. If we apply a negative voltage V to the semiconductor with respect to the metal, the semiconductor-to-metal potential is reduced to $\psi_o - V$ while ϕ_b remains unchanged (Fig. 5-2b). The reduction of the barrier potential on the semiconductor side makes it easier for electrons in the semiconductor to move to the metal. This is the forward-bias condition, and a large current can flow. On the other hand, if a positive voltage is applied to the semiconductor, the potential barrier is raised to prevent current conduction and the reverse bias is achieved (Fig. 5-2c). With uniform doping in the semiconductor, the space-charge-layer width of the Schottky barrier is identical to that of a one-sided

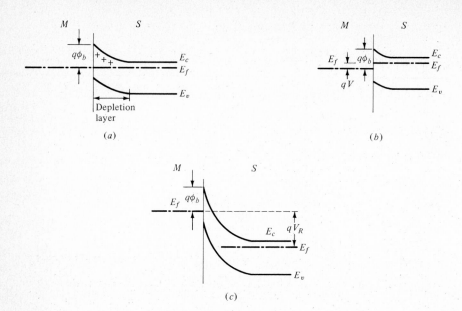

Figure 5-2 Energy-band diagram of a Schottky barrier: (*a*) without bias, (*b*) with forward bias, and (*c*) with reverse bias.

step *p-n* junction:

$$W = \left[\frac{2K\epsilon_o(\psi_o + V_R)}{qN_d}\right]^{1/2} \tag{5-3}$$

where N_d is the doping level in the semiconductor and V_R is the reverse-bias voltage. Therefore, the capacitance of the junction is given by

$$C = \frac{K\epsilon_o A}{W} = \left[\frac{qK\epsilon_o N_d}{2(\psi_o + V_R)}\right]^{1/2} A \tag{5-4}$$

which can be rewritten as

$$\frac{1}{C^2} = \frac{2}{qK\epsilon_o N_d A^2}(V_R + \psi_o) \tag{5-5}$$

As for a *p-n* junction, we can plot $1/C^2$ vs. V_R to obtain a straight-line relationship (Fig. 5-3). The built-in potential and the semiconductor doping can be calculated from the slope and intercept of such a capacitance-voltage plot. Once ψ_o is determined, the barrier height ϕ_b can be calculated using Eq. (5-2), which is rewritten as

$$\phi_b = \psi_o + V_n \tag{5-6}$$

where V_n is the potential difference between the Fermi level and E_c; its value can be deduced from the impurity concentration.

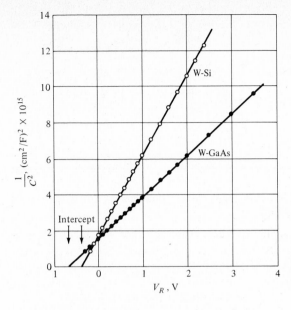

Figure 5-3 $1/C^2$ vs. applied voltage for tungsten-silicon and tungsten-gallium arsenide diodes. *(From Sze [1].)*

Example Calculate the donor concentration, built-in potential, and barrier height of the silicon Schottky diode from Fig. 5-3.

SOLUTION Using Eq. (5-5), we write

$$N_d = \frac{2}{qK\epsilon_o A^2} \frac{d(V_R + \psi_o)}{d(1/C^2)} = \frac{2}{qK\epsilon_o A^2} \frac{\Delta V_R}{\Delta(1/C^2)}$$

In Fig. 5-3, the capacitance is given in unit area so that $A = 1$. We find $1/C^2 = 6 \times 10^{15}$ at $V_R = 1$ V and $1/C^2 = 10.6 \times 10^{15}$ at $V_R = 2$ V. Thus,

$$\frac{\Delta V_R}{\Delta(1/C^2)} = \frac{1}{4.6 \times 10^{15}} = 2.17 \times 10^{-16} \text{ V-F}^2/\text{cm}^2$$

and

$$N_d = \frac{(2)(2.17 \times 10^{-16})}{(1.6 \times 10^{-19})(11.8)(8.84 \times 10^{-14})} = 2.6 \times 10^{15} \text{ cm}^{-3}$$

Using Eq. (1-8) and Table 1-2, we find

$$V_n = V_T \ln \frac{N_c}{N_d} = 0.026 \ln \frac{2.8 \times 10^{19}}{2.6 \times 10^{15}} = 0.24$$

Since $\psi_o = 0.4$ V from Fig. 5-3, we obtain

$$\phi_b = \psi_o + V_n = 0.4 + 0.24 = 0.64$$

The ideal model presented describes the Schottky-barrier characteristics well except that the barrier height ϕ_b obtained from the C-V measurement is not necessarily the difference of the work functions. In a practical Schottky

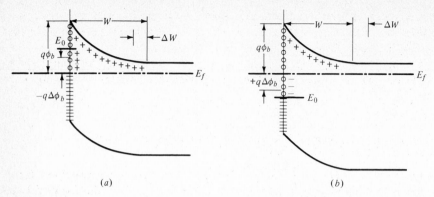

Figure 5-4 Pinning of the Fermi level by surface states. All energy states below E_f are occupied.

diode, the disruption of the crystal lattice at the interface produces a large number of energy states called *interface* or *surface states*, located within the forbidden gap. The interface states are usually continuously distributed in energy and are characterized by a *neutral level E_o* such that if the interface states are occupied up to E_o and empty above E_o, the surface is electrically neutral. In other words, the states below E_o are positively charged when empty, acting like donors; the states above E_o are negatively charged when occupied, acting like acceptors. If E_o is aligned with the Fermi level, the net charge of the surface is zero. In a practical M-S contact whenever $E_o > E_f$, the net charge of the interface states is positive, or donorlike, so that fewer ionized donors are needed in the depletion layer to reach equilibrium. As a result, the built-in potential is effectively reduced (Fig. 5-4a), and, according to Eq. (5-6), the barrier height ϕ_b is also reduced. From Fig. 5-4a we see that a smaller ϕ_b brings E_f closer to E_o. Similarly, if $E_o < E_f$, there is negative charge in the interface states and ϕ_b is increased to bring E_f closer to E_o again (Fig. 5-4b). Thus the charge in the interface states has a negative feedback effect which tends to keep E_f close to E_o. If the interface-state density is large, the Fermi level is effectively pinned down at E_o and ϕ_b becomes independent of the work functions of the metal and semiconductor. In most

Table 5.1 Schottky-barrier heights in electronvolts to n-type semiconductors (After Sze [1] and Milnes and Feucht [2])

Metal	ϕ_m, eV	Si ($\chi = 4.05$)	Ge ($\chi = 4.13$)	GaAs ($\chi = 4.07$)	GaP ($\chi = 4.0$)
Al	4.2	0.5–0.77	0.48	0.80	1.05
Au	4.7	0.81	0.45	0.90	1.28
Cu	4.4	0.69–0.79	0.48	0.82	1.20
Pt	5.4	0.9	—	0.86	1.45

practical Schottky barriers, interface states play a dominant role in determining the value of ϕ_b, and the barrier height is essentially independent of both the work-function difference and doping level in the semiconductor. Experimentally observed barrier heights are given in Table 5-1. The energy E_o is found to be near $E_g/3$ from the valence-band edge [3] for most semiconductors.

5-2 CURRENT-VOLTAGE CHARACTERISTIC

The current flow in the M-S barrier is governed by the carrier transport from the edge of the space-charge layer to the metal. In Fig. 5-2b, the forward bias reduces the electric field and potential barrier in the depletion layer. As a result, electrons diffuse through the depletion layer with a velocity v_D and are emitted from the semiconductor to the metal with a velocity v_E. If $v_E \gg v_D$, the current is controlled by the diffusion of carriers across the space-charge layer. On the other hand, if $v_E \ll v_D$, the current is governed by the emission process near the M-S interface. It turns out that at room temperature in most practical Schottky-barrier diodes, the current transport mechanism is limited by the emission process [4, 5]. The diffusion effect of carriers across the space-charge layer will be neglected in our discussion.

When the electrons are emitted from the semiconductor to the metal, they have an energy approximately $q\phi_b$ greater than that of the metal electrons. Before they collide in the metal to give up this extra energy, these electrons are considered to be *hot* because their equivalent temperature is higher than that of the electrons in the metal. For this reason the Schottky-barrier diode is sometimes called the *hot-carrier diode*. These hot carriers, however, reach equilibrium with the metal electrons in a very short time, typically less than 0.1 ns.

To understand the emission process, consider a metallic filament placed in a high-vacuum chamber. As shown in Fig. 5-5a, the potential barrier at the metal-vacuum interface is given by the work function $q\phi_m$. At a low tem-

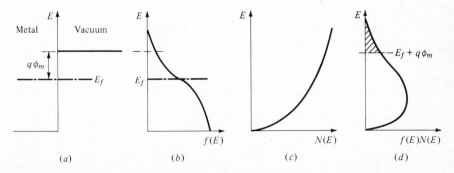

Figure 5-5 Graphical method of obtaining electrons with energy greater than $E_f + q\phi_m$.

perature, electrons in the metal cannot go beyond the metal surface because of the confinement of this barrier. If a current is passed through the filament to increase its temperature, the electrons in the metal will receive a thermal energy which enables some of them to escape from the metal into the vacuum. This phenomenon of electrons escaping from a hot surface is called *thermionic emission*. As shown in Fig. 5-5, the number of electrons occupying the energy level between E and $E + dE$ is

$$dn = N(E)f(E) \, dE \qquad (5\text{-}7)$$

where $N(E)$ is given by Eq. (1-2) with E_c set to zero and $f(E)$ is given by Eq. (1-6). Since only electrons that have a velocity component normal to the surface with an energy greater than $E_f + q\phi_m$ can escape the solid,† the thermionic current is

$$I = A \int qv_x \, dn$$

$$= \int_{E_f+q\phi_m}^{\infty} \frac{A4\pi q(2m)^{3/2}}{h^3} v_x E^{1/2} e^{-(E-E_f)/kT} \, dE \qquad (5\text{-}8)$$

where v_x is the component of velocity normal to the surface of the metal. In order to integrate Eq. (5-8), we must use the relation $p_x = mv_x$ and Eq. (1-1) to convert the energy into the momentum variable. In addition, the momentum components must be changed to rectangular coordinates. After integration, we obtain the Richardson-Dushman equation for the thermionic current (see Appendix B):

$$I_o = ART^2 \exp\left(\frac{-q\phi_m}{kT}\right) \qquad (5\text{-}9)$$

where $R = 4\pi qmk^2/h^3 = 120 \text{ A/K}^2\text{-cm}^2 = $ Richardson constant
$\quad A = $ junction area
$\quad m = $ free-electron mass

In analogy with the metal-vacuum system, electrons in the semiconductor of the M-S junction shown in Fig. 5-2a may overcome the potential barrier $q\psi_o$ to reach the metal and produce a current flow. The equation governing this current component can be derived and shown to be identical to Eq. (5-9) except that ϕ_m is replaced by ϕ_b and m is replaced by m_e, the effective electron mass. Under the thermal-equilibrium condition, however, equal numbers of electrons in the metal and semiconductor surmount the potential barrier to reach the other side, so that the net current flow across the M-S junction is zero. When a reverse-bias voltage V_R is applied, the potential barrier for electrons from the semiconductor to the metal is increased to $\psi_o + V_R$ but the potential barrier from metal to semiconductor remains unchanged, as shown in Fig. 5-2c. Under this condition, the current is given by

---

† If an electron has an energy of $E_f + q\phi_m$ but is traveling, for instance, in a direction of 45° from normal to the surface, it cannot escape into the vacuum.

I_o, the reverse saturation current. When a forward-bias voltage is applied, the potential barrier from the semiconductor to metal is reduced to $\psi_o - V$. The reduced potential barrier favors the electron transport from the semiconductor to the metal. By substituting $\phi_m = \phi_b - V$ into Eq. (5-9), we obtain the forward current as

$$I_F = AR^*T^2 \exp\left[\frac{-q(\phi_b - V)}{kT}\right] = I_o e^{V/V_T} \qquad (5\text{-}10)$$

where

$$I_o = AR^*T^2 \exp\left(\frac{-\phi_b}{V_T}\right) \qquad (5\text{-}11)$$

In addition $V_T = kT/q$, and R^* is the effective Richardson constant to indicate the use of the effective mass of electrons. Thus, the current-voltage relationship of the M-S junction for both forward and reverse bias is given by

$$I = I_o(e^{V/nV_T} - 1) \qquad (5\text{-}12)$$

where n is known as the ideality factor, which is introduced by nonideal effects. For an ideal Schottky diode, we have $n = 1$. Experimental I-V characteristics of two Schottky diodes are shown in Fig. 5-6. By extrapolating the forward I-V curve to $V = 0$, we can find the parameter I_o which, along with Eq. (5-11), can be used to obtain the barrier height. The ideality factor is calculated by using the

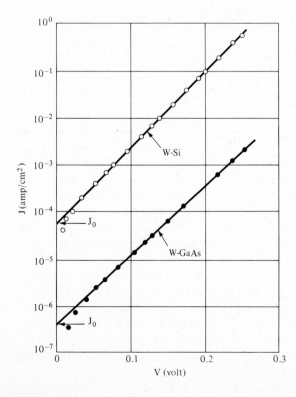

Figure 5-6 Forward current density vs. voltage of W-Si and W-GaAs Schottky diodes. *(After Sze [1].)*

slope of the semilog plot, giving $n = 1.02$ for the silicon diode and $n = 1.04$ for the GaAs diode.

Example The following parameters of a Schottky diode are given: $\phi_m = 4.7$ eV, $\chi_s = 4.0$ eV, $N_c = 10^{19}$ cm^{-3}, $N_d = 10^{16}$ cm^{-3}, and $K = 12$. Assume that the density of interface states is negligible. At 300 K calculate (a) the barrier height, built-in potential, and depletion layer width at zero bias and (b) the thermionic-emission current at a forward bias of 0.2 V.

SOLUTION (a) $\qquad \phi_b = \phi_m - \chi_s = 4.7 - 4.0 = 0.7$ V

Use Eq. (1-8) and rewrite it as

$$n = N_d = N_c e^{-(E_c - E_f)/kT} = N_c e^{-V_n/V_T}$$

Therefore $\qquad\qquad V_n = V_T \ln \dfrac{N_c}{N_d} = 0.17$ V

Thus $\qquad\qquad\qquad \psi_o = 0.7 - 0.17 = 0.53$ V

and $\qquad\qquad\qquad W = \left(\dfrac{2K\epsilon_o\psi_o}{qN_d}\right)^{1/2} = 2.6 \times 10^{-5}$ cm

(b) $\qquad\qquad J = R^* T^2 e^{-\phi_b/V_T}(e^{V/V_T} - 1)$

$\qquad\qquad\qquad\quad = 1.2 \times 10^2 (300)^2 e^{-0.7/0.026}(e^{0.2/0.026} - 1)$

$\qquad\qquad\qquad\quad = 4 \times 10^{-2}$ A/cm^2

where R^* is taken as 120 A/cm^2-K^2.

The thermionic-emission current describes the majority-carrier (electron) current across the Schottky barrier. In addition to the majority-carrier current, a minority-carrier current exists as a result of hole injection from the metal to semiconductor. The hole injection is the same as in a p-n junction; the current is given by Eq. (4-40) and is rewritten as

$$I_p = I_{po}(e^{V/V_T} - 1) \tag{5-13}$$

with $\qquad\qquad I_{po} = \dfrac{qAD_pN_vN_c}{N_dL_p} \exp\left(\dfrac{-E_g}{kT}\right) \tag{5-14}$

where Eqs. (1-13) and (1-17) have been employed. We may note that the minority-carrier current expressed in Eq. (5-13) has the same form as the majority-carrier current in Eq. (5-12). Equation (5-14) is written in a form permitting the magnitudes of I_{po} and I_o to be compared conveniently. In a covalent semiconductor such as Si, the junction barrier potential ϕ_b is always smaller than the band-gap energy E_g. As a result, the thermionic-emission current is usually much larger than the minority-carrier current. As a numerical example consider an aluminum on nSi diode with $N_d = 10^{16}$ cm^{-3}, $\phi_b = 0.69$ eV, and $L_p = 10$ μm. At room temperature we find

$$I_o = 3 \times 10^{-5} A \quad A \quad \text{and} \quad I_{po} = 3 \times 10^{-11} A \quad A \tag{5-15}$$

From this comparison, it is seen that the majority-carrier current is about 6 orders of magnitude greater than the minority-carrier current. Thus, the minority-carrier current can be neglected in most cases for Schottky barriers.

The reverse current, according to Eq. (5-11), should be a constant and equal to I_o. However, the top of the barrier at $x = 0$ in Fig. 5-2a will be slightly rounded off because of the image-force lowering (see next section). In addition, carrier generation is also taking place in the space-charge region. As a result, the reverse current will be seen as slightly voltage-dependent. For very high reverse bias, avalanche breakdown will occur just as it does in a one-sided step p-n junction. As in a p-n junction, the avalanche-breakdown voltage increases with semiconductor resistivity.

5-3 IMAGE-FORCE LOWERING OF THE BARRIER HEIGHT

In this section, we consider the effects of the image charge and the application of an electric field to the metal surface at a metal-vacuum interface. The result is then used for an M-S junction.

An electron with a charge $-q$ located at a distance x from the metal surface establishes electric field lines as shown in Fig. 5-7a. All the field lines must be normal to the surface since the metal is a good conductor. These field lines act as if an image charge $+q$ were situated at $-x$ (Fig. 5-7a). This means that the force on the electron at x is the same if the metal surface were replaced by $+q$ at $-x$. By Coulomb's force of attraction, the force on the electron is

$$F = -q\mathscr{E} = -\frac{q^2}{4\pi\epsilon_o(2x)^2} \tag{5-16}$$

Since $\mathscr{E} = -d\psi/dx$ and the potential energy is zero at $x = \infty$, we obtain from Eq. (5-16)

$$\psi(x) = \frac{1}{q}\int_x^\infty F\,dx = -\frac{q}{16\pi\epsilon_o x} \tag{5-17}$$

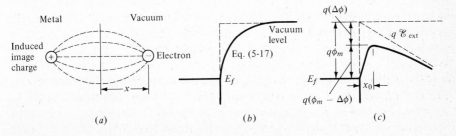

Figure 5-7 Image-force lowering of a metal-vacuum barrier.

If there is no external electric field applied to the emitting surface, the only contribution to ψ is the image force expressed in Eq. (5-17) and illustrated in Fig. 5-7b. When an external field $\mathscr{E}_{ext}$ is applied, the potential becomes

$$\psi(x) = -\frac{q}{16\pi\epsilon_o x} - \mathscr{E}_{ext}x \qquad (5\text{-}18)$$

This potential is plotted in Fig. 5-7c. It can be shown that the maximum potential is located at (Prob. 5-8)

$$x_o = \left(\frac{q}{16\pi\epsilon_o \mathscr{E}_{ext}}\right)^{1/2} \qquad (5\text{-}19)$$

and

$$\Delta\phi = \left(\frac{q\mathscr{E}_{ext}}{4\pi\epsilon_o}\right)^{1/2} \qquad (5\text{-}20)$$

where $q\,\Delta\phi$ is the amount by which the work function $q\phi_m$ is lowered when an accelerating field is applied to the surface. For example, if $\mathscr{E}_{ext} = 10^4$ V/cm, we obtain $\Delta\phi = 0.039$ eV and $x_o = 190$ Å.

At an M-S junction, the effect of barrier-height lowering is comparable with that of the metal-vacuum surface when the dielectric constant K is included in the numerator of Eq. (5-20). Therefore, the saturation current I_o is modified to

$$I_o = AR^*T^2 \exp\left[\frac{-q(\phi_b - \Delta\phi)}{kT}\right] \qquad (5\text{-}21)$$

It has been found experimentally that the foregoing equation is more accurate in describing the current-voltage characteristics of a Schottky diode, particularly in the reverse-bias direction.

5-4 METAL-INSULATOR-SEMICONDUCTOR SCHOTTKY DIODE

In practice, when a metal contact is evaporated onto a chemically prepared silicon surface, an interfacial oxide layer invariably exists between the metal and semiconductor. The oxide layer is very thin, typically on the order of 5 to 15 Å. The energy-band diagram of such a metal-insulator-semiconductor (MIS) structure is shown in Fig. 5-8. Under thermal equilibrium, there is a potential drop across the oxide layer so that the barrier height is modified. The current conduction in the MIS Schottky diode results from carrier tunneling through the oxide layer and is described by [6]

$$I = AR^*T^2 e^{-\chi^{(1/2)}\delta} e^{-q\phi_b/kT} e^{qV/nkT} \qquad (5\text{-}22)$$

where χ is the mean barrier height, in electronvolts, from the conduction-band edge and δ is the oxide thickness, in angstroms. The product $\chi^{1/2}\delta$ is normalized so that it is dimensionless. The factor n in the last exponent results from the partial drop of the applied voltage across the oxide layer so that the voltage across the semiconductor is reduced. Note that Eq. (5-22)

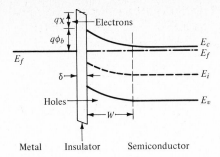

Figure 5-8 Energy-band diagram of MIS structure.

reduces to Eq. (5-10) for $\delta = 0$ and $n = 1$. In general, the thin oxide layer reduces the majority-carrier current but not the minority-carrier current if the same voltage is applied. This leads to an increase of the ratio of the minority-carrier current to majority-carrier current. It turns out that the increase in minority-carrier injection ratio is beneficial to device applications such as solar cells [7] and light-emitting diodes [8].

5-5 COMPARISON BETWEEN A SCHOTTKY-BARRIER AND A p-n JUNCTION DIODE

As described in Sec. 5-2, the current in a Schottky barrier is carried by majority carriers, whereas in a p-n junction it is by minority carriers. When a p-n junction is switched from a forward bias to a reverse bias abruptly, the minority carriers cannot be removed instantaneously, and the switching speed is limited by this minority-carrier storage effect. In a Schottky barrier, the storage time is negligible because there is no minority-carrier storage. Therefore, the frequency response is limited by the RC time constant rather than the charge storage. For this reason, Schottky-barrier diodes are ideal for high-frequency and fast-switching applications.

The saturation current in a Schottky barrier is much higher than that of a p-n junction diode for diodes of same area since the majority-carrier current is much higher than the minority-carrier current [Eq. (5-15)]. Consequently, the forward-voltage drop will be much less in a Schottky barrier than in a p-n junction for the same current. Figure 5-9 shows the current-voltage characteristics of an Al–nSi Schottky barrier and that of a p-n diode. The typical cut-in voltage or turn-on voltage (the knee of the I-V curve) of a Schottky-barrier diode is 0.3 V, and that of the Si p-n junction is 0.6 V. The low cut-in voltage makes Schottky diodes attractive for clamping and clipping applications. Under reverse bias, however, the Schottky diode has a higher reverse current which does not saturate. Furthermore, additional leakage current and soft breakdown usually exist in the Schottky diode, and special care must be taken in device fabrication. The nonideal reverse characteristics

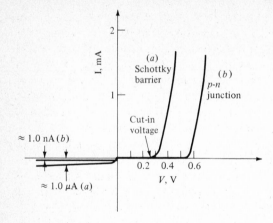

Figure 5-9 Current-voltage characteristics of a *p-n* junction and a Schottky-barrier diode.

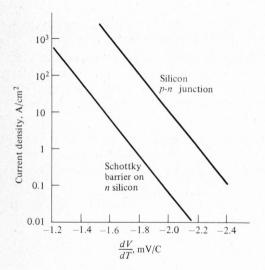

Figure 5-10 Temperature coefficient of forward voltage as a function of current density. *(After Yu [11].)*

can be eliminated by using the guard-ring or overlapped-metal structure to be discussed in a later section.

The temperature dependence of Schottky barriers and *p-n* junctions under forward bias is different. Experimental results are plotted in Fig. 5-10, where a difference of 0.4 mV/°C in the temperature coefficients is observed. This difference should be taken into account in the design of circuits utilizing both types of diodes.

5-6 OHMIC CONTACT: THE NONRECTIFYING M-S JUNCTION

An ohmic contact is defined as a contact which will not add a significant parasitic impedance to the structure on which it is used and will not sufficiently change the equilibrium-carrier densities within the semiconductor to affect the device characteristics. Let us consider a metal and semiconductor pair with $\phi_m < \phi_s$. The energy-band diagram before contact of such a pair is shown in Fig. 5-11a. When the contact is made, the exchange of carriers leading to a constant Fermi level results in the band diagram depicted in Fig. 5-11b. The potential barrier at the junction is almost nonexistent, so that carriers can freely pass in either direction. As a result, this M-S junction is nonrectifying. We can show that a metal-to-p-type semiconductor with $\phi_m > \phi_s$ is also an ohmic contact but is a rectifier if $\phi_m < \phi_s$. In practice, whether it is an n-type or p-type semiconductor, the ideal ohmic contact can only be approximated because of the charging effect at the interface states. An intimate contact between a metal and semiconductor does not generally form an ohmic junction, especially when the doping of the semiconductor is low. However, a metal-semiconductor contact is ohmic if the semiconductor is heavily doped, i.e., with an impurity density of 10^{19} cm^{-3} or more. In Fig. 5-11a, if the n-type semiconductor is heavily doped, the space-charge-layer width W becomes so thin that carriers can tunnel through instead of going over the potential barrier. Because electrons in either side of the barrier may tunnel through to the other side, an essentially symmetrical I-V curve for forward and reverse bias is realized. Consequently, the barrier is nonrectifying, and a low resistance exists

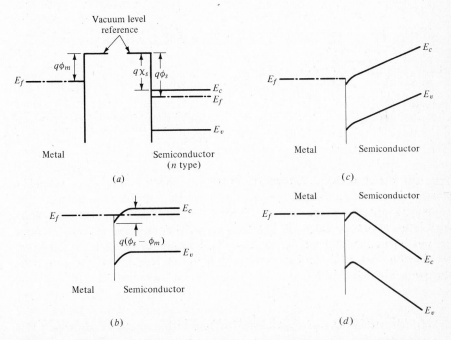

Figure 5-11 Energy-band diagram of a metal-on-n semiconductor contact with $\phi_m < \phi_s$: (a) before contact, (b) after contact at equilibrium, (c) with a negative voltage applied to the semiconductor side, and (d) with a positive voltage applied to the semiconductor side.

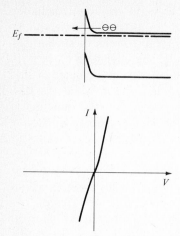

Figure 5-12 Energy-band diagram and current-voltage curve of a metal-on-n^+ semiconductor contact.

there. A practical ohmic contact can be obtained by evaporating Al, Au, or Pt on n-type Si with $N_d > 10^{19}$ cm^{-3}. Figure 5-12 shows the energy-band diagram of a nonrectifying M-S junction under small forward bias together with its I-V characteristics.

5-7 STRUCTURE OF SCHOTTKY-BARRIER DIODES

Three commonly used Schottky-barrier structures are shown in Fig. 5-13. In Fig. 5-13a, the n-type epitaxial film on an n^+Si substrate is cleansed and thermally oxidized. Subsequently, windows are opened by using the standard photoresist technique, and the metal is deposited by either evaporation or sputtering in a vacuum system. The metal geometry is defined by another photoresist step. Unfortunately, this simple structure does not give ideal Schottky-barrier characteristics because of the sharp edge and a positive fixed charge Q_{ss} that exists at the Si–SiO$_2$ interface. These conditions establish a high electric field in the depletion region in the semiconductor near the periphery, leading to excess current at the corners. This corner effect results in a soft reverse characteristic and low breakdown voltage as well as poor noise properties. The periphery effect can be eliminated by allowing the metal to overlap the oxide as shown in Fig. 5-13b. The depletion regions under the metal-oxide-semiconductor (MOS) capacitances are now rounded off, and the sharp edge which causes the soft breakdown is eliminated. The overlapped region should be small; otherwise the added capacitance would degrade the high-frequency response of the diode. The guard-ring structure shown in Fig. 5-13c uses an additional p^+ diffusion ring to reduce the edge effect in order to obtain an ideal I-V characteristics. Usually the overlapped-metal structure is preferred in integrated circuits because of its simplicity.

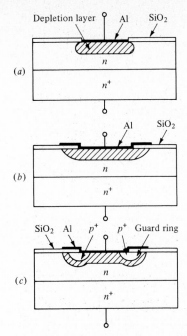

Depletion layer Al SiO$_2$

n

(a)

n^+

Al SiO$_2$

n

(b)

n^+

SiO$_2$ Al p^+ p^+ Guard ring

n

(c)

n^+

Figure 5-13 Practical Schottky-diode structures: (a) simple contact, (b) with metal overlap, and (c) with guard-ring diode.

5-8 APPLICATIONS OF SCHOTTKY-BARRIER DIODES

As a majority-carrier device, the Schottky diode does not have the minority-carrier storage effect, and it can be turned off in less than 1 ns. The simplicity in fabricating a Schottky barrier makes it possible to produce devices with a very small area for high-frequency operation, possibly up to 100 GHz (1 GHz = 10^9 Hz). In this section, we describe two useful applications; the Schottky solar cell and field-effect transistor will be discussed in later chapters.

Schottky-Barrier Detector or Mixer

In small-signal operation, the Schottky diode can be represented by the equivalent circuit shown in Fig. 5-14. In this figure C_d is the junction

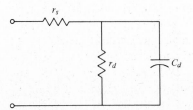

r_s

r_d C_d

Figure 5-14 Equivalent circuit of a Schottky diode.

capacitance, and r_s is the ohmic series resistance. The diode-junction resistance is defined as

$$r_d \equiv \frac{dV}{dI} \tag{5-23}$$

An efficient detector or mixer requires that the radio-frequency power be absorbed by the diode resistance r_d and that the power dissipation in r_s be small. Usually we have $r_s \ll r_d$, so that the effect of r_s at low frequency is negligible. However, increasing operating frequency reduces the junction impedance with respect to r_s. Eventually, a frequency is reached at which point the power dissipation in r_s is equal to that of the junction, namely,

$$r_s = \frac{r_d}{1 + \omega_c^2 C_d^2 r_d^2} \tag{5-24}$$

where ω_c is called the *cutoff frequency*. Since $r_d \gg r_s$, we have

$$\omega_c^2 \approx \frac{1}{C_d^2 r_d r_s} \tag{5-25}$$

For high-frequency operation C_d, r_d, and r_s should all be small. A small r_s is realizable if the semiconductor has high impurity concentration and high mobility. By using GaAs materials, operating frequency near 100 GHz appears to be possible.

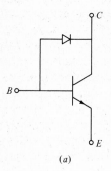

(a)

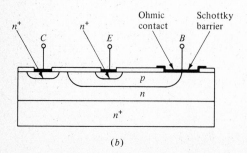

(b)

Figure 5-15 Schottky clamped transistor: (a) circuit representation and (b) integrated structure.

Schottky-Barrier Clamped Transistor

Because of its fast switching response, a Schottky barrier can be connected in parallel with the collector-base junction of an *n-p-n* transistor, as shown in Fig. 5-15*a*, to reduce the storage time of the transistor. When the transistor is saturated, the collector junction is forward-biased to approximately 0.5 V. If the forward-voltage drop in the Schottky diode (typically 0.3 V) is lower than the base-collector ON voltage of the transistor, most of the excess base current flows through the diode, which has no minority-carrier storage effect. Thus the storage time of the composite device is reduced markedly compared with that of the transistor alone. The measured storage time can be less than 1 ns. The Schottky-barrier clamped transistor is realized in integrated-circuit form by the structure shown in Fig. 5-15*b*. The aluminum forms an excellent Schottky barrier over the lightly doped collector *n* region and at the same time forms a good ohmic contact over the heavily doped base *p* region; these two contacts can be made by one single metallization step, and no extra processing is needed.

5-9 HETEROJUNCTIONS†

A junction formed by two semiconductor materials, e.g., silicon and germanium, is called a *heterojunction*. In contrast, the *p-n* junction described in Chap. 4 is a *homojunction*, in which both sides of the junction are made of the same material. Since the energy-band gaps of the two materials are different, the energy-band diagram of a heterojunction exhibits a discontinuity at the junction interface and the theory is more complicated than that of the homojunction. In this section, we shall not delve into the various complexities of the problem; instead we shall introduce some concepts that are useful for device applications. We shall find that the construction of the heterojunction energy-band diagram is similar to that of a Schottky barrier, but the current transport may follow either the *p-n* homojunction or the M-S barrier.

In the following discussion, germanium and gallium arsenide (Ge–GaAs) are chosen as the semiconductor pair for two reasons: (1) The material properties of Ge and GaAs are well understood, and the fabrication technology is well under control, and (2) the lattice constants (Table 1-2) of these materials are matched to within 1 percent. The latter criterion is of primary importance because the lattice mismatch introduces a large number of interface states and degrades the heterojunction characteristics. Since each semiconductor may be either *n* type or *p* type, there are four possible heterojunction combinations, *n*Ge–*p*GaAs, *n*Ge–*n*GaAs, *p*Ge–*n*GaAs, and *p*Ge–*p*GaAs. We shall limit our discussion to the *n*Ge–*p*GaAs and *n*Ge–*n*GaAs junctions.

† This section may be skipped without loss of continuity.

The energy-band diagrams for isolated nGe and pGaAs are shown in Fig. 5-16, where the vacuum level is used as the reference. The two semiconductors are distinctly different in band-gap energy, dielectric constant, work function, and electron affinity. The subscripts 1 and 2 refer to Ge and GaAs, respectively. The difference in energy of the conduction-band edge is represented by ΔE_c and that in the valence-band edge by ΔE_v; they are obtained from the diagram as

$$\Delta E_c = q(\chi_1 - \chi_2) \tag{5-26}$$

$$\Delta E_v = (E_{g1} - E_{g2}) - \Delta E_c \tag{5-27}$$

As the two semiconductors are brought into contact, constancy of the Fermi level must be satisfied at equilibrium. This is accomplished when electrons in the nGe are transferred to the pGaAs and holes are transferred in the opposite direction until the Fermi level is aligned. The energy-band diagram is shown in Fig. 5-17. As in homojunction, the redistribution of charges creates a

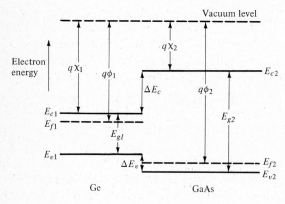

Figure 5-16 Energy-band diagram for two isolated semiconductors. *(After Anderson [10].)*

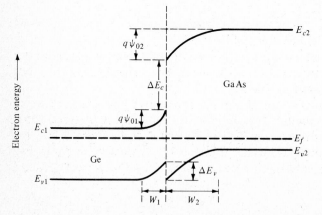

Figure 5-17 Energy-band diagram for n-p heterojunction at equilibrium. *(After Anderson [10].)*

depletion layer on each side of the junction. Within the depletion layer, the energy band bends down on the p side and up on the n side, indicating the depletion of free carriers. The bending of the energy-band edges also indicates the existence of a built-in voltage in both sides of the junction. The total built-in voltage ψ_o equals the sum of the partial built-in voltages:

$$\psi_o = \psi_{o1} + \psi_{o2} \tag{5-28}$$

where ψ_{o1} and ψ_{o2} are the portion of the built-in voltage in sides 1 and 2, respectively.

The depletion widths can be obtained by solving Poisson's equation on both sides of the heterojunction. One of the boundary conditions is that the electric displacement be continuous, i.e.,

$$K_1 \mathscr{E}_1 = K_2 \mathscr{E}_2 \tag{5-29}$$

where K and $\mathscr{E}$ are the dielectric constant and electric field, respectively. In addition, the condition of charge neutrality leads to

$$\frac{W_2}{W_1} = \frac{N_{d1}}{N_{a2}} \tag{5-30}$$

as in a p-n homojunction. The derivation of W_1 and W_2 is left as an exercise for the reader. We wish to find the portions of an applied voltage across the two sides of the depletion region. The electric fields in Ge and GaAs are

$$\mathscr{E}_1 = \frac{\psi_{o1} - V_1}{W_1} \tag{5-31}$$

$$\mathscr{E}_2 = \frac{\psi_{o2} - V_2}{W_2} \tag{5-32}$$

where V_1 and V_2 are the portions of the applied voltage appearing on the Ge side and on the GaAs side, respectively. Solving Eqs. (5-29) to (5-32), we obtain

$$\frac{\psi_{o1} - V_1}{\psi_{o2} - V_2} = \frac{K_2 N_{a2}}{K_1 N_{d1}} \tag{5-33}$$

When the doping levels on the two sides are significantly different, Eq. (5-33) indicates that the external bias voltage is across the lightly doped side, just as in the one-sided step p-n homojunction.

The relative magnitudes of the current components in a heterojunction are determined by the potential barriers involved. For the heterojunction shown in Fig. 5-17, the hole current from GaAs to Ge is expected to dominate because of the low potential barrier ψ_{o2} for hole injection and the high potential barrier $(\psi_{o1} + \psi_{o2} + \Delta E_c)$ for electron injection. The current density is therefore given by

$$J = \frac{q D_p p_{no}}{L_p} (e^{qV/kT} - 1) \tag{5-34}$$

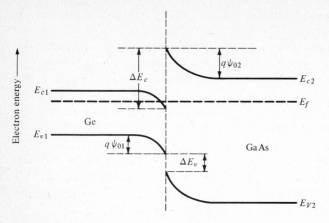

Figure 5-18 Energy-band diagram for *n-n* heterojunction at equilibrium. *(After Anderson [10].)*

where D_p, L_p, and p_{no} are, respectively, the diffusion constant, diffusion length, and equilibrium density for holes in Ge. In contrast with the homojunction, the dominant current component is not necessarily the minority-carrier current in the lightly doped side. In fact, the discontinuity in the band diagram favors the injection of majority carriers from the larger band-gap material regardless of the doping levels.

The model presented in the foregoing paragraph is a first-order approximation and has been found to be valid in Ge–GaAs and GaAs–AlGaAs *p-n* heterojunctions where the lattice mismatch is small. In other heterojunctions, such as the Ge–Si pair, a large lattice mismatch (4 percent in Ge–Si) leads to a high interface density. As a result, recombination- and tunneling-current components have to be included.

The energy-band diagram for an *n*Ge–*n*GaAs heterojunction is shown in Fig. 5-18. The barrier here acts like the metal-semiconductor contact, and the current is due to thermionic emission of electrons from GaAs to Ge.

PROBLEMS

5-1 A silicon Schottky-barrier diode has a contact area of 0.01 cm², and the donor concentration in the semiconductor is 10^{16} cm⁻³. Let $\psi_o = 0.7$ V and $V_R = 10.3$ V. Calculate (*a*) the thickness of the depletion layer, (*b*) the barrier capacitance, and (*c*) the field strength at the surface.

5-2 (*a*) Obtain the donor concentration, built-in potential, and barrier height of the GaAs Schottky diode from its capacitance-voltage plot shown in Fig. 5-3.

(*b*) Calculate the barrier height from Fig. 5-6 and compare your result with that of part (*a*).

5-3 Construct the energy-band diagram for a metal-on-*p*-type Schottky barrier with negligible surface states for (*a*) $\phi_m > \phi_s$ and (*b*) $\phi_m < \phi_s$. Indicate whether it is a rectifying or nonrectifying junction and specify the built-in potential and barrier height.

5-4 What would be the surface potential of a free silicon surface with a donor concentration of $10^{15}\,\mathrm{cm}^{-3}$ and a uniform concentration of surface states $D_{ss} = 10^{12}\,\mathrm{cm}^{-2}\,\mathrm{eV}^{-1}$ with a neutral level of $E_v + 0.3\,\mathrm{eV}$? *Hint*: First find the energy difference between the Fermi level and the neutral level. The charge residing in these surface states must be equal to the depletion-layer charge supported by the surface potential.

5-5 The following parameters of a Schottky diode are given: $\phi_m = 5.0\,\mathrm{V}$, $\chi_s = 4.05\,\mathrm{V}$, $N_c = 10^{19}\,\mathrm{cm}^{-3}$, $N_d = 10^{15}\,\mathrm{cm}^{-3}$, and $K = 11.8$. Assume that the density of interface states is negligible. At 300 K, calculate (*a*) the barrier height, the built-in potential, and the depletion-layer width at zero bias and (*b*) the thermionic, emission current density at a forward bias of 0.3 V.

5-6 In a metal-silicon contact, the barrier height is $q\phi_b = 0.8\,\mathrm{eV}$, and the effective Richardson constant $R^* = 10^2\,\mathrm{A/cm^2\text{-}K^2}$; $Eg = 1.1\,\mathrm{eV}$, $N_d = 10^{16}\,\mathrm{cm}^{-3}$, and $N_c = N_v = 10^{19}\,\mathrm{cm}^{-3}$.

(*a*) Calculate the bulk potential V_n and the built-in potential of the semiconductor at zero bias at 300 K.

(*b*) Assuming $D_p = 15\,\mathrm{cm^2/s}$ and $L_p = 10\,\mu\mathrm{m}$, calculate the ratio of injected majority-carrier current to the minority-carrier current.

5-7 Calculate the ratio of the majority-carrier current to the minority-carrier current of a gold-on-nGaAs Schottky barrier at room temperature. The donor concentration is $10^{15}\,\mathrm{cm}^{-3}$, $L_p = 1\,\mu\mathrm{m}$, and $R^* = 0.068R$.

5-8 Derive Eqs. (5-19) and (5-20).

5-9 Calculate $\Delta\phi$ and x_o for a metal-insulator barrier for $\mathscr{E}_{ext} = 10^4\,\mathrm{V/cm}$ and a dielectric constant of (*a*) $K = 4$ and (*b*) $K = 12$. Compare your results with the example given in Sec. 5-3.

5-10 (*a*) Derive an expression of dV/dT as a function of the current density in a Schottky diode. Assume that the minority-carrier current is negligible.

(*b*) Estimate the temperature coefficient if typically $V = 0.25\,\mathrm{V}$ and $\phi_b = 0.7\,\mathrm{V}$ at 300 K.

5-11 Calculate the cutoff frequency of a Schottky detector with a capacitance of 10 pF, a series resistance of 10 Ω, and a diode resistance of 100 Ω.

5-12 Construct the energy-band diagram of (*a*) a pGe–nGaAs heterojunction and (*b*) a pGe–pGaAs heterojunction.

REFERENCES

1. S. M. Sze, "Physics of Semiconductor Devices," chap. 8, Wiley, New York, 1969.
2. A. G. Milnes and D. L. Feucht, "Heterojunction and Metal-Semiconductor Junctions," Academic, New York, 1972.
3. C. A. Mead, Metal-Semiconductor Surface Barriers, *Solid-State Electron.*, **9**: 1023 (1966).
4. C. R. Crowell and M. Beguwala, Recombination Velocity Effects on Current Diffusion and Imref in Schottky Barriers, *Solid-State Electron.*, **14**: 1149–1159 (1971).

5. E. H. Rhoderick, Comments on the Conduction Mechanism in Schottky Diodes, *J. Phys. D: Appl. Phys.*, **5**: 1920–1929 (1972).

6. H. C. Card and E. H. Rhoderick, Studies of Tunnel MOS Diodes, I: Interface Effects in Silicon Schottky Diodes, *J. Phys. D: Appl. Phys.*, **4**: 1589–1601 (1971).

7. R. J. Stirn and Y. C. M. Yeh, A 15% Efficient Antireflection-coated Metal-Oxide-Semiconductor Solar Cell, *Appl. Phys. Lett.*, **27**: 95 (1975).

8. H. C. Card and B. L. Smith, Green Injection Luminescence from Au–GaP Schottky Barriers, *J. Appl. Phys.*, **42**: 5863–5865 (1971).

9. R. L. Anderson, Experiments on Ge–GaAs Heterojunctions, *Solid-State Electron.*, **5**: 341–351 (1962).

10. R. L. Anderson, Germanium-Gallium Arsenide Heterojunctions, *IBM J. Res. Dev.*, **4**: 283–286 (July 1960).

11. A. Y. C. Yu, The Metal-Semiconductor Contact: An Old Device with a New Future, *IEEE Spectrum*, **7**: 83–90 (March 1970).

ADDITIONAL READING

Milnes, A. G., and D. L. Feucht: "Heterojunction and Metal-Semiconductor Junctions," Academic, New York, 1972.

THE SOLAR CELL AND LIGHT-EMITTING DIODE

In this chapter the principles of light absorption and emission in a *p-n* junction are described. Optical energy can be absorbed in a semiconductor if the photon energy is greater than the band-gap energy. The absorbed photons generate electron-hole pairs, which produce a potential difference across the *p-n* junction. The mechanism converting optical energy into electrical energy is known as the *photovoltaic effect*. The photodiode and solar cell are the two most useful devices. The inverse mechanism, namely, converting electrical energy into optical energy, is called the *electroluminescent* effect. A *p-n* junction device that emits light upon the application of a forward-bias current is known as the *light-emitting diode* (LED). In this chapter, we shall limit our discussion to the solar cell and LEDs emitting in the visible and near-infrared regions. The topic of optical absorption in a semiconductor will be introduced first.

6-1 OPTICAL ABSORPTION IN A SEMICONDUCTOR

The unit energy of light, called a *photon*, is $h\nu$, where ν is the light frequency and h is Planck's constant. The wavelength of light λ is related to the frequency by

$$\lambda = \frac{c}{\nu} = \frac{hc}{E_{ph}} = \frac{1.24}{E_{ph}} \mu m \qquad (6\text{-}1)$$

where E_{ph} is the photon energy $h\nu$ in electron volts and c is the speed of light, that is, 3×10^{10} cm/s. When a semiconductor is illuminated, photons may or may not be absorbed, depending on the photon energy and the band-gap energy E_g. Photons with energy smaller than E_g are not readily absorbed by the semiconductor because there is no energy state available in the forbidden gap to accommodate an electron (Fig. 6-1a). Thus, light is transmitted through and the material appears transparent. If $E_{ph} = E_g$, photons are absorbed to create electron-hole pairs, as shown in Fig. 6-1b. When the photon energy is greater than E_g, an electron-hole pair is generated and, in addition, the excess energy $E_{ph} - E_g$ is dissipated as heat (Fig. 6-1c).

Let us consider the nature of absorption for a semiconductor shown in Fig. 6-2. The optical source provides monochromatic light with $h\nu > E_g$ and a flux F_{ph} or F in photons per square centimeter per second. As the light beam penetrates the crystal, the fraction of the photons absorbed is proportional to the intensity of the flux. Therefore, the absorbed photons within Δx are

$$\alpha F(x) \, \Delta x$$

where α is a proportional constant called the *absorption coefficient*. From the continuity of light in Fig. 6-2, we find

$$F(x + \Delta x) - F(x) = \frac{dF(x)}{dx} \Delta x = -\alpha F(x) \, \Delta x$$

or

$$-\frac{dF(x)}{dx} = \alpha F(x) \tag{6-2}$$

The negative sign indicates decreasing intensity of the flux along x due to

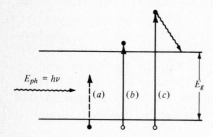

$E_{ph} = h\nu$

(a) (b) (c)

E_g

Figure 6-1 Optically generated electron-hole pairs in a semiconductor.

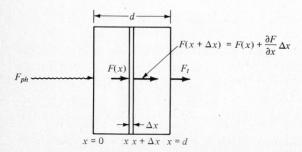

d

F_{ph}

$F(x)$

F_t

$F(x + \Delta x) = F(x) + \frac{\partial F}{\partial x} \Delta x$

Δx

$x = 0$ $x \, x + \Delta x$ $x = d$

Figure 6-2 Optical absorption in a semiconductor.

absorption. With the boundary condition $F = F_{ph}$ at $x = 0$, we obtain the solution of Eq. (6-2) as

$$F(x) = F_{ph}e^{-\alpha x} \qquad (6\text{-}3)$$

Therefore, the fraction of light transmitted through the semiconductor is

$$F_t = F(d) = F_{ph}e^{-\alpha d} \qquad (6\text{-}4)$$

where d is the thickness of the semiconductor. Since the absorption is determined by the photon energy, it is conceivable that α is a function of $h\nu$. Figure 6-3 shows the absorption spectra of a few semiconductors that are used for optoelectronic applications. In this figure, the band-gap energies are indicated, along with the wavelength. We notice that the absorption coefficient drops off sharply at the band-gap energy, indicating negligible absorption for photons with energy smaller than E_g. Thus, silicon absorbs photons with $\lambda \leqslant 1.1\ \mu\text{m}$, and GaAs absorbs photons with $\lambda \leqslant 0.9\ \mu\text{m}$.

6-2 PHOTOVOLTAIC EFFECT AND SOLAR-CELL EFFICIENCY

The process of converting optical energy into electrical energy in a p-n junction involves the following basic steps: (1) Photons are absorbed, so that

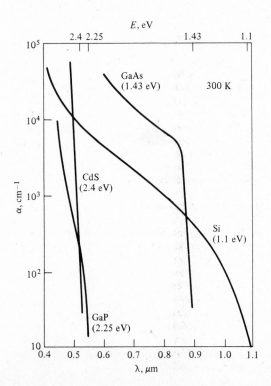

Figure 6-3 Absorption constant of various semiconductors. *(After Rappaport and Wysocki [1].)*

electron-hole pairs are generated in both the p and n sides of the junction (Fig. 6-4a). (2) By diffusion, the electrons and holes generated within a diffusion length from the junction reach the space-charge region (Fig. 6-4b). (3) Electron-hole pairs are then separated by the strong electric field; thus, electrons in the p side slide down the potential to move to the n side and holes go in the opposite direction (Fig. 6-4c). (4) If the p-n diode is open-circuited, the accumulation of electrons and holes on the two sides of the junction produces an *open-circuit voltage* (Fig. 6-4d). If a load is connected to the diode, a current will conduct in the circuit (Fig. 6-4a). The maximum current is realized when an electrical short is placed across the diode terminals, and this is called the *short-circuit current*.

The current produced by light in the load is in the same direction as the reverse saturation current of the p-n junction. Therefore, the total diode current under illumination is given by

$$I = I_L + I_o(1 - e^{V/V_T}) \tag{6-5}$$

where I_L is the *light-generated current* and the second term on the right-hand side is the reverse diode current, i.e., the negative of Eq. (4-46). For uniform absorption throughout the device, I_L is given by (Prob. 6-4)

$$I_L = qG_L(L_n + L_p)A \tag{6-6}$$

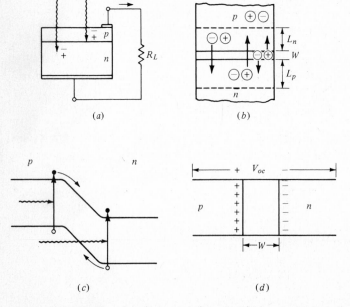

(a)   (b)   (c)   (d)

Figure 6-4 Conversion of optical energy into electrical energy: (a) a solar cell with load resistor, (b) diffusion of electrons and holes that produces current, (c) energy-band diagram of (b), and (d) establishment of the open-circuited voltage (schematic representation).

where G_L is the generation rate. The current-voltage characteristics expressed by Eq. (6-5) are depicted in Fig. 6-5 for an experimental device with the light intensity as the variable. Data were taken under air-mass 1 (AM1) illumination, defined as the sun at the zenith and the test device at sea level under a clear sky. The energy reaching the solar cell under the AM1 condition is slightly higher than $100\,\text{mW/cm}^2$. The solar spectrum just outside of the atmosphere is known as AM0, where the sun energy is $135\,\text{mW/cm}^2$. It is noted that I_L is the current at zero voltage in Fig. 6.5 and thus the short-circuited current. Setting $I = 0$ in Eq. (6-5), we obtain the open-circuited voltage as

$$V_{oc} = V_T \ln\left(1 + \frac{I_L}{I_o}\right) \tag{6-7}$$

where we have an open-circuited voltage or electric power source which supplies a current if an external load is connected across it. The conversion of optical to electrical energy is thus realized.

An equivalent circuit of the solar-cell characteristic expressed by Eq. (6-5) can be drawn as shown in Fig. 6-6. The power delivered to the load is given by

$$P = IV = I_L V - I_o V (e^{V/V_T} - 1) \tag{6-8}$$

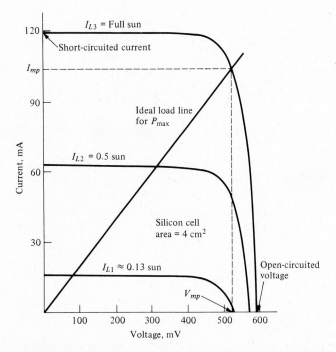

Figure 6-5 Current-voltage characteristics of a typical solar cell under air-mass 1 (AM1) illumination, i.e., sun energy at sea level under clear sky with sun at zenith.

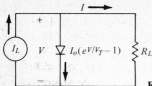

Figure 6-6 Equivalent circuit of the solar cell shown in Fig. 6-4.

Example Calculate the open-circuit voltage of a silicon n^+-p cell for substrate dopings between 10^{15} and 10^{18} cm^{-3}. Assume that $L_n = 100\ \mu$m, $D_n = 36$ cm^2/s, and $I_L/A = 35$ mA/cm^2 and that these values are independent of the doping concentration.

SOLUTION Since $N_d \gg N_a$ in an n^+-p junction, the saturation current [Eq. (4-47)] can be simplified to

$$I_o = qAn_i^2 \left(\frac{D_p}{L_p N_d} + \frac{D_n}{L_n N_a} \right) \approx \frac{qAn_i^2 D_n}{L_n N_a} = \frac{1.3 \times 10^5 A}{N_a} \qquad \text{A/cm}^2$$

Therefore, Eq. (6-7) becomes

$$V_{oc} = 26 \ln \left(1 + \frac{35 \times 10^{-3} N_a}{1.3 \times 10^5} \right) \qquad \text{mV}$$

Thus

N_a, cm^{-3}	10^{15}	10^{16}	10^{17}	10^{18}
V_{oc}, mV	504	565	625	684

In the foregoing example, the open-circuited voltage is found to increase with the substrate doping. Since the light-generated current is relatively independent of doping level, the output power of the solar cell should increase with substrate-doping concentration monotonically. In practice, however, the open-circuited voltage and output power reach a maximum in a doping less than 10^{17} cm^{-3}. Beyond this point, the space-charge-recombination dark current becomes significant, and the open-circuit voltage decreases with further increase of substrate doping [2]. The voltage corresponding to the maximum power delivery V_{mp} is obtained by taking $\partial P/\partial V = 0$. Therefore, we derive

$$\left(1 + \frac{V_{mp}}{V_T} \right) e^{V_{mp}/V_T} = 1 + \frac{I_L}{I_o} \tag{6-9}$$

The current at maximum power is denoted by I_{mp}, as indicated in Fig. 6-5. The efficiency of the solar cell is therefore

$$\eta = \frac{I_{mp} V_{mp}}{P_{in}} \times 100\% \tag{6-10}$$

where P_{in} is the input light power. To optimize the solar-cell efficiency, we should have large I_{mp} and V_{mp}. The maximum current and maximum voltage achievable in a solar cell are I_L and V_{oc}, respectively. Therefore, the ratio $V_{mp}I_{mp}/I_LV_{oc}$ is useful as a measure of the realizable power from the I-V curve. This is called the *fill factor*, and it is between 0.7 and 0.8 for a well-made cell. Typical efficiency for a commercially available silicon cell is 12 percent, although 16 percent has been reported for small-area devices (areas less than 10 cm^2).

6-3 LIGHT-GENERATED CURRENT AND COLLECTION EFFICIENCY

In the preceding section, we presented the simple theory of the solar cell. The light-generated current was obtained by assuming uniform absorption throughout the device, and the effect of photon energy on absorption was neglected. It is, however, necessary to examine the nature of the light-generated current to gain further understanding of the solar cell. Let us consider an incident photon flux F_{ph} striking the surface of a p-on-n structure. The effect of surface reflection is ignored for the moment. By using Eq. (6-3) and assuming that each absorbed photon generates an electron-hole pair, the generation rate of electron-hole pairs as a function of surface penetration is obtained:

$$G_L = \alpha F_{ph}e^{-\alpha x} \tag{6-11}$$

Adding Eq. (6-11) to the hole-continuity equation expressed in Eq. (4-32) leads to the following equation for holes in the n side of the junction under steady-state conditions:

$$D_p\frac{d^2p_n}{dx^2} - \frac{p_n - p_{no}}{\tau_p} + \alpha F_{ph}e^{-\alpha x} = 0 \tag{6-12a}$$

Here the generation term has a positive sign because it counteracts the recombination term. Similarly, the steady-state expression describing electrons in the p side is

$$D_n\frac{d^2n_p}{dx^2} - \frac{n_p - n_{po}}{\tau_n} + \alpha F_{ph}e^{-\alpha x} = 0 \tag{6-12b}$$

The electron- and hole-current components at the junction per unit area are given by

$$J_p = -qD_p\frac{dp_n}{dx}\bigg|_{x=x_j} \tag{6-13a}$$

$$J_n = qD_n\frac{dn_p}{dx}\bigg|_{x=x_j} \tag{6-13b}$$

The photon-collection efficiency is defined as

$$\eta_{col} = \frac{J_p + J_n}{qF_{ph}}$$ (6-14)

Equations (6-12) to (6-14) can be solved if the boundary conditions are given.

Example Derive expressions of light-generated minority-carrier density and current in the n side of the p^+-n cell shown in Fig. 6-7. Assume the surface recombination velocity at the back contact is S and the incoming light is monochromatic. Absorption in the p^+ layer is neglected.

SOLUTION The boundary conditions of Eq. (6-12a) are

$$p_n - p_{no} = 0 \qquad \text{at } x = 0$$

$$S(p_n - p_{no}) = -D_p \frac{dp_n}{dx} \qquad \text{at } x = W$$

The general solution of Eq. (6-12a) is

$$p_n - p_{no} = K_1 e^{x/L_p} + K_2 e^{-x/L_p} - \frac{\alpha F_{ph} \tau_p}{\alpha^2 L_p^2 - 1} e^{-\alpha x}$$

Substituting the boundary conditions into the foregoing equation yields

$$p_n - p_{no} = \frac{\alpha F_{ph} \tau_p}{\alpha^2 L_p^2 - 1} \left[\cosh \frac{x}{L_p} - e^{-\alpha x} \right. $$

$$\left. - \frac{S\left(\cosh \frac{W}{L_p}\right) + \frac{D_p}{L_p}\left(\sinh \frac{W}{L_p}\right) - (\alpha D_p - S)e^{-\alpha W}}{S\left(\sinh \frac{W}{L_p}\right) + \frac{D_p}{L_p}\cosh \frac{W}{L_p}} \sinh \frac{x}{L_p} \right]$$ (6-15)

The hole current flowing from the n side to the p^+ side is obtained using Eq. (6-13a):

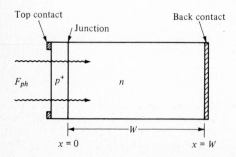

Top contact   Back contact

Junction

F_{ph} p^+ n

$\vert\!\longleftarrow\!\!\longrightarrow\!\vert$ W

$x = 0$ $x = W$ **Figure 6-7** A p-on-n cell.

$$J_p = \frac{qF_{ph}\alpha L_p}{\alpha^2 L_p^2 - 1}$$

$$\times \left[\frac{S\left(\cosh\frac{W}{L_p}\right) + \frac{D_p}{L_p}\left(\sinh\frac{W}{L_p}\right) + (\alpha D_p - S)e^{-\alpha W}}{S\left(\sinh\frac{W}{L_p}\right) + \frac{D_p}{L_p}\cosh\frac{W}{L_p}} - \alpha L_p \right] \qquad (6\text{-}16)$$

The electron current flowing from the p^+ side to the n side can be found in the same way.

Let us present some results based on the complete solution to gain some physical insight into photon collection under different wavelengths. At short wavelengths, the absorption coefficient α obtained from Fig. 6-3 is large. Therefore, the absorption of photons expressed in Eq. (6-3) decays in a short distance from the surface. In other words, most photons are converted into electron-hole pairs in a narrow layer near the surface for a short λ (5500 Å). At a longer wavelength (9000 Å), α is small, and absorption takes place mostly in the n side of the junction. The resulting minority-carrier distributions are illustrated in Fig. 6-8. If we consider that the incoming photon flux is monochromatic and the number of photons is given, we can obtain the collection efficiency in the n side for each wavelength by substituting Eq. (6-16) into Eq. (6-14). The theoretical collection efficiency at different wavelengths is calculated and plotted in Fig. 6-9. The components resulting from the absorption in the n side and p side are separated to show the individual effects.

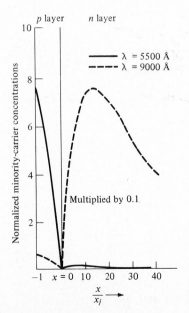

Figure 6-8 Normalized minority-carrier distributions for incident radiation at $\lambda = 5500$ and 9000 Å. Device parameters are $x_j = 2.8\ \mu$m, $W = 20$ mils, $\tau_p = 4.2\ \mu$s, $\tau_n = 10$ ns, and $S_n = 1000$ cm/s. *(After Wolf [3].)*

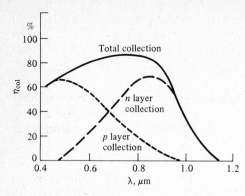

Figure 6-9 Collection efficiency vs. wavelength for solar cell of Fig. 6-8. *(After Wolf [3].)*

The collection efficiency is influenced by the minority-carrier diffusion length and the absorption coefficient. The diffusion length should be as long as possible to collect all light-generated carriers. In some solar cells, a built-in field is established by impurity gradient to improve carrier collection. As to the effect of the absorption coefficient, a large α leads to heavy absorption near the surface, resulting in a strong collection in the *skin layer*. A small α allows deep penetration of photons so that the *base* of the solar cell becomes more important in carrier collection. A typical GaAs solar cell is of the former type, and the silicon cell belongs to the latter type.

6-4 MATERIAL SELECTION AND DESIGN CONSIDERATIONS

In the preceding derivation, we have obtained the ideal conversion efficiency with which a photon generates an electron-hole pair producing a current without any loss of energy in the process. In a practical solar cell, since various factors limit the device performance, it is necessary to consider these limiting factors in solar-cell design and material selection.

Spectral Considerations

Let us consider the matching of the solar spectrum and the solar-cell absorption characteristic for terrestrial applications. The solar spectrum at sea level under a clear sky (AM1) is given in Fig. 6-10. The predominant portion of the energy in the sunlight is in the visible region, and the total power reaching the earth at sea level is approximately 100 mW/cm^2. Since only the portion of energy greater than E_g can be absorbed, photons with energy lower than E_g are not utilized. The number of photons available for electron-hole pair generation is obtained by integrating Fig. 6-10 from E_g to the maximum energy. It is found that the total number of solar photons at sea level is $4.8 \times 10^{17} \, \mathrm{cm}^{-2} \, \mathrm{s}^{-1}$ and the maximum number of photons that can be absorbed is $3.7 \times 10^{17} \, \mathrm{cm}^{-2} \, \mathrm{s}^{-1}$ in silicon and $2.5 \times 10^{17} \, \mathrm{cm}^{-2} \, \mathrm{s}^{-1}$ in GaAs. Therefore, the fraction of solar photons available for absorption in silicon is roughly 77 percent.

From the foregoing consideration, silicon would be a better material than GaAs. However, a large number of absorbed photons in silicon have energy greater than E_g. The excess energy $E_{ph} - E_g$ is dissipated as heat instead of generating more electrons and holes. For example, an energy of 1.1 eV is needed to produce an electron-hole pair in silicon. If a solar photon with $E_{ph} = 2.2 \, \mathrm{eV}$ is absorbed by silicon, half the photon energy is dissipated as heat instead of producing electricity. It turns out that if all solar photons with energy greater than 1.1 eV are considered, the total energy loss of the absorbed photons is 43 percent. This partial utilization of absorbed photon energy must be taken into account in the overall assessment of a material. In general, the smaller the energy gap, the more power is wasted near the peak of the sun spectrum. As a consequence, we find that silicon and GaAs are comparable as far as matching the solar spectrum is concerned.

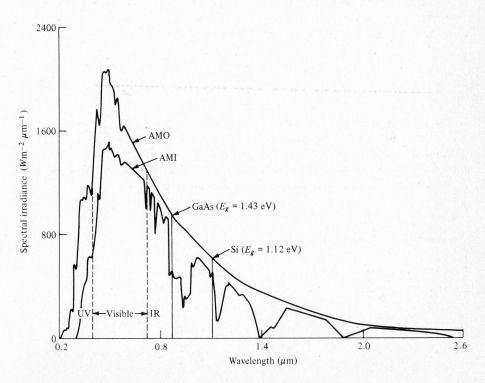

Figure 6-10 Solar spectra at AM0 and AM1 conditions with energy-cutoff points in GaAs and Si.

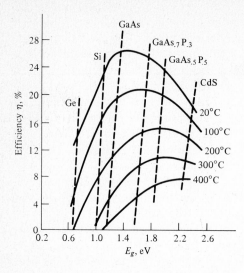

Figure 6-11 Maximum theoretical conversion efficiency vs. energy gap. *(After Wysocki and Rappaport [4].)*

Maximum-Power Considerations

The maximum power output of a solar cell is determined by the open-circuited voltage and the short-circuited current. From spectral considerations, it is found that I_L decreases with increasing E_g. The open-circuited voltage is expressed by Eq. (6-7), in which V_{oc} is inversely related to the reverse saturation current I_o. Using Eqs. (1-13), (1-17), and (4-47), we find

$$I_o \propto e^{-E_g/kT} \tag{6-17}$$

Substituting Eq. (6-17) into (6-7) yields

$$V_{oc} \propto E_g \tag{6-18}$$

which states that the open-circuited voltage is proportional to E_g. Since I_L decreases and V_{oc} increases with increasing E_g, the product $V_{oc}I_L$ exhibits a maximum. Using Fig. 6-10 and available semiconductor parameters, we plot the maximum conversion efficiency vs. E_g in Fig. 6-11 for different temperatures. It is seen that Si and GaAs are among the semiconductors most suitable for solar-cell applications.

Series-Resistance Considerations

The series resistance, the sum of the contact and sheet resistance, modifies the current-voltage characteristics as shown in Fig. 6-12. It increases the internal power dissipation and reduces the fill factor. The effect of the shunt resistance is found to be of little significance. The contact resistance can be reduced to a negligible value, but the choice of the sheet resistance is less straightforward. A small sheet resistance corresponds to a heavily doped

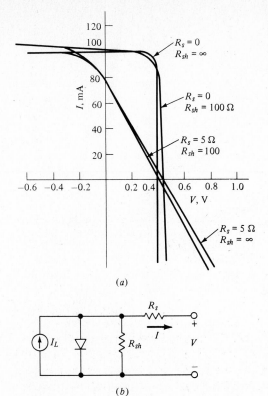

$R_s = 0$
$R_{sh} = \infty$

$R_s = 0$
$R_{sh} = 100 \, \Omega$

$R_s = 5 \, \Omega$
$R_{sh} = 100$

$R_s = 5 \, \Omega$
$R_{sh} = \infty$

(a)

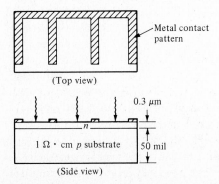

(b)

Figure 6-12 (a) Effects of series and shunt resistances on the I-V curve and (b) the equivalent circuit including parasitic resistances.

surface layer, which reduces the carrier lifetime and diffusion length of the surface layer. A compromise of doping level and junction depth is therefore necessary to arrive at the optimum design.

In addition, a small series resistance requires a large metallic contact area, which limits the area of light absorption. A practical contact in the form of a grid is shown in Fig. 6-13. This structure allows a large area of exposure but at the same time keeps the series resistance to a reasonable value.

Metal contact pattern

(Top view)

0.3 μm

n

1 Ω · cm p substrate

50 mil

(Side view)

Figure 6-13 Top and side views of a diffused n-on-p silicon cell.

Surface Reflection

The number of photons penetrating the surface is less than that of the incident photons because of reflection at the surface. The percentage of light reflected is determined by the angle of incidence and the dielectric constant of the material. If we assume normal incidence, the reflectance is given by the following law of optics [5]

$$R = \frac{(n-1)^2 + (\lambda\alpha/4\pi)^2}{(n+1)^2 + (\lambda\alpha/4\pi)^2} \qquad (6\text{-}19)$$

where $n = n_2/n_1$ and n_1 and n_2 are the refractive indices[†] of the air and semiconductor, respectively. In addition, α is the absorption coefficient of the semiconductor. In a silicon cell, the term $(\lambda\alpha/4\pi)^2$ is negligible, and $n = 3.5$. Therefore, the reflected light amounts to about 30 percent. To reduce the reflectance, we can coat the semiconductor surface with a material having a refractive index between n_1 and n_2. A practical antireflective coating for Si cells is a silicon oxide layer, which has a refractive index of 1.9. The ideal antireflective coating material should have a refractive index of $\sqrt{n_1 n_2}$.

Cost Considerations

Among the various limiting factors mentioned in the preceding paragraphs, materials synthesis to match the solar spectrum should provide the greatest improvement as far as efficiency is concerned. However, in order to capture a significant share of the energy market, cost of solar cells will be the dominant factor. Present research effort is leaning heavily toward methods for low-cost material preparation and device fabrication. Recently (1977) quoted prices for complete solar cell panels were below \$15 per peak watt. The term *peak watt* specifies a panel that produces 1 W at the peak solar energy of $100\,\text{mW/cm}^2$. To compete with existing sources of electrical energy, the price of a solar cell must be reduced to less than 50 cents per peak watt, with the ultimate goal of 10 cents per peak watt by the year 2000.

The total cost in making a solar cell consists of crystal growth and wafer preparation, device fabrication, and packaging. Packaging cost is least likely to be reduced, although automation could help somewhat. Conventional silicon crystals are grown in cylindrical rods, and they have to be sliced and polished before a junction is diffused. If crystals are grown in the form of a sheet or film, the wafer-preparation steps can be simplified. Newly proposed techniques include the edge-defined film-fed growth (EFG), the dendritic web growth, and polycrystalline sheet generation by rolling and extrusion [6].

[†] Refractive index is given by the square root of the dielectric constant. Thus, $n = \sqrt{K_s \epsilon_o / \epsilon_o} = \sqrt{K_s}$. K_s is the dielectric constant of the semiconductor. The subscript s is added here for clarification.

These semiconductor films are mostly polycrystalline, and the resulting solar cells have low efficiency. Although these film-growing methods are suitable for mass production at a very low cost, significant improvement in film quality is necessary before they can be employed. As for the device-fabrication technology, the use of ion implantation or Schottky barriers to replace or supplement the solid-state diffusion could be of practical importance in the near future. The principle of Schottky-barrier cells will be discussed in the next section.

Besides silicon cells, the use of other semiconductors such as cadmium sulfide, cuprous sulfide, and cuprous oxide provides another dimension in cost reduction. Solar cells fabricated on these materials are usually of the Schottky-barrier or heterojunction type.

A different approach to cost cutting is making use of a concentrator [7]. In this method, a large optical lens is employed to concentrate sunlight to a small area of solar cells. The light intensity can be increased up to a few hundred times. The lens can be made of plastic materials, which are much cheaper than high-quality silicon, thus lowering the overall system cost.

6-5 SCHOTTKY-BARRIER AND MIS SOLAR CELLS

One of the low-cost methods in fabricating a solar cell is using a Schottky barrier to replace the p-n junction. The cost reduction is due to the fabrication simplicity. In a typical process, a thin semitransparent metal film (50 to 100 Å thick) is evaporated onto the semiconductor, and the top contact with thick metal grid is then deposited. To minimize surface reflection at the metal-air interface, an antireflective coating is added in most devices to arrive at the final structure shown in Fig. 6-14a.

Using the energy-band diagram illustrated in Fig. 6-14b, we find two different modes of converting optical energy into electrical energy. If the incoming photon energy is greater than the barrier height but is smaller than the band-gap energy, that is, $E_g > h\nu > q\phi_b$, electrons in the metal can be excited to overcome the barrier height, resulting in a current flow. This is not a very efficient process, however, because of the requirement of momentum conservation across the M-S barrier. If the photon energy is greater than E_g, electron-hole pairs will be generated in both the depletion and bulk regions of the semiconductor. As a consequence, holes will move toward the metal and electrons toward the semiconductor, leading to a current flow. Since the bulk absorbs most of the photons, the light-generated current is primarily constituted by hole flow from the semiconductor to the metal. This second mode of operation is similar to that in a p-n junction cell and provides a good conversion efficiency.

In comparison with the p-n junction cell, the Schottky-barrier cell has a lower open-circuited voltage, resulting in a lower efficiency. According to Eq. (6-7), V_{oc} is inversely proportional to I_o. Since I_o in a Schottky-barrier diode is

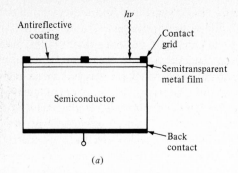

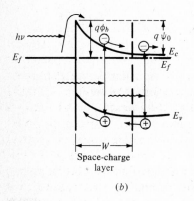

Figure 6-14 Schottky-barrier solar cell: (*a*) device structure and (*b*) energy-band diagram.

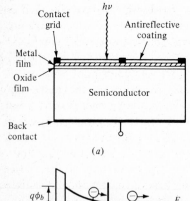

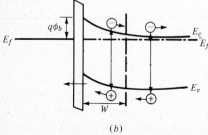

Figure 6-15 Metal-insulator-semiconductor solar cell: (*a*) device structure and (*b*) energy-band diagram.

a few orders of magnitude higher than that of the *p-n* junction (Sec. 5-3), the open-circuited voltage of the Schottky cell is significantly less than that of the *p-n* junction cell. It is noted that I_o represents the majority-carrier thermionic current which opposes the light-generated current. The thermionic current I_o can be reduced if a thin insulating layer is inserted between the metal and the semiconductor, and V_{oc} can thereby be increased [8]. The new structure is shown in Fig. 6-15, along with the energy-band diagram. In this MIS device, current conduction results from tunneling of carriers through the thin insulating layer. Efficiency as high as 12 percent for Au-Si cells and 15 percent for Au–GaAs cells has been obtained with this new structure.

6-6 GENERATION OF LIGHT WITH A *p-n* JUNCTION: THE LED

A light-emitting diode (LED) is a solid-state *p-n* junction device that emits light upon the application of a forward-biasing current. It is different from the incandescent light bulb, in which light is generated by heating a filament to a very high temperature. An LED is a cold lamp which converts electrical energy directly into optical energy without the intermediate step of thermal conversion. This luminescent mechanism, called *electroluminescence*, has an emission wavelength in the visible or infrared region. In the literature, these diodes are called *electroluminescent diodes*, and the acronym LED is reserved for devices emitting in the visible region only. In this chapter, however, the word LED is used to cover both visible and infrared emitters. LEDs are operated at low voltages and currents, typically 1.5 V and 10 mA, respectively; they can be made very small, so that it is reasonable to consider them as point sources of light. These characteristics make LEDs attractive for optical displays. In addition, the emission spectrum of LEDs is relatively narrow, and they can be switched on and off in the order of 10 ns. These properties are suitable for applications in optical data communication.

When a forward bias is applied across a *p-n* junction, carriers are injected across the junction to establish excess carriers above their thermal equilibrium values. The excess carriers recombine, and energy is released in the form of heat (phonons) or light (photons). In photon emission we derive optical energy from the biasing electrical energy. The *p-n* junction electroluminescence is illustrated graphically in Fig. 6-16. The injected electrons in the *p* side make a downward transition from the conduction band to recombine with holes in the valence band, emitting photons with an energy E_g. The corresponding emission wavelength is

$$\lambda = \frac{hc}{E_g} = \frac{1.24}{E_g} \qquad \mu m \tag{6-20}$$

where E_g is in electron volts. For example, the emission wavelength of GaAs at room temperature is 8900 Å, corresponding to a band-gap energy of 1.4 eV.

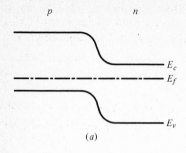

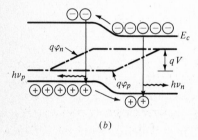

Figure 6-16 An electroluminescent p-n junction under (a) zero bias and (b) forward-bias voltage V.

Light emission in the n side of the junction follows the same pattern except that holes are now the excess carriers. In the following sections, the injection and recombination mechanisms are explained in more detail.

6-7 Minority-Carrier Injection and Injection Efficiency [9]

Depending on both the impurity profile and the external applied voltage, we can identify four current components in an LED under the forward-bias condition: (1) the electron diffusion current, (2) the hole diffusion current, (3) the space-charge-layer recombination current, and (4) the tunneling current. The tunneling current is important only in a heavily doped p-n junction at a small forward bias; thus its effect is negligible in most LEDs at light-emitting current level. The other current components have been described in Chap. 4 and are rewritten in the following equations:

$$I_n = \frac{qD_n n_i^2}{L_n N_a}(e^{qV/kT} - 1) \tag{6-21}$$

$$I_p = \frac{qD_p n_i^2}{L_p N_d}(e^{qV/kT} - 1) \tag{6-22}$$

$$I_{rec} = \frac{qn_i W}{2\tau} e^{qV/2kT} \tag{6-23}$$

Recombination inside the space-charge layer is effective if trap levels exist near the center of the forbidden gap. This process is generally nonradiative, and the component I_{rec} does not contribute to the emission of light. Furthermore, luminescence originates from the electron diffusion current in the p side of the junction in most practical LEDs for reasons beyond the scope of the present text. Consequently, we can define the current-injection efficiency as

$$\gamma = \frac{I_n}{I_n + I_p + I_{rec}} \qquad (6\text{-}24)$$

Frequently, the hole diffusion current is negligible because of the high electron-to-hole mobility ratio; e.g., in GaAs we have $\mu_n/\mu_p = 30$, and the foregoing equation can be simplified.

6-8 INTERNAL QUANTUM EFFICIENCY

The injection efficiency indicates the percentage of diode current that can produce radiative recombination in the p side of the junction. However, not all the electrons that reach the p side recombine radiatively. Electrons that survive the space-charge layer may recombine radiatively or nonradiatively, depending on the recombination paths in the p side of the junction, as shown in Fig. 6-17. The simplest recombination process is the band-to-band recombination R_1, in which a free electron and a free hole recombine directly. The second process R_2 involves a shallow impurity state, where an electron recombines with a hole trapped on a shallow acceptor state. Alternatively, the process may involve a shallow acceptor and a shallow donor state. The photon energy generated in this process is smaller than E_g. In the third possibility R_3 via deep impurity states, photons may not be generated at all; even if photons are generated, their energy is much smaller than E_g, so that it makes the emission intensity at E_g appear lower.

To simplify the complex picture, let us take an elementary case where a nonradiative recombination process via an intermediate state R_3 is competing with the band-to-band radiative recombination R_1. Following the notation in Chap. 2, we define the *radiative* and *nonradiative recombination rates*,

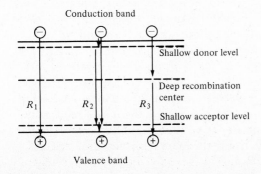

Figure 6-17 Three possible recombination paths.

respectively, as (for the p-type region)

$$U_r \equiv \frac{\Delta n}{\tau_r} \tag{6-25}$$

$$U_{nr} \equiv \frac{\Delta n}{\tau_{nr}} \tag{6-26}$$

where Δn = excess-electron density
τ_r = radiative recombination lifetime
τ_{nr} = nonradiative lifetime

The *radiative efficiency* is defined as the percentage of electrons that recombine radiatively,

$$\eta \equiv \frac{U_r}{U_r + U_{nr}} = \frac{1}{1 + \tau_r/\tau_{nr}} = \frac{\tau}{\tau_r} \tag{6-27}$$

where τ is the effective lifetime, given by

$$\frac{1}{\tau} = \frac{1}{\tau_r} + \frac{1}{\tau_{nr}} \tag{6-28}$$

Using Eq. (2-9) as the radiative lifetime and Eq. (2-27) as the nonradiative lifetime, we can write radiative efficiency in the p side of the junction as

$$\eta = \frac{1}{1 + c_n N_t / B N_a} \tag{6-29}$$

where the approximation $n_o + p_o = N_a + N_a/n_i^2 \approx N_a$ has been employed. In some practical cases, e.g., in the red GaP LED, the recombination processes involve trapping effects with R_2 and R_3 as the competing mechanisms. The radiative efficiency can be derived from the detailed balance of recombination and generation rates. The result is (Prob. 6-10)

$$\eta = \left[1 + \frac{N_t c_{p3} p}{N_a c_{n2} n} \exp\left(-\frac{E_t - E_a}{kT}\right) \right]^{-1} \tag{6-30}$$

The overall *internal quantum efficiency* can now be written as

$$\eta_i = \eta\gamma \tag{6-31}$$

The internal efficiency is influenced by the current-injection efficiency and radiative efficiency, and these two parameters depend strongly on the doping concentrations. In general, we can increase the doping concentration N_a in the p side to increase the radiative efficiency, as seen in Eqs. (6-29) and (6-30). A higher N_a also has the benefit of a smaller series resistance, thus reducing the forward-voltage drop and ohmic losses. However, very high concentration is not desirable because it increases crystal imperfections, which lead to an increase of the nonradiative centers N_t. In addition, a high doping in the p side reduces the injection efficiency. The foregoing consideration is confirmed by experimental data given in Fig. 6-18 for a GaP LED, where the externally measured efficiency peaks at $N_a = 2.5 \times 10^{17}$ cm^{-3}.

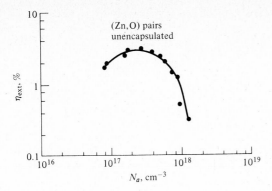

Figure 6-18 External quantum efficiency of red GaP diodes vs. net acceptor doping. *(After Bhargava [11].)*

6-9 EXTERNAL QUANTUM EFFICIENCY

The most important parameter of an LED is the *external quantum efficiency*. It may be significantly smaller than the internal quantum efficiency because of internal absorption and reflection of light. After photons are generated at the *p-n* junction, they must pass through the crystal to reach the surface. Some of the emitted photons are reabsorbed by the semiconductor. Furthermore, even after the photons have reached the surface, they may not be able to leave the semiconductor because of the large difference in the refractive indices of the semiconductor and air. According to the theory of optics, the *critical angle* θ_c (Fig. 6-19) at which total internal reflection occurs is given by the Fresnel equation

$$\sin \theta_c = \frac{1}{n} = \frac{n_1}{n_2} = \frac{1}{\sqrt{K_s}} \tag{6-32}$$

where n is the refractive index of the semiconductor with the air (n_1) as the external reference. All rays of light striking the surface at angles exceeding θ_c are reflected. Since n ranges between 3.3 to 3.8 for a typical LED material, θ_c is calculated to be between 15 and 18°. For the light striking within the critical

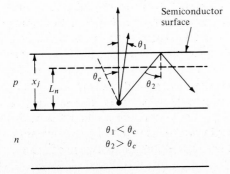

Figure 6-19 Internal reflection and critical angle in an LED.

angle, the portion that comes out is given approximately by the average transmissivity

$$T = \frac{4n}{(1+n)^2} \tag{6-33}$$

Therefore, the total light emission within the solid angle θ_c is

$$\bar{T} = T \sin^2 \frac{\theta_c}{2} \tag{6-34}$$

A simple expression relating the external quantum efficiency to the internal quantum efficiency is given by [10]

$$\eta_{ext} = \frac{\eta_i}{1 + \bar{\alpha}V/A\bar{T}} = \frac{\eta_i}{1 + \bar{\alpha}x_j/\bar{T}} \tag{6-35}$$

where $\bar{\alpha}$ = average absorption coefficient
V = diode volume
A = emitting area
In an LED, the ratio V/A may be taken as the junction depth x_j from the emitting surface. Equation (6-35) indicates that the external quantum efficiency can be increased by reducing $\bar{\alpha}$ or x_j or by increasing $\bar{T}$.

Reducing the junction depth to less than a diffusion length from the surface introduces more minority carriers to the surface. Therefore, the surface recombination centers capture a larger portion of the injected carriers, thus reducing the internal quantum efficiency. Experimental results of the junction-depth dependence of η_{ext} in a GaAs LED are shown in Fig. 6-20, where the optimized junction depth is between 15 and 25 μm.

The reduction of α can be achieved by generating a luminescence with $h\nu < E_g$ (R_2 in Fig. 6-17), as illustrated in Fig. 6-21. A high efficiency is obtained since the emitted photons have an energy below E_g. Note that the absorption coefficient is very low at the emission peak but absorption is high

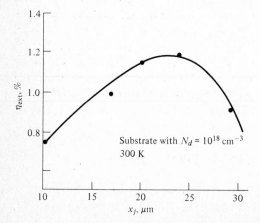

Figure 6-20 External efficiency vs. junction depth of GaAs LEDs. $L_P \sim 8\,\mu$m.

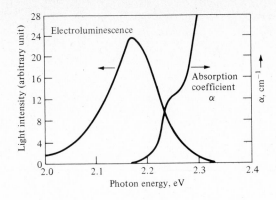

Figure 6-21 Comparison of a typical external electroluminescent spectrum of a green GaP LED and the absorption coefficient of GaP. $E_g = 2.25$ eV.

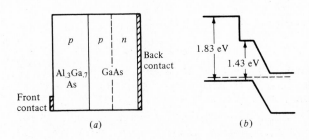

Figure 6-22 A GaAs LED with an AlGaAs layer as optical window: (*a*) structure and (*b*) energy-band diagram. Al.3Ga.7As represents a ratio of 30% Al and 70% Ga in its composition.

at E_g. Alternatively, an optical window is used, as shown in Fig. 6-22. In this device, an additional AlGaAs layer is grown on top of the GaAs diode. Since the AlGaAs material has a band gap greater than GaAs, the emitted photons are not absorbed by this added layer. At the same time, the density of recombination centers at the AlGaAs–GaAs interface is significantly lower than that of the GaAs surface without the AlGaAs layer. Therefore, the depth of the junction from the interface can be made very small.

The reduction of the internal reflection can be achieved by using a dome-shaped diode geometry (Fig. 6-23*a*) so that most of the light emitted at the junction arrives within the critical angle at the semiconductor surface. The disadvantage of this method is that a large amount of semiconductor material is needed and the machine work is not economical. A more practical technique is using an optical medium with a refractive index between that of air and the semiconductor, as shown in Fig. 6-23*b*. Hemispherical domes cast from epoxy or acrylic-polyester resin (with $n = 1.5$) are quite effective, increasing the external efficiency by a factor of 2 to 3.

6-10 EYE SENSITIVITY AND BRIGHTNESS

The response of the human eye, called *luminous efficiency*, is limited to wavelengths between 4000 and 7000 Å. The *standard-luminous-efficiency*

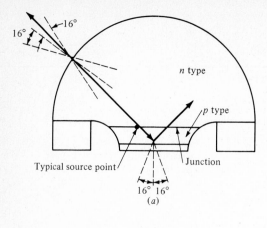

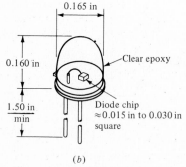

Figure 6-23 Dome-shaped LED structures: (a) with n-type semiconductor dome and (b) with clear epoxy dome.

curve is shown in Fig. 6-24. The eye is very sensitive to green or yellow color but is a poor detector in the red or violet region. Because of the large variation of eye sensitivity, the performance of an LED is appraised not only by its external quantum efficiency but also by the relative response of the eye at the wavelength of interest. An emission at 5500 Å (2.23 eV) is most desirable as far as the luminous efficiency is concerned. For this reason, we define the brightness of an LED as a measure of the visual impact of the radiation using

$$B = 1150L \frac{J}{\lambda} \frac{A_j}{A_s} \eta_{\text{ext}}, \text{ fL} \qquad (6\text{-}36)$$

where λ = emission wavelength, μm

J = current density, A/cm^2

L = luminous efficiency at λ, lumens/watt

$A_j = p\text{-}n$ junction area

A_s = observed emitting surface area

The unit of brightness is footlamberts (fL). To facilitate fair comparison of the performance of different types of LEDs, the brightness B is frequently normalized with respect to the current density. A brightness of 1500 fL at 10 A/cm^2 is typical for a commercial LED. By comparison, the brightness of a

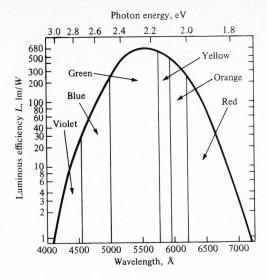

Figure 6-24 Luminous efficiency of eye vs. wavelength of incident light. $L_{max} = 680$ lm/W at 5550 Å.

40-W incandescent light bulb is about 7000 fL, but it is at a much higher current density.

6-11 MATERIAL CONSIDERATIONS

The maximum possible energy for the emitted photons in a semiconductor is determined approximately by the energy gap. For a visible display, the wavelength of the emitted light should be between 0.4 and 0.72 μm. This requirement limits our materials to those having an energy gap between 1.7 and 3.0 eV. However, larger wavelengths may be more suitable for optical communication systems because low-loss optical fibers have their minimum attenuation in the neighborhood of 1 μm. The best-known materials are GaAs, which emits in the infrared region, and GaP and GaAsP, which emit in the visible region. Other materials such as SiC, InSb, PbS, GaN, AlGaAs, and InGaP have been made in junction form to produce light to some degree, but their low efficiency or material-preparation difficulties hinder their further development. In recent years, most work in p-n junction luminescence has been concentrated on GaAs, GaP, and GaAsP. We shall limit our discussion to these three materials; readers interested in other materials are referred to the literature.

GaAs Emitters

Gallium arsenide is a direct-gap semiconductor with an energy gap of 1.4 eV at room temperature, which corresponds to an emission wavelength of

8900 Å. A typical GaAs LED is made by solid-state impurity diffusion with zinc as the p-type impurity diffused into an n-type substrate doped with tin, tellurium, or silicon. To achieve high efficiency, the concentration of both types of dopants is of the order of 10^{18} cm^{-3}. The external efficiency at room temperature is typically 5 percent. The emission spectra of a GaAs diode are shown in Fig. 6-25. The spectral line width, i.e., the width of the half-power points, is typically less than 300 Å. A GaAs diode can also be fabricated by liquid-phase epitaxy with silicon as both its n and p dopants. If a silicon atom replaces a Ga atom, it provides one additional electron; thus the resulting GaAs is an n type. If a silicon atom replaces an arsenic atom, an electron is missing, and the resulting GaAs is a p type. Because of this special property, silicon in GaAs is called an *amphoteric dopant*. In a Si-doped GaAs diode, the emission peak shifts down to 1.32 eV. The self-absorption becomes much smaller, and an external quantum efficiency as high as 20 percent has been achieved in dome-shaped devices. Since the emission is in the infrared region, GaAs light sources are suitable for applications such as the optical isolator illustrated in Fig. 6-26. The high switching speed, with a recovery time between 2 and 10 ns, makes them ideal for data transmission.

GaP LEDs

With an energy gap of 2.3 eV, gallium-phosphide diodes emit red or green light, depending on the recombination mechanism involved. In general, recombination in an indirect semiconductor such as GaP tends to take place via

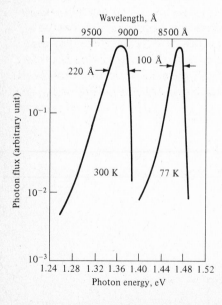

Figure 6-25 Experimental emission spectra of GaAs LED at 300 and 77 K. *(After Bergh and Dean [10].)*

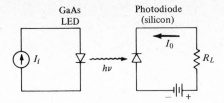

Figure 6-26 Optical isolator using a GaAs LED and a silicon photodiode.

impurity levels because this facilitates conservation of crystal momentum. We shall describe two types of GaP diodes separately, the red and the green.

Red GaP diode The radiation mechanism in a red diode is through a donor-acceptor impurity pair. When both donors and acceptors are simultaneously present in the semiconductor, the donor and acceptor states are partly occupied. It is therefore possible for an electron in a donor state to make a downward transition, recombining with a hole in an acceptor state. The radiated photon energy is given by

$$hv = E_g - (E_d + E_a) + \frac{q^2}{K_s \epsilon_o r} \tag{6-37}$$

where v = emitted frequency
E_d = donor energy
E_a = acceptor energy

The last term in Eq. (6-37) describes the Coulomb interaction energy between donors and acceptors where r is the donor-acceptor separation. In an oxygen-doped GaP diode, an emission near 7000 Å is observed. It is believed that the emission is due to a transition from a deep donor caused by oxygen to a shallow acceptor such as zinc. The zinc level is 0.04 eV above the valence-band edge, and the oxygen level is 0.803 eV below the conduction-band edge. Since the donor-acceptor separation varies among different pairs, the emission band is quite broad, having a half-width of 870 Å. Typically, red-emitting diodes have external efficiencies of 2 to 3 percent at current levels of 10 A/cm². An efficiency-vs.-current characteristic is shown in Fig. 6-27. The

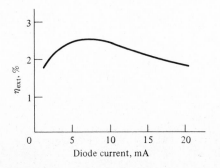

Figure 6-27 External efficiency of a red GaP diode as a function of current.

drop of efficiency at high current is caused by the saturation of impurity centers at high carrier-injection levels. Maximum efficiency in red diodes of 15 percent has been achieved in the laboratory; however, the maximum brightness is low because the emission output of these high-efficiency diodes saturates at a very low current. The switching speed is typically 100 ns.

Green GaP diode Green emission has been observed in GaP, and this has been attributed to recombination at a nitrogen atom on a phosphorus site. Because both nitrogen and phosphorus are in the same column of the periodic table, replacement of phosphorus by a nitrogen atom is described as *isoelectronic*. An isoelectronic center is a very localized potential well that can trap an electron, thus becoming charged. The resulting Coulomb field then attracts a hole which pairs with the trapped electron to form an *exciton*, i.e., a hydrogenlike bound electron-hole pair. The annihilation of this exciton by radiative recombination gives rise to green light with $\lambda = 5700$ Å at room temperature. Because of competing nonradiative recombination and thermal activation processes, the internal efficiency of green emission in GaP is less than 1 percent. However, the self-absorption is very small since the emission is below the band-gap energy, and an external efficiency of 0.7 percent has been realized. The green emission is nearly at the peak of the eye-sensitivity curve shown in Fig. 6-24. For this reason, the green GaP diode provides high brightness despite its low external quantum efficiency. In addition, its light output does not saturate with current, as seen in Fig. 6-28, where the red GaP diode is included for comparison. The current-voltage characteristic of a typical green diode is shown in Fig. 6-29.

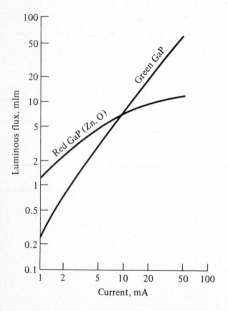

Figure 6-28 Light output vs. current in GaP LEDs showing saturation effect in the red diode. *(After Bhargava [11].)*

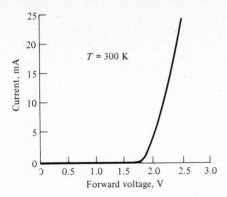

Figure 6-29 Current-voltage characteristic of a green GaP LED.

GaAs$_{1-x}$P$_x$ LEDs

Material synthesis is a practical approach to obtaining a semiconductor with a particular desired energy-band gap. Such synthesis is desirable because efficient emission in direct-band-gap semiconductors such as GaAs is not visible to the human observer and because GaP was a material difficult to handle in its early development. When GaAs is mixed with GaP, a *ternary alloy* can be formed having an energy gap between that of GaAs and GaP. The mixed alloy is called GaAs$_{1-x}$P$_x$, where x specifies the alloy composition of phosphorus. Using x as a parameter, we obtain a different energy-band gap, as shown in Fig. 6-30. It is noted that in addition to an increased E_g, the crystal changes from direct gap to indirect gap at $x = 0.46$. Since the radiative recombination is more efficient in direct-gap materials, we expect a decreasing external quantum efficiency for increasing x and E_g, as shown in Fig. 6-31 for LEDs with and without nitrogen doping. Nitrogen doping is found to enhance the radiative-recombination process and thus improve the quantum efficiency. It is also shown that the epoxy encapsulation yields an improvement of $2\frac{1}{2}$

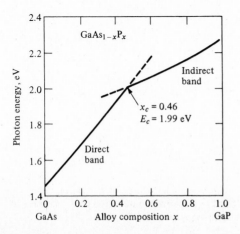

Figure 6-30 Band-gap energy E_g of GaAs$_{1-x}$P$_x$ as a function of alloy composition. *(After Casey and Trumbore [12].)*

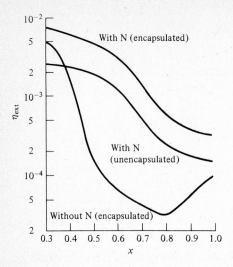

Figure 6-31 External quantum efficiency as a function of alloy composition for $GaAs_{1-x}P_x$ diodes with and without nitrogen doping. *(After Bhargava [11].)*

times. When the eye-sensitivity response is taken into account, we find that the brightness of $GaAs_{1-x}P_x$ diodes peaks approximately at $x = 0.4$ and $E_g = 1.9\,eV$ (Fig. 6-32). The corresponding emission wavelength is 6500 Å, which is in the red region. The external efficiency is less than 0.5 percent for an encapsulated diode with a brightness of $150\,fL/A\text{-}cm^2$. By increasing the GaP concentration, we can obtain orange emission, but with reduced efficiency and brightness

In most earlier commercial diodes, the GaAsP layer was grown on a GaAs substrate. Since GaAs has a lower band-gap energy, it absorbs the light

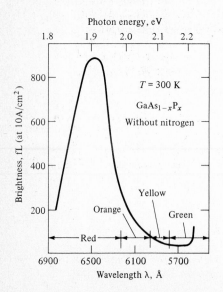

Figure 6-32 Brightness of GaAsP diodes vs. emission peak. *(After Bhargava [11].)*

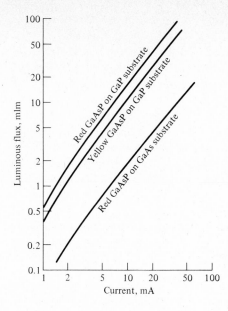

Figure 6-33 Effect of substrate material on GaAsP LED performance. *(After Bhargava [11].)*

Package physical dimensions

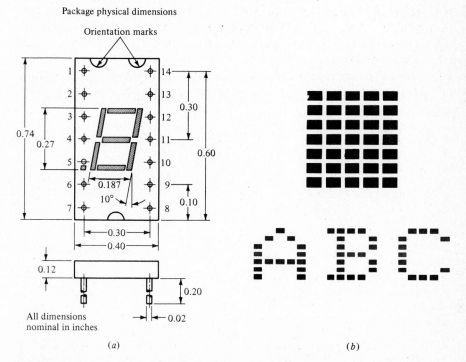

Figure 6-34 (*a*) Seven-segment LED display and (*b*) alphanumeric 5 × 7 array. *(After Casey and Trumbore [12].)*

emitted from the GaAsP junction and thus reduces the light output. Absorption by the substrate can be eliminated if the band-gap energy of the substrate is larger than the emitted-photon energy. For this reason, recent GaAsP LEDs are fabricated on a GaP substrate. The improved light-current characteristic is seen in Fig. 6-33. The output light power increases without saturation even at a high current similar to that of the GaAs diodes.

By using planar technology $GaAs_{0.6}P_{0.4}$ monolithic arrays can be fabricated for numeric and alphanumeric displays, as shown in Fig. 6-34. Individual segments are clearly visible because of the strong self-absorption, so that light does not penetrate deep into the inactive regions. With the help of a lens, $\frac{1}{2}$-in numerals can be obtained. Most commercially available red LEDs are GaAsP devices because of lower cost and ease of fabrication.

PROBLEMS

6-1 (a) Calculate the maximum wavelength λ of the light source that generates electron-hole pairs in Ge, Si, and GaAs.

(b) What is the photon energy for the light source with wavelength of 5500 and 6800 Å?

6-2 A 0.46-μm-thick sample of GaAs is illuminated with a monochromatic light source of $h\nu = 2\,eV$. The absorption coefficient α is $5 \times 10^4\,cm^{-1}$, and the incident power of the sample is 10 mW.

(a) Calculate the total energy absorbed by the sample in joules per second.

(b) Find the rate of excess thermal energy given up by the electrons to the lattice before recombination in joules per second.

(c) Calculate the number of photons per second given off from recombination.

6-3 Assume that a p^+-n diode is uniformly illuminated by a light source to produce an electron-hole generation rate of G_L. Solve the diffusion equation in the n side of the diode to show that

$$\Delta p_n = \left[p_{no}(e^{V/V_T} - 1) - G_L \frac{L_p^2}{D_p} \right] e^{-X/L_p} + \frac{G_L L_p^2}{D_p}$$

6-4 Use the result of Prob. 6-3 to derive Eq. (6-6).

6-5 (a) Carry out the derivation of Eq. (6-9).

(b) Assume that the dark current is 1.5 nA and the light-generated short-circuit current is 100 mA. Plot the I-V curve and find the load resistance graphically for maximum output power. What is the fill factor?

6-6 (a) Show that the current I_o in Eq. (6-5) for an n^+-p cell is given by

$$I_o = \frac{qAn_{po}D_n}{L_n} \frac{S \cosh(W_p/L_n) + (D_n/L_n) \sinh(W_p/L_n)}{(D_n/L_n) \cosh(W_p/L_n) + S \sinh(W_p/L_n)}$$

where S is the surface-recombination velocity at the ohmic contact, W_p is the width of the p region, and other symbols designate minority-carrier parameters.

(b) Show that

$$I_o = \begin{cases} \dfrac{qAn_{po}D_n}{L_n} \tanh \dfrac{W_p}{L_n} & \text{for } S \ll \dfrac{D_n}{L_n} \\[3mm] \dfrac{qAn_{po}D_n}{L_n} \coth \dfrac{W_p}{L_n} & \text{for } S \gg \dfrac{D_n}{L_n} \end{cases}$$

Which is the preferred condition for solar-cell application?

6-7 Assuming $J_L = 40 \text{ mA/cm}^2$, plot the open-circuit voltage as a function of acceptor concentration in an n^+-p GaAs cell with $J_o = I_o/A$ given by Prob. 6-6b for $S \ll D_n/L_n$ and $W_p = L_n = 5 \ \mu\text{m}$.

6-8 The number of photons per square centimeter per second averaged over a small bandwidth λ which has entered silicon is $Q(\lambda)$.

(a) Derive an expression to represent the loss of light-generated current $\Delta J_L(\lambda)$ at wavelength λ as a function of the reflection coefficient at the back contact, the total cell thickness, and the absorption coefficient.

(b) Estimate the loss of light-generated current at $\lambda = 9000$ Å. Assume that $Q(\lambda)$ is equal to 50 percent of the sun spectrum for a bandwidth of ± 500 Å, the average absorption coefficient is 500 cm^{-1}, and the reflection coefficient at the back contact is 0.8. The cell thickness is $10 \ \mu\text{m}$.

6-9 Show that the hole diffusion current is negligible in comparison with the electron current if $N_a \approx N_d$ in a GaAs LED. Use $\mu_n/\mu_p = 30$.

6-10 Assume $n_t = N_t \exp[-(E_t - E_f)/kT]$ and $p_a = N_a \exp[-(E_a - E_f)/kT]$. Derive Eq. (6-30) by using Fig. P6-10 and the detailed balance of recombination and generation rates.

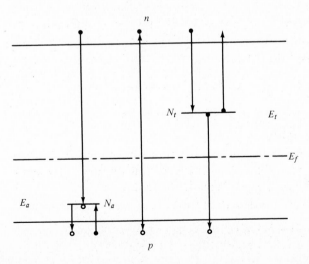

Figure P6-10.

6-11 A GaAs infrared emitter has the following device parameters: $\eta_i = 80$ percent, $\bar{\alpha} = 10^3 \text{ cm}^{-1}$, and $x_j = 10 \ \mu\text{m}$.

(a) Calculate the external quantum efficiency.

(b) Repeat (a) if a dome-shaped epoxy with a refractive index of 1.8 is used in the LED package. What is the ratio of improvement in the external quantum efficiency?

6-12 In GaAs the temperature dependence of the absorption coefficient can be approximated by $\bar{\alpha} = \alpha_o \exp(T/T_o)$, where α_o is α extrapolated to $T = 0 \text{ K}$ and T_o is approximately 100 K. The external quantum efficiency of a diode is 5 percent at 300 K. Other parameters are $x_j = 20 \ \mu\text{m}$, $\bar{T} = 0.2$, and $\bar{\alpha}(300 \text{ K}) = 10^3 \text{ cm}^{-1}$.

(a) Calculate the internal quantum efficiency at 27°C.

(b) Assuming that the internal quantum efficiency is constant for the temperature range considered here, estimate the external quantum efficiency at −23 and 77°C.

6-13 Calculate the brightness of (a) a red GaP LED with $\eta_{\text{ext}} = 5$ percent at 10 A/cm^2, (b) a green GaP with $\eta_{\text{ext}} = 0.03$ percent at 10 A/cm^2, (c) a red $\text{GaAs}_{0.6}\text{P}_{0.4}$ with $\eta_{\text{ext}} = 0.15$ percent at 20 A/cm^2. Assume that Aj/As is unity.

6-14 Estimate the range of the spacing of the donor-acceptor separation for the red GaP diode described in the text.

REFERENCES

1. P. Rappaport and J. J. Wysocki, The Photovoltaic Effect in GaAs, CdS and Other Compound Semiconductors, *Acta Electron.*, **5**: 364 (1961).
2. E. S. Rittner, An Improved Theory of the Silicon p-n Junction Solar Cell, *Int. Electron. Devices Meet., Washington, December 1976, Tech. Dig.*, pp. 69–70.
3. M. Wolf, Limitations and Possibilities for Improvements of Photovoltaic Solar Energy Converters, *Proc. IRE*, **48**: 1246 (1960).
4. J. J. Wysocki and P. Rappaport, Effect of Temperature on Photovoltaic Solar Energy Conversion, *J. Appl. Phys.*, **31**: 571 (1961).
5. J. I. Pankove, "Optical Processes in Semiconductors," Prentice-Hall, Englewood Cliffs, N.J., 1971.
6. M. Wolf, Progress in New Low-Cost Processing Methods, *11th IEEE Photovoltaic Spec. Conf.*, 1975, p. 306.
7. R. J. Schwartz and M. D. Lammert, Silicon Solar Cells for High Concentration Applications, *Int. Electron. Device Meet., Washington, December 1975*, p. 350.
8. H. C. Card and E. S. Yang, MIS-Schottky Theory under Conditions of Optical Carrier Generation in Solar Cells, *Appl. Phys. Lett.*, **29**: 51 (1976).
9. E. S. Yang, Degradation of Zinc-diffused GaAs Electroluminescent Diodes, *IEEE J. Quantum Electron.*, **QE-7**: 239 (1971).
10. A. Bergh and P. Dean, Light-emitting Diodes, *Proc. IEEE*, **60**: 156 (1972).
11. R. N. Bhargava, Recent Advances in Visible LEDs, *IEEE Trans. Electron. Devices*, **ED-22**: 691 (1975).
12. H. C. Casey, Jr., and F. A. Trumbore, Single Crystal Electroluminescent Materials, *Mater. Sci. Eng.*, **6**: 69 (1970).

ADDITIONAL READINGS

Bergh, A., and P. Dean: Light-emitting Diodes, *Proc. IEEE*, **60**: 156–223 (February 1972). Probably the most extensive review of LEDs in the literature.

Hovel, H. J.: "Solar Cells, Semiconductors and Semimetals," vol. 11, Academic, New York, 1975. The most comprehensive coverage of solar cells.

Larach, S. (ed.): "Optoelectronic Materials and Devices," Van Nostrand, Princeton, N.J., 1965.

Merrigan, J. A.: "Sunlight to Electricity," MIT Press, Cambridge, Mass., 1975. Covers both the economics and technology of solar cells.

Neuse, C. J., H. Kressel, and I. Ladany: The Future for LEDs, *IEEE Spectrum*, May 1972, pp. 28–38.

Pankove, J. I.: "Optical Processes in Semiconductors," Prentice-Hall, Englewood Cliffs, N.J., 1971. An advanced treatment of optoelectronic devices.

SEVEN

JUNCTION FIELD-EFFECT TRANSISTORS

7-1 INTRODUCTION

The junction field-effect transistor (JFET) is a three-terminal semiconductor device in which the *lateral* current flow is controlled by an externally applied *vertical* electric field. In its early development, the JFET was known as the *unipolar transistor* [1] because the current is transported by carriers of one polarity, namely, the majority carriers. This is in contrast with the bipolar-junction transistor, discussed in Chap. 9, in which both majority- and minority-carrier currents are important. An idealized JFET fabricated by the standard planar epitaxial processes is shown in Fig. 7-1a. The active region of the device consists of a lightly doped *n*-type *channel* sandwiched between two heavily doped p^+-gate regions. The lower p^+ layer is the substrate, and the upper p^+ region is formed by boron diffusion into the epitaxially grown *n*-type channel. The p^+ regions are connected either internally or externally to form the *gate* terminal. Ohmic contacts attached to the two ends of the channel are known as the *drain* and *source* terminals, through which the channel current flows. Alternatively, the JFET may be fabricated by the double-diffused technique with diffused channel and upper gate as illustrated in Fig. 7-1b. The structures shown are *n-channel* JFETs since the channel is doped with donor impurities and the channel current consists of electrons. If the channel is doped with acceptor atoms and the gate regions are n^+ type, the channel current consists of holes and the device is a *p-channel* JFET. Since electrons have higher mobility than holes, an *n*-channel device provides higher conduc-

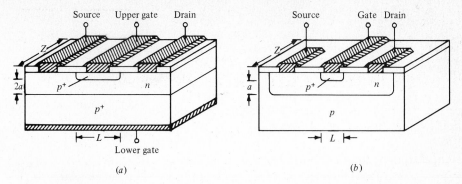

Figure 7-1 An *n*-channel JFET fabricated by (*a*) epitaxial-diffused process and (*b*) double-diffused process.

tivity and higher speed and is preferred in most applications. In the following sections, our discussion will be limited to the *n*-channel JFET.

Under normal operating conditions, a reverse bias is applied across the *p-n* gate junctions so that free carriers are depleted from the channel and space-charge regions extending into the channel are produced. Consequently, the cross-sectional area of the channel is reduced, and the channel resistance is increased. Thus the current flow between the source and the drain is modulated by the gate voltage.

Let us connect the source and gate terminals to the ground potential and set the drain voltage at V_D (Fig. 7-2). Under these conditions, the voltage across the gate junctions at $x = 0$ is zero, but the full value of V_D is across the junctions at $x = L$. As a result, the space-charge regions extend farther into the channel at the drain end, as depicted in Fig. 7-2*a*. When we increase V_D, the bottleneck in the channel becomes smaller and the channel resistance is increased. If the drain voltage is further increased, a condition will eventually be reached, as shown in Fig. 7-2*b*, in which the space-charge regions join and all free carriers are completely depleted in the joining region. This condition is called *pinchoff*. Further increase of the drain voltage beyond

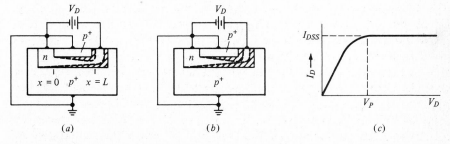

Figure 7-2 The JFET with $V_G = 0$ and (*a*) $V_D < V_P$, (*b*) $V_D = V_P$, and (*c*) idealized drain characteristic.

pinchoff would not increase the drain current significantly. Therefore, the current is *saturated*, and the channel resistance becomes very large. The current-voltage characteristic of the drain with respect to ground is shown in Fig. 7-2c, where I_{DSS} specifies the *saturation drain current* and V_P is called the *pinchoff voltage*.

7-2 THEORY OF THE JFET

With applied voltages at both the gate and the drain, the JFET shown in Fig. 7-2a is reproduced in Fig. 7-3 in an enlarged scale. The space-charge layer profile indicates that the electric field and carrier distributions are of two-dimensional nature. This is a difficult analytical problem to solve, and it must be simplified in order to obtain some physical insight into the current-conduction mechanism. Fortunately in most practical JFETs, the channel length L is 2 or more times greater than that of the channel width $2a$. In these *long* devices, the change of the channel width along the channel is small in comparison with the width of the channel. Therefore, the electric field in the space-charge layer may be considered as in the y direction, i.e., normal to the gate junctions. At the same time, the electric field in the neutral channel may be assumed to be in the x direction only. The separation of the electric field in the space-charge layer and the channel allows us to solve the one-dimensional Poisson's equation in each region. This technique was developed by Shockley and is known as the *gradual-channel approximation* [1]. Using this approach, we can obtain the important features of the JFET without resorting to the solution of two-dimensional partial differential equations

By assuming gate junctions as one-sided step junctions, the depletion-layer width in the JFET is obtained by using Eq. (4-23):

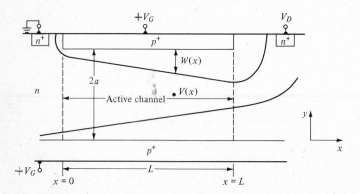

Figure 7-3 Enlarged schematic diagram of Fig. 7-2a, showing gradual change of the space-charge regions in the active channel. The n^+ regions are added to provide good ohmic contacts.

$$W(x) = \left\{ \frac{2K\epsilon_o[V(x) + \psi_o - V_G]}{qN_d} \right\}^{1/2} \quad \left(\begin{array}{c} \text{for} \\ n\text{-channel} \end{array} \right) \quad (7\text{-}1)$$

where $V(x) - V_G$ is the voltage across the reverse-biased junction at x (Fig. 7-3). The drain current can be obtained by making use of Eq. (2-63). Since the electron distribution in the neutral channel is assumed to be uniform, the gradient of electrons is zero and the diffusion current component can be neglected. Thus, the drain current consists of the electron-drift component only, and Eq. (2-63) becomes†

$$I_D = -q\mu_n nA\mathscr{E} = 2q\mu_n N_d(a - W)Z\frac{dV}{dx} \quad (7\text{-}2)$$

where I_D denotes the drain current and $2(a - W)Z$ is the cross-sectional area. Substituting Eq. (7-1) into (7-2) and integrating, we have

$$\int_0^L \frac{I_D\,dx}{2q\mu_n N_d Z} = \int_0^{V_D} \left[a - \sqrt{\frac{2K\epsilon_o}{qN_d}(V + \psi_o - V_G)} \right] dV \quad (7\text{-}3)$$

The limits of integration are defined by the length of the active region from $x = 0$ to $x = L$ with the corresponding voltage from 0 to V_D. Performing the integration yields

$$I_D = G_o\left\{ V_D - \frac{2}{3}\sqrt{\frac{2K\epsilon_o}{qa^2 N_d}}\,[(V_D + \psi_o - V_G)^{3/2} - (\psi_o - V_G)^{3/2}] \right\} \quad (7\text{-}4)$$

where

$$G_o = \frac{2qaZ\mu_n N_d}{L} \quad (7\text{-}5)$$

G_o is the channel conductance without any depletion layers. Equation (7-4) describes the drain current as a function of both the drain and gate voltages before the pinchoff condition is reached. The I-V characteristics of the drain are plotted in Fig. 7-4. Curves in Fig. 7-4a are based on the theoretical model expressed by Eq. (7-4), and it is assumed that the drain current is constant after pinchoff. We have also plotted data of an experimental device in Fig. 7-4b for comparison. The discrepancy can be accounted for when the effect of series resistance is included, and it will be explained in a later section.

The space-charge-layer width at the pinchoff point is exactly equal to the channel width. Thus, the pinchoff voltage is obtained by setting $W = a$ and $V - V_G = V_P$ in Eq. (7-1):

$$V_P + \psi_o = \frac{qa^2 N_d}{2K\epsilon_o} = V_{PO} \quad (7\text{-}6)$$

where V_P is the externally applied voltage to reach the pinchoff condition and is the pinchoff voltage. V_{PO} is the sum of the pinchoff voltage and the built-in potential, and it may be referred to as the *internal* pinchoff voltage.

---

† The drain current is defined as positive in the direction against the x axis in Fig. 7-3, thus yielding the negative sign in Eq. (7-2).

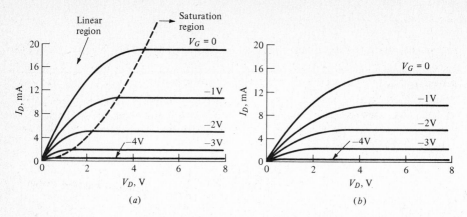

Figure 7-4 Current-voltage characteristics of a silicon n-channel JFET with $a = 1.5\ \mu m$, $Z/L = 170$, and $N_d = 2.5 \times 10^{15}\ cm^{-3}$: ($a$) theoretical plot of Eq. (7-4) with $R_s = 0$, and (b) experimental results. (*After Grove [2].*)

Example For an n-channel silicon JFET with $K = 12$, $N_d = 5 \times 10^{15}\ cm^{-3}$, $N_a = 10^{19}\ cm^{-3}$, $a = 1\ \mu m$, $L = 30\ \mu m$, $Z = 0.1\ cm$, and $\mu_n = 1350\ cm^2/V\text{-}s$ find (a) the pinchoff voltages V_{PO} and V_P and (b) the drain current at $V_D = V_P$ with both gate and source grounded.

SOLUTION (a)

$$V_{PO} = \frac{qa^2 N_d}{2K\epsilon_0} = \frac{(1.6 \times 10^{-19})(10^{-8})(5 \times 10^{15})}{(2)(12)(8.85 \times 10^{-14})} = 3.77\ V \qquad \text{from Eq. (7-6)}$$

$$\psi_o = V_T \ln \frac{N_a N_d}{n_i^2} = 0.026 \ln \frac{(5 \times 10^{15})(10^{19})}{2.25 \times 10^{20}} = 0.86\ V \qquad \text{from Eq. (4-7)}$$

$$V_P = V_{PO} - \psi_o = 3.77 - 0.86 = 2.91\ V$$

(b) $\qquad G_o = \dfrac{2qaZ\mu_n N_d}{L} = 14.4 \times 10^{-3} \quad \Omega^{-1} \qquad \text{from Eq. (7-5)}$

$$I_D = G_o \left\{ V_P - \frac{2}{3\sqrt{V_{PO}}} [(V_P + \psi_o)^{3/2} - (\psi_o)^{3/2}] \right\} = 9.6\ mA \qquad \text{from Eq. (7-4)}$$

7-3 STATIC CHARACTERISTICS

The current-voltage characteristics displayed in Fig. 7-4 can be divided into linear and saturation regions with the pinchoff condition as the boundary. In this section we examine the I-V characteristics in these two regions and discuss the significance of the gate leakage current, breakdown voltage, and series resistances.

Linear Region

Examining the current-voltage characteristics shown in Fig. 7-4, we find that the drain current is proportional to the drain voltage at small drain voltages. In addition, the slope of the I-V curves near the origin is a function of the gate voltage. This region of operation is known as the *linear region*. Mathematically, the current-voltage relationship of the linear region is obtained by first letting $V_D \ll \psi_o - V_G$ in Eq. (7-4). Using the binomial series expansion, we can write the second term in Eq. (7-4) as

$$(\psi_o - V_G)^{3/2}\left(1 + \frac{V_D}{\psi_o - V_G}\right)^{3/2} \approx (\psi_o - V_G)^{3/2}\left(1 + \frac{3}{2}\frac{V_D}{\psi_o - V_G}\right) \tag{7-7}$$

Substituting Eq. (7-7) into (7-4) and simplifying the expression, we obtain

$$I_D = G_o\left(1 - \sqrt{\frac{\psi_o - V_G}{V_{PO}}}\right)V_D \tag{7-8}$$

The foregoing equation shows indeed the linear dependence of the drain voltage. The effect of the gate voltage on the slope of the I-V curves is also apparent in Eq. (7-8).

Saturation Region†

At the point of pinchoff, the magnitudes of the biasing drain and gate voltages satisfy the condition

$$V_D - V_G = V_P \tag{7-9}$$

Thus, the drain voltage needed to reach the pinchoff condition is different for each gate voltage. Equation (7-9) is plotted in Fig. 7-4a and is known as the *pinchoff curve*. The current-voltage characteristics beyond pinchoff are called the *saturation region* because the drain current is saturated. The magnitude of the drain current in saturation I_{DS} is derived by substituting Eq. (7-9) into Eq. (7-4):

$$I_{DS} = G_o\left(\frac{2}{3}\sqrt{\frac{\psi_o - V_G}{V_{PO}}} - 1\right)(\psi_o - V_G) + \frac{G_o V_{PO}}{3} \tag{7-10}$$

Equation (7-10) expresses the saturation drain current as a function of the gate voltage. It is called the *transfer characteristic* and is plotted in Fig. 7-5, where we have also plotted the parabola

$$I_{DS} = I_{DSS}\left(1 - \frac{V_G}{V_{PO}}\right)^2 \tag{7-11}$$

where I_{DSS} denotes the drain saturation current at zero gate voltage; i.e., the

---

† The reader is cautioned here that the meaning of *saturation* in FETs is completely different from that of a bipolar transistor (Chap. 9).

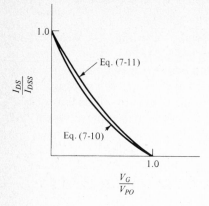

Figure 7-5 Transfer characteristic of a JFET.

gate is short-circuited to the source. Note that the simple square law expressed in Eq. (7-11) approximates Eq. (7-10) very well. It has been found experimentally [3] that even with arbitrary nonuniform impurity distribution in the y direction, the transfer characteristic of any JFET falls within the boundaries set by the two curves shown in Fig. 7-5. In amplifying applications, the JFET is usually operated in the saturation region, and the transfer characteristic is used to find the output drain current for a given input gate-voltage signal.

Gate Leakage Current

The gate leakage current is the sum of the reverse saturation, generation, and surface leakage currents. In a planar JFET, the surface leakage component is usually small. The saturation current and the generation current are given by Eqs. (4-47) and (4-55), respectively. In a typical device, the magnitude of the gate leakage current is between 10^{-9} and 10^{-12} A, yielding an input impedance greater than $10^8 \, \Omega$. However, the surface leakage current due to poor fabrication control can significantly degrade the input impedance of a JFET.

Breakdown Voltage

When the drain voltage is increased, the gate leakage current remains small until avalanche breakdown at the gate junction takes place. The breakdown occurs at the drain end of the channel because it has the highest reverse-bias voltage. This breakdown voltage is given by

$$V_B = V_{DO} + V_G \tag{7-12}$$

Similar to that of the p-n junction diode, the drain current of a JFET exhibits an abrupt increase at breakdown, as depicted in Fig. 7-6.

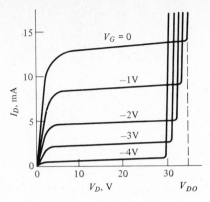

Figure 7-6 Breakdown at high V_D in a JFET.

Series Resistances

The sections between the ohmic contacts and the active region of the channel introduce ohmic series resistances R_S and R_D. These resistances, shown in Fig. 7-7, are constant since they are not affected by the gate or drain voltages. In the linear region of operation, both R_S and R_D contribute to ohmic drops. In the saturation region the effect of R_D is negligible, but R_S has a strong influence because of the negative feedback it produces (see next section). The source resistance reduces the drain current (Fig. 7-4) as well as the gain of the amplifier. For this reason, it should be kept as small as possible.

7-4 SMALL-SIGNAL PARAMETERS AND EQUIVALENT CIRCUITS

In the linear region, the drain conductance is obtained by differentiating Eq. (7-8). The result is

$$g_{dl} \equiv \left.\frac{\partial I_D}{\partial V_D}\right|_{V_G} = G_o\left(1 - \sqrt{\frac{\psi_o - V_G}{V_{PO}}}\right) \qquad \text{for } V_D \ll V_{PO} \qquad (7\text{-}13)$$

where the drain conductance is seen as a function of the applied gate voltage. This feature makes the JFET suitable for applications as a voltage-controlled

$$\frac{W}{q} = \sqrt{\frac{\psi_o}{V_{PO}}}$$

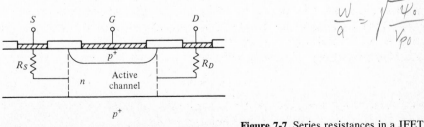

Figure 7-7 Series resistances in a JFET.

variable resistance. The transconductance of the linear region is given by

$$g_{ml} \equiv \left. \frac{\partial I_D}{\partial V_G} \right|_{V_D} = \frac{G_o}{2} \frac{V_D}{\sqrt{V_{PO}(\psi_o - V_G)}} \tag{7-14}$$

derived by using Eq. (7-8).

In the saturation region, the transconductance is derived by differentiating Eq. (7-10). The result is

$$g_m \equiv \frac{\partial I_{DS}}{\partial V_G} = G_o \left(1 - \sqrt{\frac{\psi_o - V_G}{V_{PO}}} \right) \tag{7-15}$$

Note that Eqs. (7-13) and (7-15) are identical, thus, the linear output conductance is equal to the saturation transconductance. Experimental data for the transconductance of a typical JFET are plotted in Fig. 7-8. It is seen that the experimental data agree with Eq. (7-15) when g_m is small. At large values of g_m, Eq. (7-15) does not give an accurate description of the JFET because of the series resistance R_S. Using circuit analysis, we find that the negative-feedback effect of R_S leads to an effective transconductance g'_m (Prob. 7-6):

$$g'_m = \frac{g_m}{1 + g_m R_S} \tag{7-16}$$

Thus, the effective transconductance is equal to g_m for small values of $g_m R_S$, but it is reduced when $g_m R_S$ is comparable to unity.

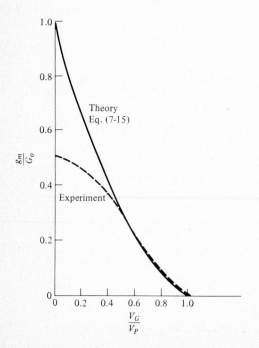

Figure 7-8 Theoretical and experimental curves of the transconductance.

The *p-n* junction between the gate and the channel has a junction capacitance under reverse bias. Let $\bar{W}$ be the average depletion-layer width; the total gate capacitance is given by

$$C_G = 2ZL\frac{K\epsilon_o}{\bar{W}} \qquad (7\text{-}17)$$

The factor of 2 accounts for the two gate junctions, each having an area of ZL. With $V_G = 0$ and at pinchoff, the average depletion-layer width is $a/2$. Therefore, the gate capacitance at pinchoff is

$$C_G = 4ZL\frac{K\epsilon_o}{a} \qquad (7\text{-}18)$$

Despite the distributed nature of the gate capacitance, for simplicity it is usually represented by two lumped capacitances, namely, the gate-to-drain capacitance C_{gd} and the gate-to-source capacitance C_{gs}. These two capacitances are voltage-dependent. In addition, a small capacitance C_{ds} is introduced by the device package between the drain and the source.

A small-signal equivalent circuit for the JFET incorporating the parameters described in the preceding paragraphs is shown in Fig. 7-9. The resistances r_{gd} and r_{gs} represent the gate leakage current. They are usually very large and can be neglected for most practical purposes. The resistance r_{ds} is the finite drain resistance introduced by the channel-length modulation (discussed later). Its typical value is between 100 kΩ and 2 MΩ. A simplified equivalent circuit is given in Fig. 7-10. This simple representation is sufficient

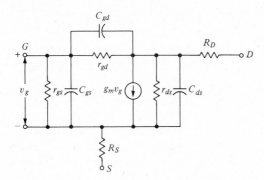

Figure 7-9 A small-signal equivalent circuit.

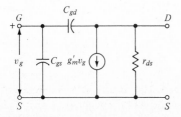

Figure 7-10 A simplified small-signal equivalent circuit.

for most applications. In addition, the capacitances may be ignored for low-frequency operation.

7-5 CUTOFF FREQUENCY

The maximum operating frequency f_{co} is defined as the condition when the JFET is no longer amplifying the input signal. Let us use the equivalent circuit shown in Fig. 7-10 with output short-circuited and replacing g'_m by g_m. The unity-gain condition is reached when the current through the input capacitance is equal to the output drain current. The input current is given by

$$I_i = 2\pi f_{co}(C_{gs} + C_{gd})v_g = 2\pi f_{co}C_G v_g \qquad (7\text{-}19)$$

and the output current is

$$I_o = g_m v_g \qquad (7\text{-}20)$$

Equating Eqs. (7-19) and (7-20), we obtain the cutoff frequency as

$$f_{co} = \frac{g_m}{2\pi C_G} \leqslant \frac{qa^2\mu_n N_d}{4\pi K\epsilon_o L^2} \qquad (7\text{-}21)$$

The last expression in Eq. (7-21) is obtained by using Eqs. (7-18) and (7-15) with $g_m \leqslant G_o$. If Eq. (7-17) is used with $\bar{W} = a$ for minimum gate capacitance, the derived cutoff frequency will double the value given by Eq. (7-21). We notice that the term $a^2 N_d/K\epsilon_o$ in Eq. (7-21) is equal to $2V_{PO}$ [Eq. (7-6)] and therefore is determined by the pinchoff voltage. Usually, this term cannot be adjusted for maximum frequency consideration. The other variables in Eq. (7-21) are the mobility and channel length. To achieve the best high-frequency performance, we should have large mobility and short channel length. Silicon JFETs operating at 4 GHz have been realized [4].

7-6 THE SCHOTTKY-BARRIER FET

In the last section, it was mentioned that the carrier mobility and channel length are the most important factors limiting the high-frequency performance of a JFET. Gallium arsenide is more suitable for high-frequency applications since its electron mobility is approximately 6 times that of silicon. In addition, a shorter channel length can be realized if a Schottky barrier is used to replace the p-n gate junction. Such a Schottky FET is depicted in Fig. 7-11. The active channel layer is grown epitaxially on a highly insulating substrate to reduce parasitic effects. The Schottky barrier is formed on the top surface, together with evaporated ohmic contacts for the drain and the source. This structure is also known as the MESFET, an acronym for metal semiconductor field-effect transistor. Devices with maximum operating frequency as high as 18 GHz have been fabricated, and the projected maximum cutoff frequency is

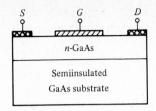

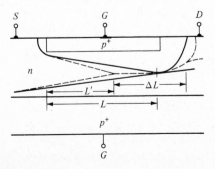

Alloyed ohmic contact

Metal gate

Figure 7-11 A GaAs Schottky-barrier FET.

in the range of 30 GHz [4]. In digital applications, the expected bit rate using MESFETs is 10^{10} bits/s. Another attractive feature of the MESFET is its noise performance. With a noise figure as low as 2.7 dB at 10 GHz, the MESFET is surpassed only by parametric amplifiers in the high-frequency operating region.

7-7 BEHAVIOR OF THE JFET BEYOND PINCHOFF

The pinchoff condition specifies that the two space-charge regions meet at the center of the channel, as shown by the solid line in Fig. 7-12. When the drain voltage is increased further, more free carriers are depleted from the channel. As a result, the length of the depleted region is increased, and the length of the neutral channel is decreased. This is called *channel-length modulation*, and the change of the depletion-layer profile is drawn as the dotted line in Fig. 7-12. In the center of the channel, the applied drain voltage is now shared by the depletion and neutral region, with the neutral-channel region supporting the potential V_P and the depleted-channel region supporting the potential $V_D - V_P$. Since the reduced neutral-channel length supports the same V_P, the drain current will increase slightly for any drain voltage beyond pinchoff. For this reason, the drain current beyond pinchoff is not saturated, and the drain resistance is finite.

Figure 7-12 Channel-length modulation beyond pinchoff.

With this physical picture in mind, let us derive the drain resistance in the saturation region. The drain current beyond pinchoff is obtained by modifying Eq. (7-10) to

$$I'_{DS} = G'_o \left(\frac{2}{3} \sqrt{\frac{\psi_o - V_G}{V_{PO}}} - 1 \right) (\psi_o - V_G) + \frac{G'_o V_{PO}}{3} \tag{7-22}$$

where

$$G'_o = \frac{2qaZ\mu_n N_D}{L'} \tag{7-23}$$

The new channel length L' supports the pinchoff voltage V_P. According to Eq. (4-23), the drain voltage beyond pinchoff increases the length of the depleted channel by

$$\Delta L = \left[\frac{2K\epsilon_o(V_D - V_P)}{qN_d} \right]^{1/2} \tag{7-24}$$

Assuming that the depleted channel extends equally toward the source and the drain, we obtain L' as

$$L' = L - \tfrac{1}{2}\Delta L = L - \frac{1}{2} \left[\frac{2K\epsilon_o(V_D - V_P)}{qN_d} \right]^{1/2} \tag{7-25}$$

The small-signal drain resistance at pinchoff is approximately given by the slope of the current-voltage characteristics of the drain. Therefore, we have

$$r_{ds} = \frac{\Delta V}{\Delta I} = \frac{V_D - V_P}{I'_{DS} - I_{DS}} \tag{7-26}$$

Since the current is not linearly related to the drain voltage, the drain resistance r_{ds} must be calculated for each drain voltage to obtain the variation of the drain resistance. The procedure of calculating r_{ds} is demonstrated in the following example.

Example Consider the JFET in the example of Sec. 7-2. Find the drain resistance at $V_D = V_P + 2$ V and $V_G = 0$.

SOLUTION Using Eq. (7-10), we may rewrite Eq. (7-22) as

$$I'_{DS} = I_{DS} \frac{L}{L'} = I_{DS} \frac{L}{L - \Delta L/2}$$

and

$$\Delta L = \left[\frac{(2)(12)(8.85 \times 10^{-14})\,\Delta V}{(1.6 \times 10^{-19})(5 \times 10^{15})} \right]^{1/2}$$

Let us take $V'_D = V_P + 1$ V and $V''_D = V_P + 3$ V as the two points on the V-I curve for our calculation. We find

$$\Delta L' = 0.52 \ \mu\text{m} \qquad I'_{DS} = I_{DS} \frac{30}{29.74} \qquad \text{at } V'_D$$

$$\Delta L'' = 0.9 \ \mu\text{m} \qquad I'_{DS} = I_{DS} \frac{30}{29.55} \qquad \text{at } V''_D$$

Equation (7-26) is rewritten as

$$r_{ds} = \frac{V_D'' - V_D'}{I_{DS}'' - I_{DS}'} = \frac{2}{(30/29.55 - 30/29.74)(9.6 \times 10^{-3})} = 32 \, k\Omega$$

where $I_{DS} = 9.6 \, mA$ is used.

The model presented here is valid for long JFETs, i.e., with a channel length-to-width ratio greater than 4, and is applicable to most commercial general-purpose JFETs. In a *short* device ($L/a < 2$), however, the saturation mechanism is more involved and is beyond the scope of this text [5].

PROBLEMS

7-1 A silicon n-channel JFET has the structure of Fig. 7-1a and following parameters: $N_a = 10^{18} \, cm^{-3}$, $N_d = 10^{15} \, cm^{-3}$, $a = 2 \, \mu m$, $L = 20 \, \mu m$, and $Z = 0.2 \, cm$. Calculate (a) the built-in potential ψ_o, (b) the pinchoff voltages V_{PO} and V_P, (c) the conductance G_o, and (d) the actual channel conductance with zero bias at the gate and drain terminals.

7-2 Derive the current-voltage relationship of the JFET having the n channel with a cross section of $2a$ by $2a$ surrounded by the p^+ region. The length of the device is L.

7-3 Derive the current-voltage relationship of the junction field-effect tetrode in which the two gates are separated. The applied voltages at the two gates are V_{G1} and V_{G2}, respectively. Assume one-sided step junctions.

7-4 Calculate and plot the transfer characteristics of the JFET in Prob. 7-1 at 25, 150, and $-50°C$. Use the electron-mobility data given in Chap. 1. Use 0.5-V increments for the gate voltage.

7-5 (a) Calculate and plot the small-signal saturation transconductance of the JFET of Prob. 7-1 at 25°C.

(b) Repeat (a) if $R_s = 50 \, \Omega$.

7-6 Derive Eq. (7-16) by using the simplified small-signal equivalent circuit shown in Fig. 7-9. Neglect the capacitances and r_{gd}. Assume $r_{ds} \gg R_S + R_D$.

7-7 (a) Estimate the cutoff frequency of the JFET of Prob. 7-1.

(b) Repeat (a) if $L = 2 \, \mu m$.

(c) Repeat (b) if n-type GaAs is used.

7-8 Calculate the drain resistance r_{ds} of the JFET of Prob. 7-1 at $V_D = V_P + 5 \, V$ and $V_G = -1 \, V$.

REFERENCES

1. W. Shockley, A Unipolar Field-Effect Transistor, *Proc. IRE*, **40**: 1365 (1952).
2. A. S. Grove, "Physics and Technology of Semiconductor Devices," chap. 8, Wiley, New York, 1967.

3. R. D. Middlebrook and I. Richer, Limits on the Power-Law Exponent for Field-Effect Transistor Transfer Characteristics, *Solid-State Electron.*, **6**: 542 (1963).
4. C. A. Liechti, Recent Advances in High-Frequency Field-Effect Transistors, *Int. Electron. Device Meet., Washington, 1975, Tech. Dig.,* p. 5.
5. E. S. Yang, Current Saturation Mechanisms in Junction Field-Effect Transistors, *Adv. Electron. Electron. Phys.* **31**: 247 (1972).

ADDITIONAL READINGS

Cobbold, R. S. C.: "Theory and Application of the Field-Effect Transistors," Wiley, New York, 1970.
Hamilton, D. J., F. A. Lindholm, and A. H. Marshak: "Principles and Applications of Semiconductor Device Modeling," Holt, New York, 1971.
Hauser, J. R.: Unipolar Transistors, sec. II in R. M. Burger and R. P. Donovan (eds.), "Fundamentals of Silicon Integrated Device Technology," vol. II, Prentice-Hall, Englewood Cliffs, N.J., 1968.
Sevin, L. J. Jr.: "Field Effect Transistors," McGraw-Hill, New York, 1965.

EIGHT

METAL-OXIDE-SEMICONDUCTOR TRANSISTORS

The operation of field-effect transistors described in Chap. 7 is based on the control of channel current by two reverse-biased p-n junctions. Instead of using p-n junctions, it is possible to control the channel current by applying a voltage at a gate electrode through an insulator. Since the gate is insulated from the channel in this new structure, it is called the insulated-gate field-effect transistor (IGFET). Figure 8-1 shows a typical IGFET, where SiO_2 serves as the insulator and the aluminum metal plate serves as the gate electrode. The structure shown has a p-type substrate and n-type source and drain regions. This device is more commonly known as the n-channel MOSFET because of the metal-oxide-semiconductor sandwich structure. A p-channel device is obtained by interchanging the n and p regions.

The principle of operation of the MOSFET will be examined briefly here. With no voltage applied to the gate, the two back-to-back p-n junctions between the drain and source of the MOS structure prevent current flow in either direction. When a positive voltage is applied to the gate, mobile negative charge is induced in the semiconductor below the semiconductor-oxide interface. The negative carriers provide a conduction channel between the source and the drain. The conductance of this *induced* channel increases with the magnitude of the gate voltage, yielding current-voltage characteristics similar to those of a JFET.

Before we can understand the detailed operation of the MOSFET we must first find the surface charge condition and the mechanism of channel formation. In this chapter, we begin our discussion on the formation of the

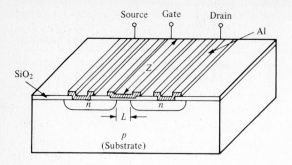

Figure 8-1 An *n*-channel MOS transistor.

surface space-charge region and energy-band diagram for an idealized MOS system. Subsequently, the topics of MOS capacitance, channel conductance, and threshold voltage are covered for both idealized and practical MOS structures. Finally, the theory of the MOSFET together with structural designs will be presented.

8-1 THE SURFACE SPACE-CHARGE REGIONS OF AN IDEALIZED MOS STRUCTURE

The properties of the surface region in a semiconductor can be understood better from the metal-oxide-semiconductor structure shown in Fig. 8-2a. The metal electrode and the silicon wafer are separated by a thin SiO_2 layer, which is a good insulator preventing flow of direct current. It is important to note that, without the flow of direct current, the surface space-charge region will

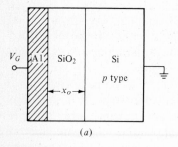

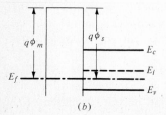

Figure 8-2 The MOS capacitor: (*a*) structure and (*b*) idealized energy-band diagram.

be in thermal equilibrium even with an externally applied voltage. As a result, the Fermi level will be constant throughout the surface space-charge region, and the condition $np = n_i^2$ is valid in the MOS device. To simplify our discussion, we assume (1) that there is no charge located in the oxide or at the interface between the oxide and the semiconductor and (2) that the work-function difference between metal and semiconductor is zero, as depicted in Fig. 8-2b. With the two assumptions stated here, we have eliminated localized space-charge regions and the built-in potential difference in this section. These assumptions will be removed later.

The application of a voltage across the MOS capacitor establishes an electric field $\mathscr{E}_o$ between the plates. As a result, a displacement of mobile carriers near the surface of each plate takes place, giving rise to two space-charge regions. The density of the induced charge Q_s on each plate is given by Gauss' law

$$-Q_s = K_o \epsilon_o \mathscr{E}_o = K_s \epsilon_o \mathscr{E}_s \qquad (8\text{-}1)$$

where ϵ_o = free-space permittivity

K_o = dielectric constant of oxide

$\mathscr{E}_s$ = field at semiconductor surface

K_s = dielectric constant of semiconductor

The potential distribution is depicted in Fig. 8-3 for a p-type semiconductor for two different acceptor densities. In order to avoid the discontinuity of the field at the surface, the position coordinate x has been changed to x/K, with K as the appropriate dielectric constant in the material concerned. The penetration of the field into the semiconductor produces a potential barrier beneath the surface with the depth of penetration inversely proportional to the doping density. With negligible voltage drop in the metal plate, the applied voltage is shared by the voltage across the oxide V_o and *surface potential* ψ_s. Thus,

$$V_G = V_o + \psi_s \qquad (8\text{-}2)$$

The electric field that exists in the semiconductor modifies the band diagram and establishes a space-charge region beneath the surface. Depending on the

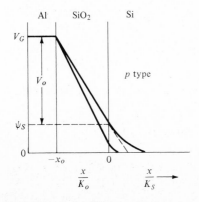

Figure 8-3 Potential distribution in an MOS structure with applied voltage V_G. The potential at the SiO_2–Si interface, ψ_s, is called the surface potential.

polarity of the applied voltage and its magnitude, it is possible to realize three different surface conditions: (1) carrier accumulation, (2) carrier depletion, and (3) carrier inversion.

The carrier densities under thermal equilibrium are given by Eqs. (1-21) and (1-22), repeated here for reference:

$$n = n_i e^{(E_f - E_i)/kT} \tag{8-3}$$

$$p = n_i e^{(E_i - E_f)/kT} \tag{8-4}$$

From these equations we find that $n > p$ for $E_f > E_i$ and the semiconductor is n type. Similarly, we have $n < p$ for $E_f < E_i$ in a p-type semiconductor. Equations (8-3) and (8-4) will be used to determine the carrier concentration at the surface under different biasing voltages on the MOS structure.

Carrier Accumulation Fig (8-4 a)

When the surface charge density just below the silicon surface is greater than the equilibrium-carrier density in the bulk, we have the condition of carrier accumulation. This condition is realized by applying a negative voltage at the metal electrode. The resulting negative surface potential ψ_s produces an upward bending of the energy-band diagram, as shown in Fig. 8-4a. Since E_f remains constant, the band bending leads to a larger $E_i - E_f$ near the surface. According to Eqs. (8-3) and (8-4), we have a higher hole density and a lower electron density at the surface compared with that of the bulk. Consequently, holes are accumulated at the surface, and the surface conductivity is increased.

Carrier Depletion Fig(8-4) b

When a small positive V_G is applied, the surface potential is positive and the energy bands bend downward, as shown in Fig. 8-4b. The Fermi level near the silicon surface is now farther away from the valence-band edge, indicating a smaller hole density. In other words, the value of $E_i - E_f$ is decreased, and holes are depleted from the vicinity of the oxide-silicon interface, establishing a space-charge region consisting of stationary acceptor ions. The total charge per unit area Q_s is given by

$$Q_s = Q_B = -qN_a x_d \tag{8-5}$$

where x_d is the width of the depletion layer, as shown in Fig. 8-4b. The symbol Q_B is defined as the *bulk charge* in the semiconductor, and the negative sign specifies the polarity of charge. The relationship between ψ_s and x_d can be obtained by solving Poisson's equation using the depletion approximation. The result is

$$\psi_s = \frac{qN_a x_d^2}{2K_s \epsilon_o} \tag{8-6}$$

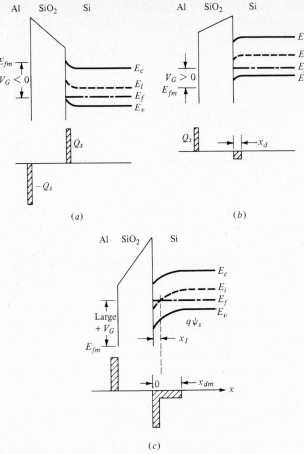

Figure 8-4 Energy band and charge distribution under $(a) - V_G$, (b) small $+ V_G$, and (c) large $+ V_G$.

The potential distribution in the semiconductor is given by

$$\psi = \psi_s \left(1 - \frac{x}{x_d}\right)^2 \tag{8-7}$$

These equations are identical to those of the one-sided step junction having a lightly doped p side [compare with Eqs. (4-16) and (4-17)].

Carrier Inversion $Fig \left(8-4 \; c\right)$

If a large positive V_G is applied to the gate, the downward band bending would be more significant than that in carrier depletion. The large band bending may even cause the midgap energy E_i to cross over the constant Fermi level at or near the silicon surface. When this happens, as shown in Fig.

8-4c, an inversion layer is formed in which the electron density is greater than the hole density. Making use of Eqs. (8-3) and (8-4), we find that the right side of x_I in Fig. 8-4c remains p type but the left side of x_I becomes n type. Therefore, a p-n junction is *induced* under the metal electrode.

The surface is inverted as soon as E_f becomes greater than E_i. However, the density of electrons remains small until E_f is considerably above E_i. This region of *weak inversion* will be discussed in a later section. For most MOSFET operation, it is desirable to define a condition after which the charge due to electrons in the inversion layer becomes very large. This condition, called *strong inversion*, is reached when the electron density per unit volume at the surface is equal to the hole density in the bulk. From Eqs. (8-3) and (8-4), strong inversion demands that the midgap potential be below E_f at the surface by as much as it is above E_f in the bulk. The energy-band diagram in the semiconductor at the onset of strong inversion is shown in Fig. 8-5. Since the bands are flat at zero bias, the required surface potential for strong inversion is

$$\psi_{si} = 2\phi_f \tag{8-8}$$

where $q\phi_f$ is defined as the difference between E_i and E_f in the bulk of the semiconductor. The corresponding width of the surface depletion region of the induced junction is derived by using Eq. (8-6) and setting $\psi_s = \psi_{si}$:

$$x_{dm} = \sqrt{\frac{2K_s\epsilon_0\psi_{si}}{qN_a}} \tag{8-9}$$

Let us examine what would happen if the applied voltage is so large that $\psi_s > \psi_{si}$. Since the increase of ψ_s is added to the difference of E_f and E_i, a small increase of ψ_s produces a large increase of electrons at the surface, according to the exponential nature of Eq. (8-3). Therefore, the surface inversion layer is acting like a narrow n^+ layer, and the induced junction resembles an n^+-p junction for a large positive gate voltage. However, all induced charge will be in the inversion layer after strong inversion, and the charge inside the depletion layer will remain constant. Thus, the space-charge-layer width remains at x_{dm} as we further increase the gate voltage. With a

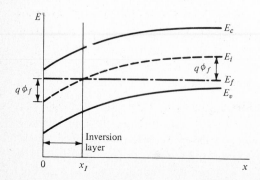

Figure 8-5 Energy-band diagram at strong inversion.

depletion-layer width of x_{dm}, the bulk charge Q_B becomes

$$Q_B = -qN_a x_{dm} \qquad (8\text{-}10)$$

Beyond strong inversion, the charge condition in the surface region is given by

$$Q_s = Q_I + Q_B = Q_I - qN_a x_{dm} \qquad (8\text{-}11)$$

where Q_I is the charge density per unit area in the inversion layer. It is important to note that Q_I is a function of the applied gate voltage, and these induced mobile charges become the current carriers in an MOSFET.

8-2 THE MOS CAPACITOR

The MOS structure shown in Fig. 8-2 is basically a capacitor with the SiO_2 as the dielectric material. If the silicon were a perfect conductor, the parallel-plate capacitance per unit area would have been given by the oxide capacitance

$$C_o = \frac{K_o \epsilon_o}{x_o} \qquad (8\text{-}12)$$

where x_o is the oxide thickness. However, the MOS capacitor is more complicated because of the voltage dependence of the surface space-charge layer in silicon. The space charge of the depletion layer acts as another capacitor C_s in series with C_o, giving an overall capacitance of

$$C = \frac{C_o C_s}{C_o + C_s} \qquad (8\text{-}13)$$

The space-charge-layer capacitance is

$$C_s = \frac{K_s \epsilon_o}{x_d} \qquad (8\text{-}14)$$

Under the condition of carrier accumulation, there is no depletion layer under the silicon surface, and the overall capacitance is equal to C_o. Beyond strong inversion, the maximum space-charge width x_{dm} becomes a constant, and C_s is also a constant. With biasing voltage between the condition of carrier accumulation and strong inversion, the space-charge-layer width x_d is a function of the bias voltage V_G. Replacing $\mathscr{E}_o$ by V_o/x_o in Eq. (8-1) and solving it with Eq. (8-2), we obtain

$$V_G = -\frac{Q_s}{C_o} + \psi_s \qquad (8\text{-}15)$$

where Eq. (8-12) has been used. Substituting Eqs. (8-5) and (8-6) into Eq. (8-15) and solving for x_d gives

$$x_d = \frac{K_s \epsilon_o}{C_o} + \frac{K_s \epsilon_o}{C_o} \sqrt{1 + \frac{2V_G}{2K_s \epsilon_o N_a} C_o^2} \qquad (8\text{-}16)$$

Substitution of Eqs. (8-14) and (8-16) into Eq. (8-13) leads to

$$C = \frac{C_o}{[1 + (2C_o^2/qN_aK_s\epsilon_o)V_G]^{1/2}} \qquad (8\text{-}17)$$

The overall capacitance as a function of the gate voltage is plotted in Fig. 8-6, which also shows the two capacitors in series.

The experimental capacitance-voltage characteristic of an MOS structure is well represented by Fig. 8-6 as long as the measurement frequency is high. However, if the measurement frequency is low enough, the C-V curve will be changed; the curves at different frequencies are shown in Fig. 8-7. The mechanism responsible for the low-frequency C-V characteristics is carrier generation within the space-charge layer [1]. To understand this effect, let us examine the incremental change of charges upon the application of a positive voltage, shown in Fig. 8-8. The incremental positive gate voltage increases the silicon surface potential ψ_s so that holes are depleted and the depletion layer is widened. Therefore, more negative fixed charge is established at the edge of the neutral p-type semiconductor, as shown in Fig. 8-8a. This leads to two capacitors in series with an overall capacitance, expressed by Eq. (8-13). If we assume now that electron-hole pairs can be generated fast enough, the generated holes will replenish the depleted holes at the edge of the depletion

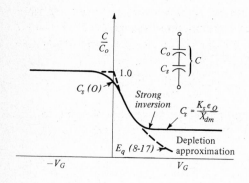

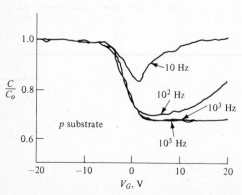

Figure 8-6 Capacitance-voltage curve for an MOS structure with a p substrate. The solid curve incorporates the limit of X_{dm} (Eq. 8-9). In addition, it is found that $C_s(0) = \sqrt{2}K_s\epsilon/L_D$ at $V_G = 0$,[3] where L_D is the extrinsic Debye length (Eq. 4-9). $C_s(0)$ is known as the flat-band capacitance.

Figure 8-7 Frequency dependence of the C-V characteristics. *(After Grove et al. [4].)*

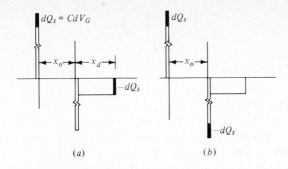

Figure 8-8 Charge distribution in an MOS structure at (a) high and (b) low frequencies.

region. At the same time, the generated electrons will be drawn by the field and accumulate at the silicon-oxide interface (Fig. 8-8b). The overall capacitance now is just C_o. As a result, the capacitance-voltage characteristics are frequency-dependent, and the generation rate determines the frequency range of transition.

8-3 CHANNEL CONDUCTANCE

In addition to the capacitance characteristics of the MOS structure, the conductance of the inversion layer is very important in the operation of an MOS transistor. By applying a positive voltage at the gate of the MOS transistor shown in Fig. 8-1, an inversion layer is formed under the gate electrode. This inversion layer provides a conducting path between the source and the drain and it is known as the *channel*. The channel conductance is given by†

$$g_I = \frac{Z}{L} \int_0^{x_I} q\mu_n n_I(x) \, dx \tag{8-18}$$

where $n_I(x)$ is the electron density in the channel and x_I is the channel width depicted in Fig. 8-5. Integrating $n_I(x)$ over the channel width yields

$$\int_0^{x_I} qn_I(x) \, dx = -Q_I \tag{8-19}$$

where Q_I is the total charge density per unit area in the inversion layer and the negative sign indicates negative charge. Therefore, the channel conductance becomes

$$g_I = -\frac{Z}{L} \mu_n Q_I \tag{8-20}$$

Let us now define a *threshold voltage* V_{TH} as the gate voltage required to bring about strong inversion, i.e.,

$$V_{TH} \equiv -\frac{Q_B}{C_o} + \psi_{si} \tag{8-21}$$

† The effective surface mobility is found experimentally to be roughly $\frac{1}{2}$ of the bulk mobility for both electrons and holes.

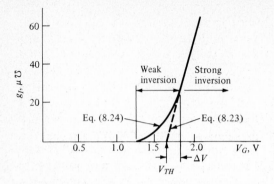

Figure 8-9 Channel conductance as a function of gate voltage. *(After Swanson and Meindl [2].)*

Physically, the threshold voltage in Eq. (8-21) supports a bulk charge Q_B and at the same time introduces band bending at the surface to reach the strong inversion potential ψ_{si}. Substituting Eqs. (8-15) and (8-21) into (8-11) yields

$$Q_I = -C_o(V_G - V_{TH}) \tag{8-22}$$

Thus, Eq. (8-20) becomes

$$g_I = \frac{Z}{L} \mu_n C_o(V_G - V_{TH}) \tag{8-23}$$

Experimental results are shown in Fig. 8-9, along with the straight-line approximation given by Eq. (8-23). Note that Eq. (8-23) describes the channel conductance accurately at large gate voltages, i.e., for $V_G > V_{TH} + \Delta V$, where ΔV is indicated in Fig. 8-9. This is known as the *region of strong inversion*, and most MOS transistors operate under this condition. For $V_G < V_{TH} + \Delta V$, the surface is inverted, but the surface band bending is less than $2\phi_f$. This is known as the *weak-inversion region*. Recently, considerable research efforts have been focused on MOS transistors for use in circuits with very low power dissipation. The low-power devices are found to be operated in the weak-inversion region, and the total charge under weak inversion is given by [2]

$$-Q_I = C_o \, \Delta V \, \exp \frac{V_G - V_{TH} - \Delta V}{\Delta V} \tag{8-24}$$

which is reproduced from the literature without proof to avoid lengthy derivation. This equation shows that Q_I is an exponential function of V_G, which is consistent with the general nature of the conductance plot in Fig. 8-9. On the other hand, the inversion-layer charge varies in a straight-line manner with V_G in Eq. (8-22) under strong inversion.

8-4 FLAT-BAND AND THRESHOLD VOLTAGES

In the previous discussions on the idealized MOS structure, we assumed that the energy-band diagram (Fig. 8-2b) is flat when the gate voltage is zero. In practice, this condition is not realized because of unavoidable work-function difference and charges in the oxide and surface states.

The work function of a material was defined in Chap. 5 as the energy required to bring an electron from the Fermi level to the vacuum level. In an MOS structure, the appropriate energies to be considered are the *modified work functions* from the respective Fermi level in the metal and semiconductor to the oxide conduction-band edge. In general, the modified work function of the metal is not equal to that of the semiconductor. Similar to the metal-semiconductor contact described in Chap. 5, a band bending exists in the semiconductor in order to satisfy the requirement of constant Fermi level under thermal equilibrium. This characteristic is illustrated in Fig. 8-10a for an Al–SiO$_2$–Si structure. The gate voltage required to eliminate the band bending so that no electric field exists in the silicon is clearly the difference of the modified work functions

$$V_{G1} = \phi'_{ms} = \phi'_m - \phi'_s \qquad (8\text{-}25)$$

The result is depicted in Fig. 8-10b. The modified work function for several metals is tabulated in Table 8-1. At room temperature, the experimental value of the modified electron affinity χ' for silicon is 3.25 eV and $E_g = 1.1$ eV. Therefore, the modified work function according to Fig. 8-10 is given by

$$\phi'_s = 3.25 + \frac{1.1}{2} + \phi_f = 3.8 + \phi_f \qquad \text{V} \qquad (8\text{-}26)$$

The gate voltage necessary to achieve the flat-band condition is further modified by charges residing within the oxide. Let us consider a thin layer of charge $+Q_o$ per unit area located at x from the metal as shown in Fig. 8-11a. With zero gate voltage, Q_o induces charges in both the metal and semiconductor, and the sum of the induced charges equals $-Q_o$. The corresponding

Table 8-1 Modified work function V for metal–SiO$_2$ system (After Sze [3])

Metal	ϕ_m, V	$\phi'^{\dagger}_m$, V
Al	4.1	3.2
Ag	5.1	4.2
Au	5.0	4.1
Cu	4.7	3.8
Mg	3.35	2.45
Ni	4.55	3.65

† $\phi'_m = \phi_m - 0.9$ V (SiO$_2$ electron affinity).

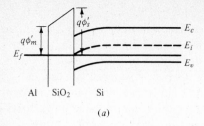

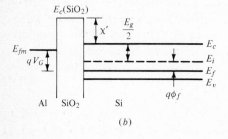

Figure 8-10 Energy-band diagrams for an Al–SiO$_2$–Si structure with (a) $V_G = 0$, (b) $V_G = -V_{FB} = +q(\phi'_m - \phi'_s)$.

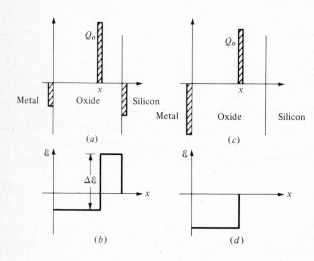

Figure 8-11 Effect of oxide charge on electric field distribution: (a) Q_o induces charges in metal and silicon at $V_G = 0$, (b) electric field for (a), (c) charge distribution for the flat-band condition, and (d) the electric field for (c).

electric field distribution is obtained by integrating the charge and the result is plotted in Fig. 8-11b. The induced charge in the silicon produces a nonzero field at the semiconductor surface so that the energy-band diagram is not flat. The flat-band condition can be reestablished if the charge at the silicon surface is reduced to zero. To achieve the flat-band condition, we apply a negative gate voltage to lower the electric field distribution until the charge is reduced to zero at the silicon surface. The electric field increment is obtained by integrating the charge distribution, and the result is

$$\Delta \mathcal{E} = \int_0^{x^+} \frac{\rho \, dx}{K_o \epsilon_o} = \frac{Q_o}{K_o \epsilon_o} \tag{8-27}$$

Both the charge and field distributions are plotted in Fig. 8-11c and d. The gate voltage necessary to realize the flat-band condition (i.e. the electric field is zero at $x+$) is therefore

$$V_{G2} = -\int_0^{x+} \Delta \mathscr{E} \, dx = -\frac{xQ_o}{K_o\epsilon_o} = -\frac{x}{x_o}\frac{Q_o}{C_o} \qquad (8\text{-}28)$$

This equation shows that the necessary gate voltage to reach the flat-band condition depends on the density and location of the oxide charge. If Q_o is located at $x = 0$, the corresponding flat-band voltage will be zero. On the other hand, the same charge residing at x_o requires the largest gate voltage to achieve the flat-band condition. In most cases, the charge at the silicon-oxide interface resulting from surface states dominates, and the flat-band voltage is

$$V_{G2} = -\frac{Q_o}{C_o} \qquad\qquad Q_o = \frac{1}{q}\int_0^x \rho \, dx \qquad (8\text{-}29)$$

The effects of oxide charge and work-function difference can be combined, and the gate voltage required to achieve the flat-band condition, known as the *flat-band voltage*, is given by

$$V_{FB} = V_{G1} + V_{G2} = \phi'_{ms} - \frac{Q_o}{C_o} \qquad (8\text{-}30)$$

The flat-band voltage can be estimated from the C-V plot shown in Fig. 8-12. The shift of the C-V curve away from $V = 0$ is a reasonable measure of the flat-band voltage.

The threshold voltage must also be modified to include the effects of work-function difference and oxide charge. The new threshold voltage is the sum of (1) the flat-band voltage, (2) the voltage to support a depletion-region charge Q_B, and (3) the voltage to produce band bending for strong inversion.

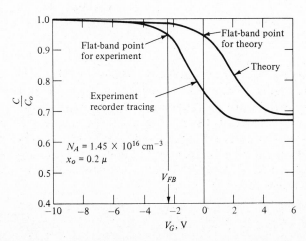

Figure 8-12 The combined effects of metal-semiconductor work-function difference and charges within the oxide on the capacitance-voltage curves. (*After Grove et al. [4].*)

Thus, Eq. (8-21) becomes

$$V_{TH} = V_{FB} + \psi_{si} - \frac{Q_B}{C_o} = \phi'_{ms} + \psi_{si} - \frac{Q_o}{C_o} - \frac{Q_B}{C_o} \tag{8-31}$$

If a voltage V is established in the channel by external voltages at the source and drain, Q_B under strong inversion can be rewritten as

$$Q_B = -[2qK_s\epsilon_o N_a(V + \psi_{si})]^{1/2} \tag{8-32}$$

[handwritten: p-channel: use Nd and positive]

Note that V is the voltage of the n-inversion-layer side of the junction with respect to the p-side substrate.

Example An aluminum-gate MOS transistor is fabricated on an n-type $\langle 111 \rangle$ silicon substrate with $N_d = 10^{15}$ cm^{-3}. The thickness of the gate oxide is 1200 Å, and the surface charge density at the oxide-silicon interface is 3×10^{11} cm^{-2}. Calculate the threshold voltage with $Q_B = 10^{11}$ cm^{-2}.

SOLUTION We obtain first the oxide capacitance

$$C_o = \frac{K_o\epsilon_o}{x_o} = 2.9 \times 10^{-8} \text{ F/cm}^2$$

[handwritten: $\text{Å} = 10^{-10}$ m; p-channel; $\phi_f = -V_T \ln \frac{Nd}{n_i}$]

Letting $n = N_d$, we obtain ϕ_f by using Eq. (8-3)

$$\phi_f = V_T \ln \frac{N_d}{n_i} = 0.29$$

The work-function difference is calculated from Table 8-1 and Eqs. (8-25) and (8-26):

$$\phi'_{ms} = 3.2 - (3.8 - 0.29) = -0.31$$

The threshold voltage from Eq. (8-31) is

$$V_{TH} = -0.31 - 2 \times 0.29 - \frac{(1.6 \times 10^{-19})(3 \times 10^{11})}{2.9 \times 10^{-8}} - \frac{(1.6 \times 10^{-19})(10^{11})}{2.9 \times 10^{-8}}$$

$$= -0.31 - 0.58 - 1.65 - 0.55 = -3.1 \text{ V}$$

8-5 CHARGES IN SURFACE STATES AND OXIDE

Let us digress for a moment to examine the origins of charges in the MOS system. In the previous discussion we implicitly assumed that Q_o includes the effects of (1) the charge in the oxide, (2) fixed surface-state charge, and (3) fast surface states. These charges have different characteristics and deserve a few words of explanation.

The charge inside the oxide originates from ionic contamination or ionized traps induced by radiation. In particular, sodium ions were the source of instability in early MOS transistors because they carry a positive charge and migrate within the oxide under an electric field. For this reason, the flat-band voltage changes with time, and the C-V curves are unstable. However, sodium contamination can be eliminated by appropriate precautions in device fabrication. Any radiation-induced charge can be annealed out at a temperature above 300°C.

The fixed *surface-state charge* Q_{ss} appears to have originated from excess ionic silicon which has moved into the oxide during oxidation. Usually, it is located within 200 Å of the oxide-silicon interface and cannot be removed by varying the surface band bending. The density of Q_{ss} depends strongly on the crystal orientation and conditions of oxidation and annealing. It is found that the (111) surface has the highest fixed surface charge (5×10^{11} cm^{-2}) and (100) the lowest surface charge (9×10^{10} cm^{-2}). The fixed surface charge displaces the C-V curve by a fixed voltage without changing the shape of the curve. A final heat treatment at a temperature above 600°C in a nitrogen ambient results in a density of less than 2×10^{11} cm^{-2} for a (111) wafer.

The disruption of the periodic lattice at the semiconductor surface results in a large number of atoms with missing covalent bonds, thus introducing energy states in the forbidden gap near the surface. These energy states are known as *fast surface states* N_{FS} because they exchange charge with the conduction or valence band rapidly at room temperature. From the picture of missing bonds at the surface, it is reasonable to assume that there is one fast surface state for every surface atom. Since the silicon atom density is 5×10^{22} cm^{-3}, the surface atom density is approximately 10^{15} cm^{-2}, yielding the same number of fast surface states. This is indeed the case in a very clean surface cleaved under high-vacuum environment. In a thermally oxidized

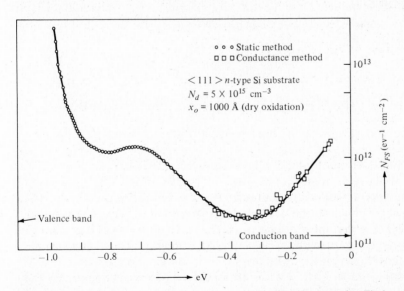

Figure 8-13 Distribution of fast surface states in the Si–SiO$_2$ system. *(After Ziegler and Klausmann [5].)*

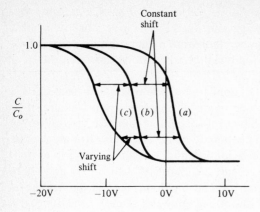

Figure 8-14 Effect of fast surface states on the C-V curve: (*a*) ideal, (*b*) experiment without fast surface states, and (*c*) experiment with fast surface states. (*After Deal et al. [6].*)

Si–SiO$_2$ system, however, N_{FS} is typically in the order of 10^{11} to 10^{12} cm^{-2}. The distribution of these states in the forbidden gap has been measured to be that depicted in Fig. 8-13. The upper portion of states near the conduction-band edge consists of acceptor states, and the lower portion near the valence-band edge consists of donor states; they are roughly equal in number. Depending on the position of the Fermi level at the surface, the surface charge can be either positive or negative. Therefore, the charge in the fast surface states will vary with the surface potential, and the C-V curve will be displaced by an amount which itself changes with the surface potential. For this reason, the C-V curve is displaced as well as distorted, as shown in Fig. 8-14. As for their influence on the surface recombination, only states which are close to the middle of the band gap are effective, and their density is less than 5×10^{10} eV^{-1} cm^{-2}.

8-6 STATIC CHARACTERISTICS OF THE MOS TRANSISTOR

A schematic diagram of an n-channel MOSFET is shown in Fig. 8-15, in which bias voltages are included. To simplify our analyses, we have grounded the source and substrate terminals. The static characteristics of the MOSFET have two distinct regions of operation similar to that of a JFET. At low drain voltages, the drain-to-source characteristics are basically ohmic, and the drain current is proportional to the drain voltage in a nearly linear fashion. This operation region is the linear region. At high drain voltages, the drain current remains almost constant with increasing drain voltage. This second region is the saturation region since the current is saturated. We derive first the current-voltage characteristics of the transistor in the linear region and then extend the results to the onset of saturation.

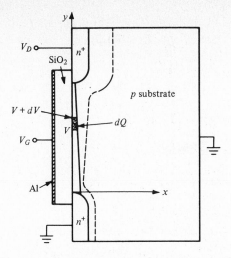

Figure 8-15 The n-channel MOS transistor.

Linear Region of Operation

In the following analysis, we employ the gradual-channel approximation, in which the vertical and lateral electric fields are assumed to be independent of each other. In Fig. 8-15 the electric field in the x direction induces an inversion layer, whereas the electric field in the y direction produces a drain current flowing along the silicon surface.

Let us consider a small section at y of the transistor shown in Fig. 8-15 under the condition that the gate voltage is greater than the threshold voltage so that mobile carriers are induced in the inversion layer. The relationship between the mobile charge Q_I and the gate voltage is given by Eq. (8-22) if we take the channel voltage as zero. As a result of the drain-to-source bias, however, a potential V is established at y. Therefore, the induced channel charge should be modified to

$$Q_I = -C_o(V_G - V_{TH} - V) \tag{8-33}$$

The channel current, being a majority-carrier current, can now be written as

$$I_D = Z\mu_n Q_I \mathscr{E}_y \tag{8-34}$$

which is obtained by using the transport equation expressed in Eq. (2-63) and by neglecting the diffusion component. Substituting $\mathscr{E}_y = -dV/dy$ and Eq. (8-33) into Eq. (8-34), we obtain

$$I_D\, dy = Z\mu_n C_o(V_G - V_{TH} - V)\, dV \tag{8-35}$$

Integration of this equation from $y = 0$ to $y = L$ and from $V = 0$ to $V = V_D$ yields

$$I_D = C_o\mu_n \frac{Z}{L}[(V_G - V_{TH})V_D - \tfrac{1}{2}V_D^2] \tag{8-36}$$

Example The MOS structure of the example in Sec. 8-4 is used as an MOSFET. The following parameters are given: $L = 10\ \mu\text{m}$, $Z = 300\ \mu\text{m}$, and $\mu_p = 230\ \text{cm}^2/\text{V-s}$. Calculate I_{DS} for $V_G = -4\ \text{V}$ and $V_G = -8\ \text{V}$.

SOLUTION Since $C_o = 2.9 \times 10^{-8}\ \text{F/cm}^2$ in Sec. 8-4, we have

$$C_o\mu_p\frac{Z}{L} = (2.9 \times 10^{-8})(230)(30) = 2 \times 10^{-4}\ \text{F/V-s}$$

Substituting this value into Eq. (8-36) and setting $V_D = V_G - V_{TH}$ yields

$$I_{DS} = \frac{2 \times 10^{-4}}{2}(V_G - V_{TH})^2$$

Since $V_{TH} = -3.1\ \text{V}$

$$I_{DS} = \begin{cases} 0.8 \times 10^{-4} = 80\ \mu\text{A} & \text{for } V_G = -4\ \text{V} \\ 24 \times 10^{-4} = 2.4\ \text{mA} & V_G = -8\ \text{V} \end{cases}$$

In the derivation of Eq. (8-36), we have assumed that V_{TH} is independent of V. In other words, V is assumed to be zero in Eq. (8-32). This approximation can lead to erroneous results. In fact, V_{TH} increases toward the drain due to the increase of Q_B, as shown in Eq. (8-32). If we use Eq. (8-32) without simplification, the drain-current equation becomes

$$I_D = C_o\mu_n\frac{Z}{L}\left\{\left(V_G - \phi'_{ms} - \psi_{si} + \frac{Q_o}{C_o} - \frac{V_D}{2}\right)V_D \right.$$
$$\left. - \frac{2}{3}\frac{\sqrt{2qK_s\epsilon_oN_a}}{C_o}[(V_D + \psi_{si})^{3/2} - \psi_{si}^{3/2}]\right\} \qquad (8\text{-}37)$$

Equations (8-36) and (8-37) are plotted in Fig. 8-16 for a p-channel device with

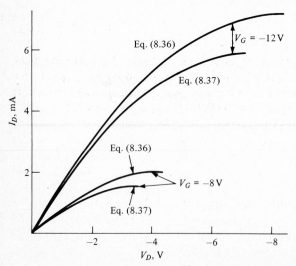

Figure 8-16 Comparison of Eqs. (8-36) and (8-37).

typical device parameters. It is shown that the simplified expression overestimates the drain current and the difference is small here, but it often proves significant, particularly if the substrate doping is high. Nevertheless, the simplicity of Eq. (8-36) provides better physical insight into the device operation, and it is useful in obtaining a first-order design, particularly in digital-circuit applications. For circuit design, a good empirical fit to actual device characteristics can be obtained by arbitrarily adjusting $C_o\mu_n$ so that Eq. (8-36) fits the observed characteristics in the I-V range of interest. We shall use the simplified expression in subsequent analyses; interested readers should consult the references at the end of the chapter.

Saturation Region

The foregoing analysis is based on the premise that an inversion layer is formed throughout the semiconductor surface between the source and the drain. If we increase the drain voltage so that the gate voltage is neutralized, the inversion layer will disappear at the drain side of the channel. The channel is now pinched off, and further increases of the drain voltage would not increase the drain current significantly. Therefore, the drain current is saturated, and the equations derived for the linear region are rendered inoperative.

The condition for the onset of the current saturation is given by setting Q_I equal to zero in Eq. (8-33). Therefore,

$$V = V_G - V_{TH} = V_{DS} \tag{8-38}$$

where V_{DS} is the drain voltage, which is equal to V for the channel adjacent to the drain; it is known as the *drain saturation voltage.* Substitution of Eq. (8-38) into Eq. (8-36) yields

$$I_{DS} = \frac{\mu_n C_o Z}{2L} (V_G - V_{TH})^2 \tag{8-39}$$

This equation is valid at the onset of saturation. Beyond this point the drain current can be considered as constant so that the foregoing equation is also applicable to the drain current for higher drain voltages. The complete current-voltage characteristics of the MOSFET are plotted in Fig. 8-17, in which the dotted curve specifies the onset of current saturation.

Cutoff Region

If the gate voltage is smaller than the threshold voltage, no inversion layer is formed. As a result, the MOSFET behaves like two *p-n* junctions connected back to back to prevent current flow in either direction. The transistor acts as an open circuit in this region of operation.

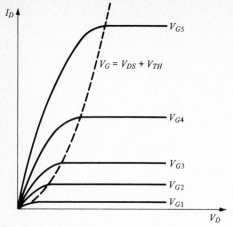

Figure 8-17 Current-voltage characteristics of an n-channel MOSFET.

8-7 SMALL-SIGNAL PARAMETERS AND EQUIVALENT CIRCUIT

The small-signal conductance is defined by

$$g_d \equiv \frac{\partial I_D}{\partial V_D}\bigg|_{V_G = \text{const}} \tag{8-40}$$

In the linear region where V_D is small, the conductance is obtained by differentiation of Eq. (8-36):

$$g_d = \mu_n C_o \frac{Z}{L}(V_G - V_{\text{TH}} - V_D) \approx \mu_n C_o \frac{Z}{L}(V_G - V_{\text{TH}}) \tag{8-41}$$

Note that the drain conductance is ohmic and linearly dependent on the gate voltage. Experimental data for an n-channel MOSFET are plotted in Fig. 8-18. We observe that g_d follows the gate voltage linearly except for high gate voltages, for which the decrease of mobility at high surface-carrier concentration is responsible. We also find that extrapolation of the data to the V_{GS} axis gives the experimental value of the threshold voltage. Besides the room-temperature data, two additional curves are included for the high- and low-temperature extremes. The decrease of slope with increasing temperature is caused by the reduction of mobility at high temperatures. The resistance in the linear region, frequently referred to as the ON resistance, is given by

$$R_{\text{on}} = \frac{1}{g_d} = \frac{L}{\mu_n C_o Z(V_G - V_{\text{TH}})} \tag{8-42}$$

An important parameter in a MOSFET is the transconductance, defined by

$$g_m \equiv \frac{\partial I_D}{\partial V_G}\bigg|_{V_D = \text{const}} \tag{8-43}$$

In the linear region, the transconductance is obtained by differentiation of Eq.

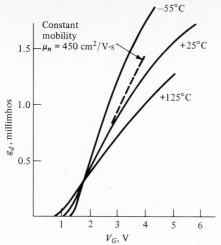

Figure 8-18 Channel conductance vs. V_G in an n-MOSFET.

(8-36), yielding

$$g_m = C_o \mu_n \frac{Z}{L} V_D \tag{8-44}$$

The transconductance in the saturation region is derived by differentiating Eq. (8-39), which leads to

$$g_m = \frac{\mu_n C_o Z}{L} (V_G - V_{\text{TH}}) \tag{8-45}$$

Note that the expressions of the saturation g_m and the linear g_d are identical, which is valid only when Q_B is assumed to be constant.

According to the idealized theory, the drain current remains constant for any drain voltage beyond pinchoff. In other words, the drain resistance is infinite for $V_D > V_{DS}$. However, all experimental MOSFETs show a finite slope in their drain current-voltage characteristics beyond pinchoff. The property is similar to that of the JFET, as described in Sec. 7-7. Therefore, we define a drain resistance in the saturation region as

$$r_{ds} = r_d(\text{sat}) \equiv \frac{\partial V_{DS}}{\partial I_{DS}} \bigg|_{V_G = \text{const}} \tag{8-46}$$

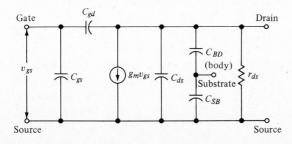

Figure 8-19 Small-signal equivalent circuit of an MOS transistor. C_{BD} and C_{SB} are the substrate-to-drain and substrate-to-source capacitances.

The drain resistance in the saturation region can be obtained graphically from the drain characteristics.

An equivalent circuit of the MOSFET is shown in Fig. 8-19. The gate-to-drain capacitance usually dominates the high-frequency characteristics because of the Miller effect.

Cutoff Frequency

Following the same reasoning for a JFET, we find that the cutoff frequency of an MOSFET is given by

$$f_o = \frac{g_m}{2\pi C_G} = \frac{\mu_n V_D}{2\pi L^2} \qquad \text{if } V_D \leq V_{DS} \tag{8-47}$$

where C_G is the total gate capacitance. Note that a short channel and high mobility are desirable for high-frequency operation.

8-8 EFFECT OF SUBSTRATE (BODY) BIAS

In the previous analyses we have assumed that the source is connected to the substrate and both terminals are grounded. Let us keep the source at ground potential but apply a negative voltage V to the p-type substrate (body) (Fig. 8-15). A reverse-bias voltage V_{SB} will be across the induced junction between the channel and body so that the space-charge layer is widened. As a result, the negative fixed charge Q_B in the space-charge layer is increased. When V_{SB} is zero, the charge in the space-charge layer is given by Eq. (8-5) and is rewritten as

$$Q_B = -qN_a x_d = -(2qK_s\epsilon_o N_a \psi_o)^{1/2} \tag{8-48}$$

where x_d is given by Eq. (4-18) with x_n and N_d replaced by x_d and N_a, respectively. For an arbitrary reverse-bias voltage V_{SB} we have

$$Q_B = -[2qK_s\epsilon_o N_a(V_{SB} + \psi_o)]^{1/2} \tag{8-49}$$

Therefore, the incremental charge is

$$\Delta Q_B = -(2qK_s\epsilon_o N_a)^{1/2}[(V_{SB} + \psi_o)^{1/2} - \psi_o^{1/2}] \tag{8-50}$$

To reach the condition of strong inversion, the applied gate voltage must be increased to compensate for ΔQ_B. Therefore,

$$\Delta V_{\text{TH}} = -\frac{\Delta Q_B}{C_o} = \frac{(2qK_s\epsilon_o N_a)^{1/2}}{C_o}[(V_{SB} + \psi_o)^{1/2} - \psi_o^{1/2}] \tag{8-51}$$

Figure 8-20 shows experimental data on the change of threshold voltage with V_{SB}, and it agrees with Eq. (8-51). Equations derived in Secs. 8-6 and 8-7 are valid even with substrate bias, provided that the change of threshold voltage is taken into account.

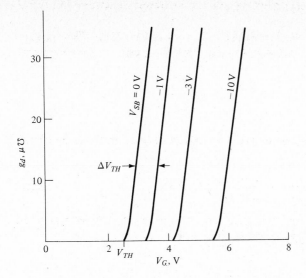

Figure 8-20 Effect of substrate bias on the threshold voltage.

8-9 FABRICATION TECHNOLOGIES

The industrial standard for the fabrication of MOS devices in the past was the p-channel process. Currently, two additional techniques, the silicon gate and ion implantation, are gaining wider acceptance, and significant progress has been made in the n-channel process. From announcements of new products in 1977, for example, the Intel MCS-85 microcomputer line, it is not unreasonable to say that the n-channel process has replaced the p-channel process as the most important MOS technology. Because there are variations of each technology, we confine our discussion to the basic steps.

The p-Channel Process

A typical p-channel metal-gate MOS transistor is shown in Fig. 8-21, where both the top view and the cross-sectional view are depicted, together with the major processing steps. The top view is the layout diagram, in which outlines of four different masks are superimposed. The thin oxide under the metal gate is called the *gate oxide*, and its formation and protection are the most critical in the entire process. Frequently, charged ions contaminate the gate oxide and the threshold voltage is modified drastically. In the early development of the MOSFET, sodium ions were a serious problem because the ion migration causes the threshold voltage to change with time. In the past few years, the use of electron-beam evaporation and very clean processes enable manufacturers to produce stable MOS devices.

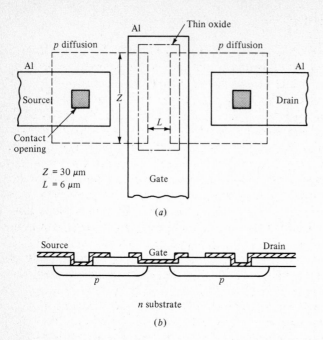

$Z = 30\ \mu m$
$L = 6\ \mu m$

(a)

(b)

Figure 8-21 A p-channel MOS transistor.

Instead of a single layer of SiO_2, the gate dielectric may be composed of SiO_2 and Si_3N_4, as shown in Fig. 8-22. This is called an MNOS device. The nitride has a higher dielectric constant, which produces an increased gate capacitance. In addition, it prevents sodium-ion migration and stabilizes the threshold voltage. According to Eqs. (8-31) and (8-45), a larger gate capacitance results in a higher transconductance and a lower threshold voltage. One may ask: Why not eliminate the SiO_2 layer altogether? Unfortunately, this is not possible because a large density of traps exists at the silicon–silicon nitride interface when the SiO_2 layer is removed. The interface charges would degrade the device performance.

The threshold voltage of the MOS transistor is affected by the crystal orientation of the substrate material. Typical values of the surface-state densities are $5 \times 10^{11}\ cm^{-2}$ in a (111) substrate and $0.9 \times 10^{11}\ cm^{-2}$ in a (100) substrate.

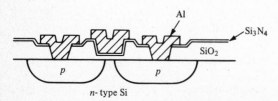

Figure 8-22 The MNOS structure. *(After Penney [8].)*

The Silicon-Gate Technology

Reduction of the threshold voltage can be accomplished by employing amorphous or polycrystalline silicon as the gate to replace the aluminum electrode. This process is called the *silicon-gate technology* [7].

The processing steps of a *p*-channel silicon-gate device are shown in Fig. 8-23. First, a thick oxide, approximately 1 μm thick, is thermally grown over the *n* substrate. Next, a window is opened by photomasking to define the source, channel, and drain regions of the final device. Subsequently, thin layers of silicon dioxide and silicon nitride† are formed by thermal oxidation and deposition, respectively, followed by the deposition of a thin amorphous silicon layer. A second photomasking step is used to remove the silicon and silicon nitride layers except in the specified gate area. The thin oxide is also removed by an oxide etch. Impurities are then diffused to produce the *p*-type drain and source regions as well as making the gate silicon highly *p* type. Next, a thick oxide layer is deposited over the entire wafer, and a mask is used to define the contact windows. Finally, aluminum is evaporated and etched to provide interconnections.

The threshold-voltage reduction can be explained in the following way. If the material of the gate silicon is identical to that of the substrate, the gate-to-substrate work-function difference in Eq. (8-25) becomes zero. There-

---

† Presently, silicon nitride is rarely used as gate dielectric except in MNOS memory devices, described later.

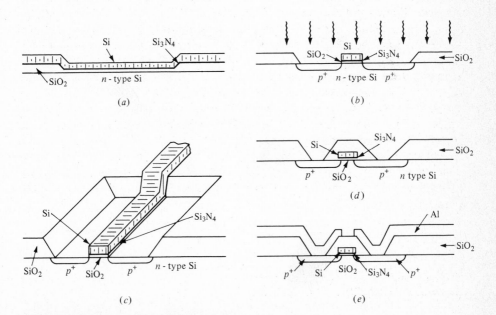

Figure 8-23 The silicon-gate technology. *(After Vadasz et al. [7].)*

fore, the threshold voltage is reduced. Furthermore, it is possible to dope the silicon gate so that the polarity of the work-function term is reversed, thus further reducing the threshold voltage. An additional attractive feature of this process is the self-alignment between gate electrode and p-diffusion source and drain regions. Since the gate silicon is also used as a mask for the p diffusion, the only overlapping between the gate and the drain region is the lateral diffusion of impurities in the drain region. The gate-source and gate-drain capacitances are reduced significantly.

Ion-Implantation Process

Self-alignment between the gate electrode and the source and drain regions can also be obtained by ion implantation with the gate as the mask, although this is not a common practice. Figure 8-24 shows the device structure in which ion implantation is employed. The feature of self-aligned gate enables us to fabricate devices with small channel length which exhibit good high-frequency response.

Ion implantation can be used to adjust the threshold voltage of an MOS transistor. For example, as shown in Fig. 8-25, boron ions are implanted through the gate oxide to compensate the doping of the channel, thus reducing V_{TH}. The implantation is performed just before the gate conductor is deposited.

The n-Channel Process

The value of the electron mobility in silicon is approximately 3 times that for holes. As a result, when other factors are equal, the transconductance is 3

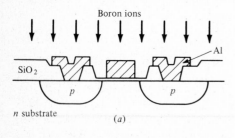

(a)

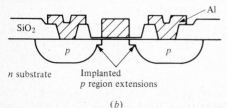

(b)

Figure 8-24 Ion-implanted MOSFET. *(After Penney [8].)*

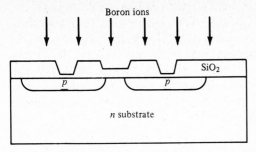

Figure 8-25 Control of threshold voltage by ion implantation.

times larger and the speed is 3 times higher in an n-channel device than in a p-channel device. The advantage of n-channel structure, however, has been difficult to realize because of manufacturing problems. Basically, the process sequence of an n-channel device is the same as the p-channel device. The difficulty arises from the charge distribution of the resulting MOS system. It is found experimentally that charge in oxide and surface states in the silicon–silicon dioxide system is invariably positive. Consequently, negative charge is induced in the silicon under the oxide, forming an inversion layer in the p substrate. Therefore, a conducting path exists between the source and the drain even at zero gate voltage; thus, the device is permanently turned on. A permanent channel can be avoided if higher doping is incorporated in the substrate. A high doping in the substrate, however, makes it difficult to produce satisfactory devices because of lower breakdown voltages, higher depletion-layer capacitances, and the increased effect of bulk charges on V_{TH}. An alternative approach is to use the ion-implantation technique to achieve charge compensation in the desired areas. The n-channel process has now been mastered and without any doubt n-channel devices will eventually replace the p-channel devices in importance because of the advantage in the mobility ratio.

8-10 OTHER FETs

In the preceding sections we have described the operation of the MOSFET in which no conducting channel exists without the applied gate voltage. This MOSFET is called the *enhancement* or normally OFF MOSFET. If a conducting channel is incorporated by providing an n-doped layer between the n-type source and drain, as shown in Fig. 8-26, the transistor will exhibit a high conductance without a gate voltage. It is then necessary to apply a negative voltage to the gate to deplete the free electrons in the channel and turn the device off. The transistor behaves like a JFET and is called a *depletion* or normally ON MOSFET.

Another technology that may have an impact on FETs is the use of sapphire substrate to support the silicon devices, as shown in Fig. 8-27. It is

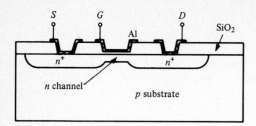

Figure 8-26 Structure of an *n*-channel depletion-mode MOS transistor.

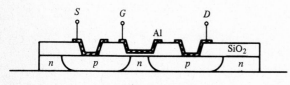

Sapphire substrate
(insulator)

Figure 8-27 A *p*-channel silicon-on-sapphire (SOS) transistor.

called the *silicon-on-sapphire* (SOS) technology [9]. Because of the insulating property of sapphire, parasitic capacitances are greatly reduced. However, economic considerations of high-quality sapphire substrate manufacturing at present make it unlikely for SOS devices to be competitive with other technologies mentioned in Sec. 8-9.

PROBLEMS

8-1 Sketch the energy-band diagrams and the charge distribution in an MOS structure under biasing conditions corresponding to carrier accumulation, depletion, and strong inversion. Use an *n*-type substrate and neglect the effects of surface states and work-function difference. (Specify the polarity of the gate voltage with respect to the substrate.)

8-2 Derive expressions of the bulk charge, surface potential, and surface field to show their dependence on the substrate doping density N_a at strong inversion. Plot the bulk charge, surface potential, and field as a function of N_a from 10^{14} to 10^{18} cm^{-3}.

8-3 An idealized MOS capacitor has an oxide thickness of 0.1 μm and $K_o = 4$ on a *p*-type silicon substrate with acceptor concentration of 10^{16} cm^{-3}. What is the capacitance at (*a*) $V_G = +2$ V and $f = 1$ Hz, (*b*) $V_G = +20$ V and $f = 1$ Hz, and (*c*) $V_G = +20$ V and $f = 1$ MHz?

8-4 Using superposition, show that the change in the flat-band voltage corresponding to a charge distribution $\rho(x)$ in the oxide is given by

$$\Delta V_{FB} = -\frac{q}{C_o} \int_0^{x_o} \frac{x\rho(x)}{x_o} \, dx$$

$$V_{FB} = \phi'_{ms} - \frac{1}{C_{ox}} \int_0^{x_o} \frac{x\rho}{x_o} \, dx$$

8-5 Calculate the flat-band voltage of an MOS device with $x_o = 1000$ Å, $q\phi_m = 4$ eV, $q\phi_s = 4.5$ eV, and a uniform positive oxide charge of 10^{16} cm^{-3}. Assume $K_o = 4$. Use the expression in Prob. 8-4. $\rho(x) = q \cdot 10^{16}$

8-6 Compare the following cases by using the result of Prob. 8-4.

(a) A density of 1.5×10^{12} charges/cm^2 is uniformly distributed in the oxide in an MOS structure. Calculate the flat-band voltage due to this charge if the oxide thickness is 1500 Å. $Q_o = q \int_0^{x_o} \rho \, dx$

(b) Repeat (a) if all the charge is located at the silicon–silicon oxide interface.

(c) Repeat (a) if the charge forms a triangular distribution which peaks at $x = 0$ and zero at $x = x_o$.

8-7 An aluminum-gate MOS transistor is fabricated on a p-type Si $\langle 111 \rangle$ substrate with $N_a = 10^{15}$ cm^{-3}. The thickness of the gate oxide is 1200 Å, and the surface charge density is 3×10^{11} cm^{-2}. Calculate the threshold voltage.

8-8 An MOS structure consists of an n-type substrate with $N_d = 5 \times 10^{15}$ cm^{-3}, an oxide of 1000 Å, and an aluminum contact. The measured threshold voltage is -2.5 V. Calculate the surface-charge density.

8-9 (a) Verify the following expression of the temperature dependence of the threshold voltage for an MOS structure with an n substrate:

$$\frac{dV_{TH}}{dT} = \frac{1}{T}\left(2 - \frac{Q_B}{2C_o\phi_f}\right)\left(\phi_f - \frac{E_g}{2q}\right)$$

(b) Does V_{TH} increase or decrease with temperature? What is the effect of the variation of oxide thickness and substrate doping?

8-10 A p-channel aluminum-gate MOS transistor has the following parameters: $x_o = 1000$ Å, $N_d = 2 \times 10^{15}$ cm^{-3}, $Q_{ss} = 10^{11}$ cm^{-2}, $L = 10$ μm, $Z = 50$ μm, and $\mu_p = 230$ cm^2/V-s. Calculate I_{DS} for V_{GS} equal to -4 and -8 V, and plot the current-voltage characteristics.

8-11 Calculate the saturation current for $V_G = 4$ V of an n-channel MOS transistor with the following parameters: $K_o = 4$, $x_o = 1000$ Å, $Z/L = 10$, $\mu_n = 1000$ cm^2/V-s, and $V_{TH} = +0.5$ V.

8-12 In the MOS transistor of Prob. 8-10, let $V_G - V_{TH} = 1$ V and (a) calculate the oxide capacitance and the cutoff frequency.

(b) Repeat (a) if $Z = 10$ μm and $L = 50$ μm.

8-13 (a) Derive an equation describing the source-to-drain V-I characteristics for n-channel MOS enhancement mode FET if the source and substrate are grounded and the gate is short-circuited to the drain. Assume V_{TH} is constant.

(b) Plot the V-I curves using data from Prob. 8-11.

(c) Calculate R_{on} at $V_G - V_{TH} = 1$ V.

(d) Repeat (c) if $L/Z = 1.0$.

8-14 (a) A p-channel MOS transistor is fabricated on an n substrate of 10^{15} cm^{-3} with a gate oxide thickness of 1000 Å. Calculate the threshold voltage if $\phi_{ms} = -0.6$ eV and $Q_{ss} = 5 \times 10^{11}$ cm^{-2}.

(b) The threshold voltage of the MOSFET in part (a) is reduced by using boron ion implantation. What is the required boron doping density in order to obtain a threshold voltage of -1.5 V?

8-15 Sketch a set of masks for the fabrication of an MOS transistor using the silicon-gate technology.

REFERENCES

1. A. S. Grove, E. H. Snow, B. E. Deal, and C. T. Sah, Simple Physical Model for the Space-Charge Capacitance of Metal-Oxide-Semiconductor Structures, *J. Appl. Phys.*, **35**: 2458 (1964).
2. R. M. Swanson and J. D. Meindl, Ion-implanted CMOS Transistors in Low-Voltage Circuits, *IEEE J. Solid-State Circuits*, **SC-7**: 146 (1972).
3. S. M. Sze, "Physics of Semiconductor Devices," p. 469, Wiley, New York, 1969.
4. A. S. Grove, B. E. Deal, E. H. Snow, and C. T. Sah, Investigation of Thermally Oxidized Silicon Surface Using Metal-Oxide-Semiconductor Structures, *Solid-State Electron.*, **8**: 145 (1965).
5. K. Ziegler and E. Klausmann, Static Technique for Precise Measurements of Surface Potential and Interface State Density in MOS Structures, *Appl. Phys. Lett.*, **26**: 40 (April 1, 1975).
6. B. E. Deal, M. Sklar, A. S. Grove, and E. H. Snow, Characteristics of the Surface-State Charge of Thermally Oxidized Silicon, *J. Electrochem. Soc.*, **114**: 266 (1967).
7. L. L. Vadasz, A. S. Grove, T. A. Rowe, and G. E. Moore, Silicon Gate Technology, *IEEE Spectrum*, **6**: 28 (October 1969).
8. W. M. Penney (ed.), "MOS Integrated Circuits," chap. 3, Van Nostrand Reinhold, New York, 1972.
9. A. Karl Rapp and E. C. Ross, Silicon-on-Sapphire Substrates Overcome MOS Limitations, *Electronics*, Sept. 25, 1972, p. 113.
10. H. R. Camenzind, "Electronic Integrated System Design," Van Nostrand Reinhold, New York, 1972.

ADDITIONAL READINGS

Goetzberger, A., E. Klausman, and M. J. Schulz, Interface States on Semiconductor/Insulator Surfaces, *CRC Critical Reviews in Solid State Sciences*, January 1976.
Grove, A. S.: "Physics and Technology of Semiconductor Devices," chaps. 9–12, Wiley, New York, 1967.
Ihantola, H. K. J., and J. L. Moll: Design Theory of a Surface Field-Effect Transistor, *Solid-State Electron.*, **7**: 423 (1964).
Many, A., Y. Goldstein, and N. B. Grover: "Semiconductor Surfaces," Wiley, New York, 1965.
Penney, W. M. (ed.): "MOS Integrated Circuits," Van Nostrand Reinhold, New York, 1972.
Sze, S. M.: "Physics of Semiconductor Devices," chaps. 9 and 10, Wiley, New York, 1969.

NINE

BIPOLAR JUNCTION TRANSISTORS†

The most important electronic device presently available is the *bipolar junction transistor* (BJT), an active three-terminal device that can be used as an amplifier or switch. It is the fundamental component in most integrated circuits.

In this chapter, the amplifying principle is first described qualitatively. The carrier distribution is derived by solving the continuity equation of minority carriers in the base. On the basis of carrier distribution, we obtain the current components, current gains, and current-voltage characteristics. The Ebers-Moll and hybrid-pi models are shown to be useful in describing transistor behavior. In addition, the frequency response and switching properties are examined, and limitations due to base resistance and junction breakdown are discussed.

If an additional *p-n* junction is fabricated on a bipolar transistor structure, we get a four-layer *p-n-p-n* device. The *p-n-p-n* transistors are bistable devices whose operation depends on an internal feedback mechanism resulting in high- and low-impedance stable states. They are available in a wide range of voltage and current ratings. The low-power devices are designed to be used primarily for switching and logic circuitry, while the high-power devices find wide applications in industry as ac switches, phase-control devices, power inverters, and dc choppers.

† Readers are urged to review Chap. 4 on *p-n* junction theory before studying this chapter.

9-1 THE TRANSISTOR ACTION

A transistor consists of a layer of *p*-type silicon sandwiched between two layers of *n*-type silicon. Alternatively, a transistor may consist of a layer of *n*-type between two layers of *p*-type material. In the former case, the transistor is referred to as an *n-p-n* transistor and in the latter case a *p-n-p* transistor. The standard fabrication technique is the planar technology described in Chap. 3. An *n-p-n* transistor can be made by introducing an additional *n*-type impurity diffusion subsequent to the step shown in Fig. 3-3*e*. First, an oxide layer is grown to cover the entire surface. Then a window is opened by means of photoresist-masked etching, and a phosphorus diffusion is made to form a second junction. The final structure after contact metallization and lead attachment, shown in Fig. 9-1*a*, is known as the silicon *double-diffused planar-epitaxial* transistor.

In order to study the basic characteristics of the transistor, we take a one-dimensional section (shown between the dotted lines in Fig. 9-1*a*) to represent the transistor and redraw it in Fig. 9-1*b*. The three layers are known as *emitter, base,* and *collector.* In the following discussion, our attention is directed toward this simplified model because it demonstrates all the important features of transistor action. The circuit representation of the *n-p-n* transistor is shown in Fig. 9-1*c*, where the arrow on the emitter lead specifies the direction of current flow when the emitter-base junction is forward-biased. It should be noted that the external circuit currents I_E, I_B, and I_C are assumed to be positive when currents flow *into* the transistor. The symbols V_E (for V_{BE}) and V_C (for V_{BC}) are the base-to-emitter and base-to-collector voltages, respectively.

Under normal operation, the emitter junction is forward-biased and the collector junction is reverse-biased. The energy-band diagram is shown in Fig. 9-2*a*. The forward biasing of the emitter junction lowers the emitter-base potential barrier by V_E, whereas the reverse biasing of the collector junction increases the collector-base potential barrier by V_C. The lowering of the emitter-base barrier results in an injection of electrons into the base and holes into the emitter. The injected holes constitute the emitter hole current I_{pE}. The injected electrons, which are the minority carriers in the base, diffuse across the base layer and give rise to a current I_{nE}. Some injected electrons recombine with holes in the base, but the majority of the injected electrons survive recombination and reach the collector junction. The electrons which reach the collector junction fall down the potential barrier and are *collected* by the collector as I_{nC}. Consequently, a large current may flow in a *reverse-biased* (collector) junction resulting from carrier injection of a nearby (emitter) junction. This is the basis of the transistor action, and it is realized only when the two junctions are close enough physically to interact in the manner described. The difference between I_{nE} and I_{nC} represents the holes supplied by the base to recombine with electrons in the base layer. These current components are illustrated in Fig. 9-2*b*. Note that the arrows in this figure

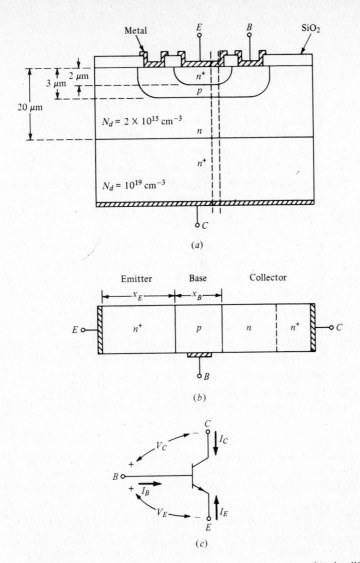

Figure 9-1 The n-p-n bipolar transistor: (a) planar structure, (b) simplified model, and (c) circuit symbol.

represent the direction of carrier motion. Therefore, the direction of electron current is in the direction opposite the arrows, whereas the hole current is in the direction shown. In addition, the emitter space-charge-layer recombination current I_{rg}, the collector generation current I_G, and reverse-saturation currents I_{nCO} and I_{pCO} are included. In practice, the components I_G, I_{nCO}, and I_{pCO} are very small and are usually combined and designated as I_{CO}, the *collector reverse-saturation current.*

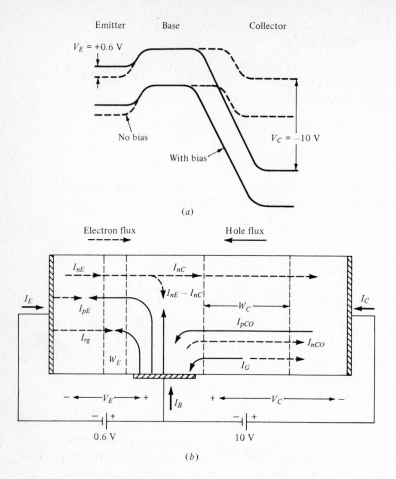

Figure 9-2 (*a*) Energy-band diagram and (*b*) current components.

The emitter, base, and collector currents are obtained by examining Fig. 9-2*b*, noting that the dotted lines represent the negative direction of electron-current components:

$$-I_E = I_{pE} + I_{nE} + I_{rg} \tag{9-1}$$

$$I_B = I_{pE} + I_{rg} + (I_{nE} - I_{nC}) - I_{CO} \tag{9-2}$$

$$I_C = I_{nC} + I_{CO} \tag{9-3}$$

Furthermore, the sum of all currents entering the transistor must be zero, so that

$$I_E + I_B + I_C = 0 \tag{9-4}$$

These equations will be used in later sections in relation to the current gains and the current-voltage characteristics.

9-2 CURRENT GAINS AND CURRENT-VOLTAGE CHARACTERISTICS

We now define the *emitter injection efficiency* γ and the *base transport factor* β_T† as

$$\gamma \equiv \frac{I_{nE}}{|I_E|} = \frac{I_{nE}}{I_{nE} + I_{pE} + I_{rg}} \qquad (9\text{-}5)$$

$$\beta_T \equiv \frac{I_{nC}}{I_{nE}} \qquad (9\text{-}6)$$

In addition, the ratio of the negative of the collector-current increment to the emitter-current change from zero to I_E is defined as the *current gain* α:

$$\alpha \equiv -\frac{I_C - I_{CO}}{I_E} \qquad (9\text{-}7)$$

Using Eqs. (9-1), (9-3), and (9-5) to (9-7), we obtain

$$\alpha = \frac{I_{nC}}{I_{nE} + I_{pE} + I_{rg}} = \gamma \beta_T \qquad (9\text{-}8)$$

which states that α is the product of the transport factor and emitter efficiency. It is an extremely important parameter in transistor theory. Since $|I_{nC}| < |I_{nE}| < |I_E|$, α is always smaller than unity, although it approaches unity for well-fabricated transistors.

Equation (9-7) can be rewritten as

$$I_C = -\alpha I_E + I_{CO} \qquad (9\text{-}9)$$

which relates the output collector current to the input emitter current with the base as the common terminal. Therefore, α is also known as the *common-base* current gain. Equation (9-9) is valid if the emitter is forward-biased and the collector is reverse-biased. Under these conditions, the transistor is said to be in its *active region* of operation, where the collector current is essentially independent of collector voltage. If both forward- and reverse-bias conditions for the collector junction are considered, however, I_{CO} in Eq. (9-9) has to be replaced by the diode equation (4-46) with I_o replaced by $-I_{CO}$ and V by V_C. The complete expression for I_C becomes

$$I_C = -\alpha I_E + I_{CO}(1 - e^{V_C/V_T}) \qquad (9\text{-}10)$$

Note that if V_C is negative and has a large magnitude compared with V_T, Eq. (9-10) reduces to Eq. (9-9). The current-voltage characteristics of the transistor can be obtained by plotting I_C vs. V_{CB} from Eq. (9-10) with I_E as a parameter, the voltage V_{CB} being the negative of V_C. The current-voltage

---

† The symbol β is commonly used to represent the base transport factor in the literature. This is unfortunate since β is also used for the common-emitter current gain, described later. We have added the subscript T here to avoid confusion.

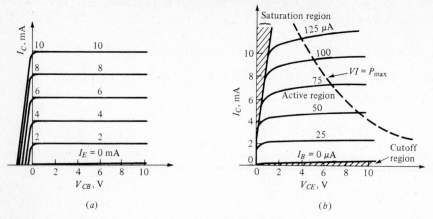

Figure 9-3 Collector current-vs.-voltage characteristics for (a) the common-base configuration and (b) the common-emitter configuration.

characteristics of a typical planar transistor are shown in Fig. 9-3a. Note that the collector current remains constant except when the collector junction is forward-biased, i.e., $V_{CB} < 0$, and the result is consistent with Eq. (9-10).

In most circuit applications, the emitter is used as the common terminal, with the base and collector as the input and output terminals, respectively. The relationship between the collector and base currents can be obtained by substituting I_E from Eq. (9-4) into Eq. (9-9):

$$I_C = \frac{\alpha}{1-\alpha} I_B + \frac{I_{CO}}{1-\alpha} = h_{FE} I_B + I_{CEO} \tag{9-11}$$

where we define

$$h_{FE} \equiv \frac{\alpha}{1-\alpha} \tag{9-12}$$

$$I_{CEO} \equiv \frac{I_{CO}}{1-\alpha} \tag{9-13}$$

The symbol h_{FE}, also known as β, is called the *common-emitter current gain.* Experimental current-voltage characteristics of the common-emitter configuration are shown in Fig. 9-3b, where three regions of operation are specified. The *cutoff* region corresponds to the condition that both the emitter and collector junctions be reverse-biased. The *normal active* region indicates that the emitter junction is forward-biased and the collector junction is reverse-biased. In the *saturation* region, both junctions are forward-biased. As an amplifier, the transistor is operated within the normal active region. However, a transistor switch usually traverses between the saturation (ON) and cutoff (OFF) regions.

The maximum power P_{max} that can be dissipated in a transistor without overheating is given in transistor specifications. The constant product $VI = P_{max}$ is plotted in Fig. 9-3b, and the transistor should not be operated beyond this curve.

9-3 DERIVATION OF CURRENT COMPONENTS AND GAIN EXPRESSIONS

The injected-current components shown in Fig. 9-2b can be derived by solving the continuity equations in Sec. 4-4. To simplify our analysis, we assume that the doping concentration is uniform in the base and emitter and that the injection level is low. The electron current and continuity equations (4-33) and (4-34) are rewritten here for the dc case:

$$I_n = qAD_n \frac{dn_p}{dx} \tag{9-14}$$

$$D_n \frac{d^2 n_p}{dx^2} - \frac{n_p - n_{po}}{\tau_n} = 0 \tag{9-15}$$

Equation (9-15) has a general solution

$$n_p - n_{po} = K_1 e^{-x/L_n} + K_2 e^{x/L_n} \tag{9-16}$$

where K_1 and K_2 are constants to be determined by the boundary conditions. With a forward-biased emitter junction, the boundary condition at the edge of the emitter depletion layer in the base side is, according to Eq. (4-29),

$$n_p(0) = n_{po} e^{V_E/V_T} \tag{9-17}$$

The boundary condition at the depletion-layer edge of the reverse-biased collector junction is usually assumed to be†

$$n_p(x_B) = 0 \tag{9-18}$$

This is a reasonable assumption since the electric field of the collector junction sweeps carriers into the collector so that the collector is nearly a perfect sink. Substitution of the boundary conditions into Eq. (9-16) yields (Prob. 9-2)

$$n_p - n_{po} = \frac{n_{po}(e^{V_E/V_T} - 1) \sinh [(x_B - x)/L_n]}{\sinh (x_B/L_n)} \tag{9-19}$$

Under normal bias conditions, we have $V_E/V_T \gg 1$, so that, Eq. (9-19) becomes

$$n_p = \frac{n_{po} e^{V_E/V_T} \sinh [(x_B - x)/L_n]}{\sinh (x_B/L_n)} \tag{9-20}$$

† Strictly speaking, x_B is a function of the collector voltage which controls the extension of the collector depletion-layer width into the base. Known as the Early effect, this represents a feedback from the collector to the base. The Early effect is not important in double-diffused transistors in which the collector doping is lower than the base doping (see Fig. 9-9).

For most practical transistors, the base width x_B is much smaller than L_n, and Eq. (9-20) can be further simplified as[†]

$$n_p = n_{po} e^{V_E/V_T}\left(1 - \frac{x}{x_B}\right) \tag{9-21}$$

Equation (9-20) is plotted in Fig. 9-4 for different x_B/L_n. It is seen that the curve for $x_B/L_n = 0.1$ approaches the linear relationship expressed by Eq. (9-21).

Let us substitute Eq. (9-19) into Eq. (9-14) and evaluate the electron current at the emitter and collector junctions. The results are

$$I_{nE} = \frac{qAD_n n_{po}}{L_n}(e^{V_E/V_T} - 1)\coth\frac{x_B}{L_n} \qquad \text{at } x = 0 \tag{9-22}$$

$$I_{nC} = \frac{qAD_n n_{po}}{L_n}(e^{V_E/V_T} - 1)\operatorname{csch}\frac{x_B}{L_n} \qquad \text{at } x = x_B \tag{9-23}$$

The base transport factor defined by Eq. (9-6) is therefore given by

$$\beta_T = \operatorname{sech}\frac{x_B}{L_n} \quad = \quad \cosh\frac{x_B}{L_n} \tag{9-24}$$

For $x_B/L_n \ll 1$, Eqs. (9-22) and (9-24) become, after series expansion,[‡]

$$I_{nE} = qAD_n \frac{n_i^2}{N_a x_B}(e^{V_E/V_T} - 1) \tag{9-25}$$

[†] By using $\sinh y = y + y^3/6 + \cdots \approx y$ for $y \ll 1$.
[‡] Use $\coth y = 1/y + y/3 - y^3/45 \approx 1/y$ and $\operatorname{sech} y = 1 - y^2/2 + 5y^4/24 + \cdots \approx 1 - y^2/2$.

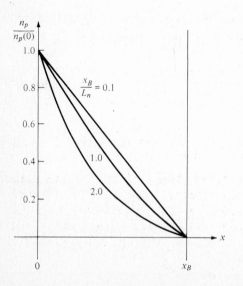

Figure 9-4 Minority-carrier distribution in the base for different x_B/L_n. Typically, $x_B = 1\,\mu\text{m}$ and $L_n = 20\,\mu\text{m}$, so that $x_B/L_n < 0.1$.

and
$$\beta_T = 1 - \frac{1}{2}\frac{x_B^2}{L_n^2} \tag{9-26}$$

Note that the emitter electron current increases with decreasing base width. In addition, the transport factor approaches unity for small x_B/L_n, meaning negligible loss of electrons during the transport across the base region.

The hole current I_{pE} can be derived in a manner similar to the solution of the electron current. The result is

$$I_{pE} = qAD_p \frac{n_i^2}{N_{dE}x_E}(e^{V_E/V_T} - 1) \tag{9-27}$$

where N_{dE} and x_E are the donor concentration and the width of the emitter, respectively. If x_E is large compared with L_p, the hole diffusion length in the emitter, x_E, should be replaced by L_p.

The recombination current in the space-charge layer of the forward-biased emitter was derived in Sec. 4-6 and is repeated here:

$$I_{rg} = \frac{qAn_iW_E}{2\tau_o}e^{V_E/2V_T} \tag{9-28}$$

where W_E is the emitter depletion-layer width. The base current according to Eq. (9-2) is sometimes written, using Eqs. (9-25) through (9-28),

$$I_B = qAn_i^2\left[\left(\frac{D_p}{N_{dE}x_E} + \frac{D_nx_B}{2N_aL_n^2}\right)(e^{V_E/V_T} - 1) + \frac{W_E}{2\tau_on_i}e^{V_E/2V_T}\right] \propto e^{V_E/\eta V_T} \tag{9-29}$$

where η is a factor with a value varying between 1 and 2 in a silicon transistor. An experimental current-voltage plot on semilog paper is shown in Fig. 9-5 for both I_B and I_C. The effect of I_{rg} is clearly observed in the base-current data in the low-current region with $\eta = 1.8$. On the other hand, I_C follows the exponential with a slope of $1/V_T$ for more than **6 decades of current**, according to Eq. (9-23).

In a typical transistor, β_T approaches unity, and $I_C = I_{nC} \approx I_{nE}$. Therefore, the reciprocal of the common-emitter current gain expressed in Eq. (9-11) with $I_{CEO} \approx 0$ can be written as

$$\frac{1}{h_{FE}} = \frac{I_B}{I_C} \approx \frac{I_{pE}}{I_{nE}} + \frac{I_{nE} - I_{nC}}{I_{nE}} + \frac{I_{rg}}{I_{nE}} \tag{9-30}$$

Using Eqs. (9-25) to (9-28), we obtain

$$\frac{1}{h_{FE}} = \frac{N_ax_BD_p}{N_{dE}x_ED_n} + \frac{x_B^2}{2L_n^2} + \frac{N_ax_BW_E}{2D_nn_i\tau_o}e^{-V_E/2V_T} \tag{9-31}$$

The contribution of the last term on the right-hand side of Eq. (9-31) leads to a nonlinear relationship between the applied emitter voltage and current gain. Thus, the current gain is not constant and is a function of the current in the transistor. In general, the larger the I_{rg} the lower the h_{FE}. At a sufficiently high current level, however, the effect of I_{rg} becomes negligible due to the factor $2V_T$ in the exponent. _Assumptions_ $X_E \ll L_{pE}$

$X_B \ll L_{nB}$

$X_C \gg L_{pc}$

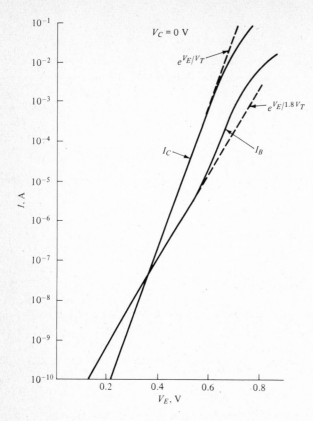

$V_C = 0$ V

e^{V_E/V_T}

$e^{V_E/1.8 V_T}$

I_C

I_B

I, A

V_E, V

Figure 9-5 Static current-voltage characteristics of an n-p-n transistor.

Using the data given in Fig. 9-5, we plot the current-gain variation with collector current in Fig. 9-6. The plot is in good agreement with Eq. (9-31) except in the high-current region. When the collector current is greater than 10 mA, h_{FE} starts to level off and then decreases with further increase of current. The drop of gain at high currents is due to the fact that the injected minority-carrier density becomes comparable to the majority-carrier density in the base. As a result, the fraction of total emitter current carried by holes injected from base to emitter increases, leading to a decrease in emitter efficiency and a decline in h_{FE}. The equations obtained in this section are no longer applicable. The theory of high-level injection is not covered in the text, and the current gain at high current is given here without proof [1]:

$$\frac{1}{h_{FE}} \approx \frac{N_a x_B D_p}{N_{dE} x_E D_n} \left(1 + \frac{I_E x_B}{2 q D_n A N_a} \right) + \frac{x_B^2}{4 L_n^2} \qquad (9\text{-}32)$$

Note that the current gain h_{FE} is inversely proportional to the emitter current, so that the current gain is reduced as we increase the operating emitter current.

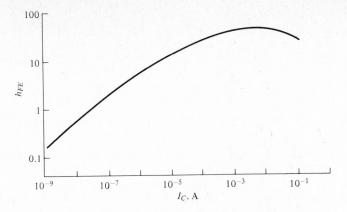

Figure 9-6 Dependence of current gain on the collector current.

9-4 THE EBERS-MOLL EQUATIONS

In the previous section we considered the transistor in the normal active region, i.e., forward-biased emitter junction and reverse-biased collector junction. We have pointed out that it is possible for the collector junction to be forward-biased. The boundary condition given by Eq. (9-18) must now be changed to

$$n_p(x_B) = n_{po}e^{V_C/V_T} \tag{9-33}$$

With Eqs. (9-17) and (9-33) as the boundary conditions, the constants K_1 and K_2 in Eq. (9-16) are not difficult to find. Usually, however, the base width x_B is small compared with L_n, and the electron current at $x = 0$ can be shown to be

$$I_{nE} = \frac{qAD_n n_i^2}{x_B N_a} [(e^{V_E/V_T} - 1) - (e^{V_C/V_T} - 1)] \tag{9-34}$$

Let us now consider the equivalent circuit shown in Fig. 9-7, where the space-charge-layer recombination currents are taken as external-current components. The current I_E' is the sum of Eqs. (9-34) and (9-27):

$$I_E' = -(I_{pE} + I_{nE}) = a_{11}(e^{V_E/V_T} - 1) + a_{12}(e^{V_C/V_T} - 1) \tag{9-35}$$

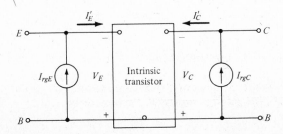

Figure 9-7 Equivalent circuit with I_{rg} removed from the intrinsic transistor.

where
$$a_{11} = -qAn_i^2\left(\frac{D_n}{N_a x_B} + \frac{D_p}{N_{dE} x_E}\right) \qquad a_{12} = \frac{qAD_n n_i^2}{N_a x_B} \tag{9-36}$$

In a similar manner we obtain

$$I_C' = a_{21}(e^{V_E/V_T} - 1) + a_{22}(e^{V_C/V_T} - 1) \tag{9-37}$$

It can be shown that

$$a_{21} = \frac{qAD_n n_i^2}{N_a x_B} \qquad a_{22} = -qAn_i^2\left(\frac{D_n}{N_a x_B} + \frac{D_p}{N_{dC} L_{pC}}\right) \tag{9-38}$$

where N_{dC} and L_{pC} are, respectively, the doping concentration and hole diffusion length in the collector. Note that $a_{12} = a_{21}$, a result that is valid for a transistor possessing any geometry. Equations (9-35) and (9-37), which are valid for either positive or negative values of V_E and V_C, are known as the *Ebers-Moll equations* [2].

If V_E is eliminated from Eqs. (9-35) and (9-37), we have

$$I_C' = \frac{a_{21}}{a_{11}} I_E' + \left(a_{22} - \frac{a_{21}a_{12}}{a_{11}}\right)(e^{V_C/V_T} - 1)$$

$$= -\alpha_N I_E' - I_{CO}(e^{V_C/V_T} - 1) \tag{9-39}$$

where α_N is known as the *normal* or *forward alpha* and I_{CO} is the *collector reverse saturation current*. Equation (9-39) is identical to Eq. (9-10) except that α_N has replaced α. Note that α_N can be obtained from Eq. (9-8) with $I_{rg} = 0$ since I_{rg} in Fig. 9-7 is taken as a parasitic current component.

If V_C is eliminated from Eqs. (9-35) and (9-37), the result is

$$I_E' = \frac{a_{12}}{a_{22}} I_C' + \left(a_{11} - \frac{a_{12}a_{21}}{a_{22}}\right)(e^{V_E/V_T} - 1)$$

$$= -\alpha_I I_C' - I_{EO}(e^{V_E/V_T} - 1) \tag{9-40}$$

where α_I is known as the *inverse alpha* and I_{EO} is the *emitter reverse saturation current*. Physically, the inverse alpha is the current gain when the collector junction is *forward*-biased and the emitter junction is *reverse*-biased. In other words, the collector is emitting electrons, and the emitter is collecting electrons in the inverse mode of operation. From the definition given in Eqs. (9-39) and (9-40), we find that the parameters α_N, α_I, I_{CO}, and I_{EO} are related by the condition [2]

$$\alpha_N I_{EO} = \alpha_I I_{CO} \tag{9-41}$$

Thus, the transistor is specified if three of these parameters are measured. However, practical devices often violate Eq. (9-41) because of major nonideal components of emitter and collector saturation currents.

Equations (9-39) and (9-40) have a simple interpretation in terms of an equivalent circuit known as the *Ebers-Moll* model (Fig. 9-8). It has two ideal diodes connected back to back with reverse saturation currents I_{EO} and I_{CO} and two dependent current sources shunting the ideal diodes. The I_{rg}

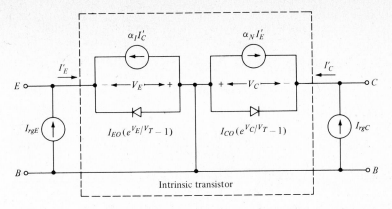

Figure 9-8 The Ebers-Moll model.

terms are included to complete the equivalent representation of the transistor. The Ebers-Moll model can be used to describe the transistor dc characteristics. In addition, it is employed to characterize the transient response when the frequency dependence of α_N and α_I are known. Frequency- and time-dependent response of the current gain will be studied in a later section.

9-5 GRADED BASE TRANSISTORS

In the previous sections, our analyses were based on the assumption that the impurity distribution is homogeneous in the base and emitter regions. In a practical transistor such as the structure shown in Fig. 9-1, the impurity profile is nonuniform, as depicted in Fig. 9-9. The graded emitter doping has little effect on the transistor performance and will be ignored. The graded impurity distribution in the base, according to Sec. 2-7, introduces a built-in field

$$\mathscr{E} = \frac{V_T}{N_a(x)} \frac{dN_a(x)}{dx} \tag{9-42}$$

This equation is obtained by using Eq. (2-50) with N_d replaced by N_a, the impurity distribution in the base, and removing the negative sign.

This electric field, which keeps the holes in their place, is in such a direction that it aids the transport of injected electrons in most of the base region. The electrons are now moved by both diffusion and drift across the base layer, so that the transport factor is increased. Let us now derive an expression of the transport factor for the graded base transistor.

We obtain the following equation by substituting Eq. (9-42) into Eq. (1-51) and making use of Einstein's relationship:

$$\frac{dn}{dx} + \frac{n}{N_a} \frac{dN_a}{dx} = \frac{I_n}{qAD_n} \tag{9-43}$$

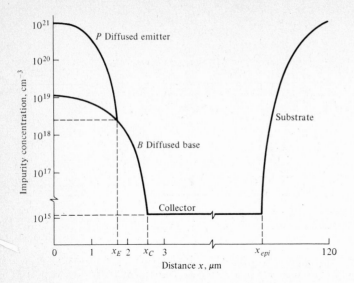

Figure 9-9 Impurity profile of 2N3866.

Multiplying both sides of Eq. (9-43) by N_a and integrating yields

$$N_a n(x) = \int \frac{I_n N_a}{qAD_n} \, dx + C \qquad (9\text{-}44)$$

where C is an integration constant. In most practical planar transistors, the base recombination is negligible, so that I_n is constant in Eq. (9-44). We can now substitute the boundary condition $n_p(x_B) = 0$ [Eq. (9-18)] into Eq. (9-44), leading to

$$n(x) = \frac{I_n}{qAD_n N_a} \int_x^{x_B} N_a \, dx \qquad (9\text{-}45)$$

The current I_n is obtained by making use of Eq. (9-17) at $x = 0$ with n_{po} replaced by $n_i^2/N_a(0)$:

$$I_n = \frac{qAD_n n_i^2}{\displaystyle\int_0^{x_B} N_a \, dx} e^{V_E/V_T} \qquad (9\text{-}46)$$

The integral in the denominator of Eq. (9-46) represents the number of impurity atoms in the base and is known as the *Gummel number* [3]. Notice that Eq. (9-46) reduces to Eq. (9-25) for a base with uniform doping. In addition a larger electron current is realized with a smaller Gummel number, which corresponds to a narrower base width.

The base transport factor can be estimated by first taking the total base recombination current as

$$\frac{qA}{\tau_n} \int_0^{x_B} n(x)\, dx$$

Thus, from the definition of the base transport factor, we obtain

$$\beta_T = \frac{I_n}{I_n + \dfrac{qA}{\tau_n} \displaystyle\int_0^{x_B} n(x)\, dx} = -\frac{1}{1 + \dfrac{qA}{\tau_n I_n} \displaystyle\int_0^{x_B} n(x)\, dx}$$

$$\approx 1 - \frac{qA}{I_n \tau_n} \int_0^{x_B} n(x)\, dx \qquad (9\text{-}47)$$

where we have used the approximation $(1 + y)^{-1} \approx 1 - y$ for $y \ll 1$. Substituting Eq. (9-45) into (9-47) and setting $D_n \tau_n = L_p^2$, we have

$$\beta_T = 1 - \frac{1}{L_n^2} \int_0^{x_B} \left(\frac{1}{N_a} \int_x^{x_B} N_a\, dx \right) dx \qquad (9\text{-}48)$$

This is the general expression of the base transport factor for arbitrary base impurity distribution. For a uniform base distribution, Eq. (9-48) reduces to Eq. (9-26).

9-6 BASE SPREADING RESISTANCE AND CURRENT CROWDING

The base current is a majority-carrier current flowing in the direction perpendicular to the minority-carrier current injected from the emitter. As illustrated in Fig. 9-10, this base current produces a lateral potential drop in both the passive and active base regions. In the passive region, the potential drop can be represented by an ohmic base resistance. In the active region, the potential drop reduces the forward-bias voltage across the emitter junction in such a way that the forward-bias voltage is larger at the periphery near the contact than at the center of the active region. As a result, the minority-

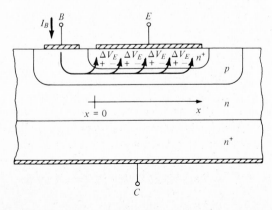

Figure 9-10 Lateral base current and ohmic drop in the base, leading to largest forward bias at $x = 0$.

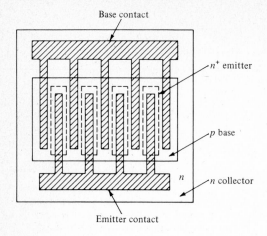

Base contact

n^+ emitter

p base

n n collector

Emitter contact

Figure 9-11 Top view of an interdigitated geometry of a medium-power bipolar transistor.

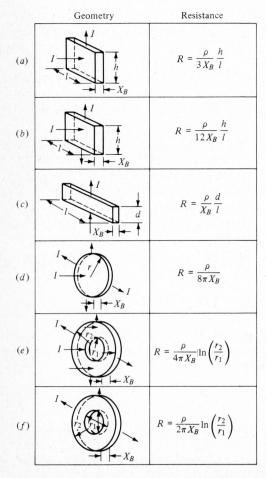

Geometry	Resistance
(a)	$R = \dfrac{\rho}{3 X_B}\dfrac{h}{l}$
(b)	$R = \dfrac{\rho}{12 X_B}\dfrac{h}{l}$
(c)	$R = \dfrac{\rho}{X_B}\dfrac{d}{l}$
(d)	$R = \dfrac{\rho}{8\pi X_B}$
(e)	$R = \dfrac{\rho}{4\pi X_B}\ln\left(\dfrac{r_2}{r_1}\right)$
(f)	$R = \dfrac{\rho}{2\pi X_B}\ln\left(\dfrac{r_2}{r_1}\right)$

Figure 9-12 A summary of resistance formulas for a number of geometrical elements. These can be used in appropriate combinations to generate approximate expressions for transistor base resistance. (*After Warner and Fordemwalt [5].*)

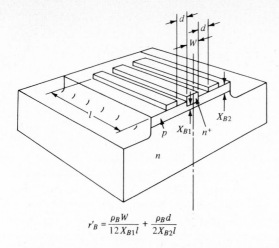

$$r'_B = \frac{\rho_B W}{12 X_{B1} l} + \frac{\rho_B d}{2 X_{B2} l}$$

Figure 9-13 An approximate expression for the base spreading resistance of the stripe transistor geometry. *(After Warner and Fordemwalt [5].)*

carrier injection falls off from the edge of the base inward. The nonuniform carrier injection gives rise to a nonuniform current distributed along the emitter junction. This results in a higher current density near the edges and is known as the *current-crowding effect* [4]. Current crowding reduces the effective area of the active transistor region and is not desirable, particularly in power transistors. To minimize this effect, power transistors are usually designed with a high periphery-to-area ratio, as in the interdigitated geometry shown in Fig. 9-11.

Fortunately, over a considerable range of current, it is possible to represent the active base by a constant resistance. The sum of the active and passive base resistances is known as the *base spreading resistance* $r_{bb'}$. We can calculate the base spreading resistance if the transistor geometry is known. Usually, the base of the transistor is divided into sections with the resistances given in Fig. 9-12. An example of the use of these formulas is shown in Fig. 9-13, in which the first term in the $r_{bb'}$ expression represents the active resistance and the second term represents the passive resistance.

9-7 FREQUENCY RESPONSE OF TRANSISTORS

In this section, we consider the transistor small-signal characteristics in the normal active mode of operation. The small-signal common-base and common-emitter current gains are defined as

$$\alpha \equiv \frac{dI_C}{dI_E}\bigg|_{V_{CB}=\text{const}} \qquad h_{fe} \equiv \frac{dI_C}{dI_B}\bigg|_{V_{CE}=\text{const}}$$

At a low frequency, the current gains are independent of the operating frequency. However, the magnitude of the gains decreases after a certain critical frequency is reached. A sketch of the typical frequency response of

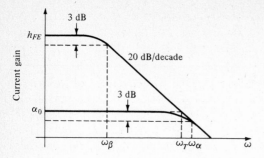

Figure 9-14 Current gains as a function of frequency.

the current gains is given in Fig. 9-14. The various frequencies shown in this figure are defined as (1) the *common-base cutoff frequency* ω_α, (2) the *common-emitter cutoff frequency* ω_β, and (3) the *gain-bandwidth product* ω_T, that is, the frequency at which h_{fe} becomes unity. Since the slope of α vs. frequency is 20 dB/decade, it can be described by

$$\alpha = \frac{\alpha_o}{1 + j\omega/\omega_\alpha} \tag{9-49}$$

where α_o is the current gain at low frequency. At $\omega = \omega_\alpha$, the magnitude of α is $0.707\alpha_o$ (3 dB down), so that ω_α is also known as the 3-dB frequency. Various factors limit the cutoff frequency, and the four most important ones are described below. Each of these factors introduces a time delay when the signal propagates from the emitter to collector.

Base Transit Time

The most severe limitation on the transistor frequency response is the carrier transport through the base layer. It is possible to derive the equations describing the transistor frequency characteristic by solving the ac diffusion equation for minority carriers in the base following the procedures outlined in Sec. 4-9. Instead of carrying out the rigorous derivation, however, we present here a simple technique known as the *transit-time analysis* [6], which yields the same information as the solution of the diffusion equation. Let us define the effective minority-carrier velocity $v(x)$ in the base by the current equation

$$I_n = qAn_p(x)v(x) \tag{9-50}$$

Since $dx = v(x)\,dt$, the time required for an electron to traverse the base is obtained by using Eq. (9-50), and, by integration,

$$\tau_B = \int_0^{x_B} \frac{dx}{v(x)} = \int_0^{x_B} \frac{qAn_p(x)}{I_n}\,dx \tag{9-51}$$

For most transistors, we have $x_B/L_n \ll 1$, and the electron distribution is given by Eq. (9-21). Substituting Eq. (9-21) into Eq. (9-51) and integrating leads to (Prob. 9-14)

$$\tau_B = \frac{x_B^2}{2D_n} \tag{9-52}$$

A small τ_B means a short delay of signal or high frequency of operation. Therefore, transistors are designed with a small base width to achieve better frequency response. In a transistor with a built-in field such as the double-diffused transistor, the base transit time is found to be one-half of that given by Eq. (9-52), so that the maximum frequency is roughly double that of a transistor with uniform doping in the base [7].

Emitter Transition-Capacitance Charging Time

The forward-biased emitter-junction transition capacitance C_{TE} is a function of the bias voltage, and it is difficult to measure because of the shunting of the diffusion capacitance (see Sec. 9-8). A rough estimate of this capacitance is 4 times the zero-bias capacitance given by Eq. (4-82) with $V_R = 0$. This capacitance is in parallel with a junction resistance r_e, and the charging-time constant is

$$\tau_E = r_e C_{TE} = \frac{4V_T}{I_E} C_{TE}(0) \tag{9-53}$$

The resistance r_e is derived by taking dI_E/dV_E from the diode equation (4-46) with I replaced by I_E and V by V_E.

Collector Depletion-Layer Transit Time

With a high reverse bias across the collector junction, the depletion-layer width is significant, and it takes time for carriers to get through. Since the electric field is very high there, the carriers may be assumed to have reached the saturation velocity v_{th}. Therefore, the transit time for carriers to traverse the depletion layer is

$$\tau_d = \frac{x_m}{v_{th}} \tag{9-54}$$

where x_m is the total thickness of the collector depletion layer.

Collector Capacitance Charging Time

The collector junction is reverse-biased, so that the shunting resistance across the junction capacitance is very large. As a result, the charging-time constant is determined by the capacitance C_{TC} and the collector series resistance r_{sc}:

$$\tau_C = r_{sc} C_{TC} \tag{9-55}$$

The collector resistance of the planar epitaxial transistor in Fig. 9-1 is small because of the heavily doped epitaxial substrate, and τ_C is negligible. However, it should be included in the calculation for integrated transistors.

The cutoff frequency ω_α is equal to the reciprocal of the total time delay incurred in signal propagation from the emitter to collector. Therefore, we have

$$\frac{1}{\omega_\alpha} = \tau_E + \tau_B + \tau_d + \tau_C \tag{9-56}$$

By using the relationship between α and h_{fe}, we find

$$h_{fe} = \frac{\alpha}{1 - \alpha} = \frac{h_{FE}}{1 + j\omega/\omega_\beta} \tag{9-57}$$

where
$$\omega_\beta = \omega_\alpha(1 - \alpha_o) \quad \text{and} \quad h_{FE} = \frac{\alpha_\sigma}{1 - \alpha_o} \tag{9-58}$$

From these equations we find that the common-emitter cutoff frequency is much lower than ω_α. But the gain-bandwidth product is

$$\omega_T = h_{FE}\omega_\beta \tag{9-59}$$

which is equal to ω_α. In practical transistors, however, ω_T is always smaller than ω_α because an excess phase shift exists in the transport of carriers from the emitter to the collector.

Let us examine the dependence of the cutoff frequency on the operating current. As the emitter current increases, the emitter-junction time constant τ_E becomes smaller, so that ω_α in Eq. (9-56) is increased. As a result, an improvement in frequency response can be achieved with increasing operating current. However, if the current is allowed to increase indefinitely, the cutoff frequency will eventually decrease. This is known as the Kirk effect [8] and is most evident in the planar-epitaxial transistor shown in Fig. 9-9. In this diagram, the doping level of the epitaxial n layer is lower than the base doping. Therefore, the collector depletion layer extends mostly into the epitaxial layer, and the depletion layer contains positive charge (fixed ions). As the emitter current is increased, the large number of injected electrons, upon arriving at the collector junction, neutralize these positively charged ions to form a neutral region. Consequently, the base-collector junction is relocated farther away from the emitter junction. At very high emitter current, the effective base width becomes

$$x'_B = x_{epi} - x_E \tag{9-60}$$

where x_{epi} is the thickness of the epitaxial layer. Within the region of x'_B, the electric field is very small, and electrons are transported via the diffusion mechanism [9]. As a result, τ_B becomes very large and ω_α is lowered. The Kirk effect is particularly important in high-frequency and high-power transistors.

9-8 HYBRID-Pi EQUIVALENT CIRCUIT

It is frequently necessary to represent the transistor by a simple equivalent circuit in order to calculate the small-signal circuit response. The most widely used equivalent circuit is the hybrid-pi model, shown in Fig. 9-15. It represents a transistor operated in the normal active mode in the common-emitter configuration. A very important feature of the hybrid-pi model is that all the parameters are related to the physical processes in the transistor.

Basically, the hybrid-pi model represents the dynamic incremental variation of the stored charge in the base upon the incremental change of the applied emitter voltage. As illustrated in Fig. 9-16, a change of $+\Delta V_E$ gives rise to an increase of ΔQ_B and thus ΔI_C. The transconductance g_m is defined by

$$g_m \equiv \frac{dI_C}{dV_E} \tag{9-61}$$

From the discussion in Sec. 9-3, we have

$$n_p(0) = n_{po} e^{V_E/V_T} \tag{9-62}$$

and
$$|I_C| = \frac{qAD_n n_p(0)}{x_B} \quad \text{for } \frac{x_B}{L_n} \ll 1 \tag{9-63}$$

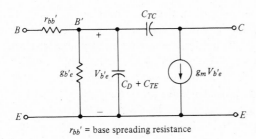

$r_{bb'}$ = base spreading resistance

Figure 9-15 The hybrid-pi equivalent circuit.

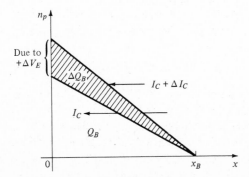

Figure 9-16 Schematic representation of the relationship between V_E, Q_B, and I_C.

Thus, we derive

$$g_m = \frac{|I_C|}{V_T} \tag{9-64}$$

The parameter $g_{b'e}$ is defined by

$$g_{b'e} \equiv \frac{dI_B}{dV_E} \tag{9-65}$$

If the space-charge recombination current is ignored in Eq. (9-29), I_B is given by

$$I_B = qAn_i^2\left(\frac{D_p}{x_E N_{dE}} + \frac{D_n x_B}{2N_a L_n^2}\right)(e^{V_E/V_T} - 1) \tag{9-66}$$

Therefore, we find

$$g_{b'e} = \frac{I_B}{V_T} = \frac{I_C}{h_{FE}V_T} = \frac{g_m}{h_{FE}} \tag{9-67}$$

The *diffusion capacitance* is defined as

$$C_D \equiv \frac{dQ_B}{dV_E} \tag{9-68}$$

Since the total charge stored in the base is

$$Q_B = \frac{q}{2}\, n_p(0)x_B A \tag{9-69}$$

we obtain the following expression by using Eqs. (9-62) to (9-64)

$$C_D = \frac{qAx_B n_p(0)}{2V_T} = g_m \frac{x_B^2}{2D_n} \tag{9-70}$$

The hybrid-pi equivalent circuit is completed when the transition capacitances are added. It can be shown (Prob. 9-17) that the cutoff frequency of the common-emitter short-circuit current gain is

$$\omega_\beta = \frac{g_m}{(C_D + C_{TE} + C_{TC})h_{FE}} \tag{9-71}$$

and the gain-bandwidth product for $C_D \gg C_{TE} + C_{TC}$ is

$$\omega_T = h_{FE}\omega_\beta = \frac{2D_n}{x_B^2} \tag{9-72}$$

Note that the gain-bandwidth product is identical to the inverse of the base transit time derived in the last section when the transition capacitances are neglected.

9-9 THE TRANSISTOR AS A SWITCH

In addition to its application as an amplifier, the transistor is used as an ON-OFF switch. We consider the transistor circuit shown in Fig. 9-17a. Driven

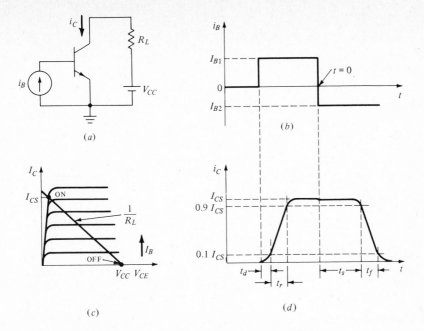

Figure 9-17 Switching operation of a bipolar transistor: (a) circuit diagram, (b) base-current drive, (c) output V-I characteristics, and (d) the output-current waveform.

by a current pulse waveform shown in Fig. 9-17b, the transistor is made to operate in the cutoff region and in saturation as depicted in Fig. 9-17c. The transistor is considered to be ON in saturation because the collector current is large and its impedance is low. It is considered to be OFF in the cutoff region since there is no current flow under this condition. In saturation, the collector current is limited by the load resistance so that

$$I_{CS} = I_C(\text{sat}) = \frac{V_{CC} - V_{CE}(\text{sat})}{R_L} \tag{9-73}$$

where $V_{CE}(\text{sat})$ is the collector-to-emitter voltage in saturation (typically 0.2 V for silicon transistors). The minimum base current I_{B1} needed to drive the transistor into saturation is

$$I_{B1} \geqslant \frac{I_C(\text{sat})}{h_{FE}} \approx \frac{V_{CC}}{h_{FE}R_L} \tag{9-74}$$

where $V_{CE}(\text{sat})$ is neglected.

Physically, the saturation condition is satisfied if both emitter and collector junctions are forward-biased, and the cutoff condition is satisfied if both junctions are reverse-biased. The corresponding minority carriers in the base and collector for the planar epitaxial transistor are shown in Fig. 9-18. The minority-carrier storage in the collector is usually small compared with the

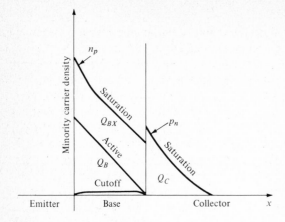

Figure 9-18 Charge storage in the base and collector at saturation. Also shown are the base charge at cutoff and active region.

total charge storage in the base and will be neglected in our consideration. Switching between the ON and OFF state is accomplished by a change of the carrier distribution; as with the p-n junction diode, these carrier densities cannot be changed instantaneously. The transition time from one state to the other, known as the *switching time*, corresponds to the establishment or removal of the minority carriers involved. A typical switching waveform of the collector current is shown in Fig. 9-17d. The definition of the switching times is given in the following paragraphs.

Delay Time

The turn-on delay time t_d is the time lapse between the application of the input step pulse to the time the output current reaches 10 percent of its final value $0.1I_C(\text{sat})$. It is limited by (1) the charging time of the junction transition capacitances from reverse bias to the new voltage levels and (2) the carrier transit times through the base and the collector depletion layer.

Rise Time and Fall Time

The rise time t_r is the time required for the current to rise from 10 to 90 percent of $I_C(\text{sat})$. It corresponds to the establishment of minority carriers in the base to reach 90 percent of the collector saturation current and is affected by the output time constant $C_{TC}R_L$. The turnoff fall time t_f specifies the interval during which the collector current falls from 90 to 10 percent of its maximum value. This is the reverse process of the rise time and is limited by the same factors.

Storage Time

The storage time t_s is measured between the negative step in the base current and the time when the collector current reaches $0.9I_C(\text{sat})$. This is the most important parameter in limiting the switching speed of a switching transistor. When the transistor is in saturation, both the emitter and collector junctions are forward-biased. As a result, excess carriers above and beyond those necessary to maintain the active mode of operation, are stored in the base and collector, as depicted in the shaded area in Fig. 9-18. The transistor output current cannot change until the excess carriers are removed. Thus, the storage time corresponds to the time required to remove the excess minority carriers.

The complete switching-time response can be calculated with the Ebers-Moll model given in Fig. 9-8. The procedure [10] involves (1) replacing α_N and α_I by their frequency-dependent form expressed in Eq. (9-49), (2) analyzing the Ebers-Moll equations in the frequency domain with the input base current as a step function, and (3) using the Laplace transform to obtain the time response. Although the procedure is straightforward, the details of such an analysis are quite complicated. For this reason, we consider the charge-control analysis [11] first introduced in Sec. 4-11. Since the storage time is the dominant parameter, we shall limit our discussion to its derivation.

The charge-control equation for the base region in the normal active mode is given by Eq. (4-92), with $I_p(0)$ replaced by i_B, Q_s by Q_B, and τ_p by τ_n:

$$i_B = \frac{dQ_B}{dt} + \frac{Q_B}{\tau_n} \tag{9-75}$$

where

$$Q_B = qA\frac{n_p(0)x_B}{2} \tag{9-76}$$

It should be pointed out that the collector current is related to the charge storage by Eqs. (9-63) and (9-76), so that the collector current is specified if Q_B is known. Under steady-state conditions, the time-dependent term in Eq. (9-75) is zero, and the base current is given by

$$I_B = \frac{Q_B}{\tau_n} \tag{9-77}$$

Let us designate the base current that brings the transistor to the onset of saturation as I_{BA}. Using Eq. (9-74), we find

$$I_{BA} = \frac{V_{CC}}{h_{FE}R_L} \tag{9-78}$$

When the saturation sets in, the total charge is given by $Q_B + Q_{BX}$, and the charge-control equation becomes (assuming $Q_C = 0$ in Fig. 9-18)

$$i_B = \frac{Q_B}{\tau_n} + \frac{Q_{BX}}{\tau_s} + \frac{dQ_B}{dt} + \frac{dQ_{BX}}{dt} \tag{9-79}$$

where Q_{BX} is the excess amount of charge storage and τ_s is the time constant associated with the removal of Q_{BX}. Let us now abruptly change the base current from I_{B1} to $-I_{B2}$ (Fig. 9-17b); the excess charge starts to decrease, but the active charge Q_B remains the same between $t = 0+$ and t_s. During this time interval, we may set

$$\frac{dQ_B}{dt} = 0 \tag{9-80}$$

and

$$\frac{Q_B}{\tau_n} = I_{BA} \tag{9-81}$$

The second equation is obtained since the ratio Q_B/τ_n corresponds to the onset of saturation. At $t = 0-$, the excess charge is obtained by setting the time-dependent terms in Eq. (9-79) to zero and using Eq. (9-81)

$$Q_{BX} = \tau_s(I_{B1} - I_{BA}) \tag{9-82}$$

This is the initial condition for Eq. (9-79), and the solution for Q_{BX} is given by

$$Q_{BX} = \tau_s(I_{B1} - I_{B2})e^{-t/\tau_s} + \tau_s(I_{B2} - I_{BA}) \tag{9-83}$$

At $t = t_s$, all the excess minority carriers are removed, and $Q_{BX} = 0$. Thus, we find

$$t_s = \tau_s \ln \frac{I_{B1} - I_{B2}}{I_{BA} - I_{B2}} \tag{9-84}$$

τ_s is related to the minority-carrier lifetime in the base and is usually given in the specifications of a switching transistor.

9-10 BREAKDOWN VOLTAGES

The basic limitation of the maximum voltage in a transistor is the same as that in a p-n junction diode, i.e., the avalanche or Zener breakdown. However, the voltage breakdown depends not only on the nature of the junction involved but also on the external circuit arrangement. For example, the breakdown voltage for the common-emitter circuit differs from that of the common-base circuit, and the external base impedance influences the maximum operating voltage before breakdown. In the following paragraphs, we consider the breakdown voltages under different operating conditions. The mechanism of *punch-through* breakdown, unique to transistors, is also discussed.

Common-Base Configuration

The maximum allowable reverse-bias voltage between the collector and the base terminals of the transistor with the emitter lead open-circuited is designated by BV_{CBO}. This voltage is determined by the avalanche breakdown voltage of the collector-base junction. The avalanche multiplication factor

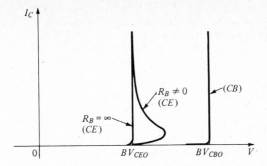

Figure 9-19 Breakdown voltage for common-emitter and common-base circuits.

given in Eq. (4-109) can be expressed empirically by

$$M = \frac{1}{1 - (V_{CB}/BV_{CBO})^n} \tag{9-85}$$

for the common-base circuit. The collector current-voltage characteristic for $I_E = 0$ in the breakdown region is sketched in Fig. 9-19. The abrupt increase of I_C at BV_{CBO} is the distinct feature of the breakdown phenomenon. The collector current is related to the emitter current by

$$\alpha^* = M\alpha \tag{9-86}$$

indicating that the effective current gain is increased by the M factor. The breakdown is satisfied when M approaches infinity.

Common-Emitter Configuration

Since $h_{FE} = \alpha/(1 - \alpha)$, it follows that the common-emitter current gain incorporating the avalanche effect h_{FE}^* is

$$h_{FE}^* = \frac{\alpha^*}{1 - \alpha^*} = \frac{M\alpha}{1 - M\alpha} \tag{9-87}$$

Accordingly, the new current gain becomes infinite when the condition $M\alpha = 1$ is reached. Since α is very close to unity, the common-emitter breakdown condition is satisfied when M is not much greater than unity. The breakdown voltage is denoted BV_{CEO} when the base is open-circuited. By setting $V_{CB} = BV_{CEO}$ in Eq. (9-85) and equating M to $1/\alpha$, we can solve for BV_{CEO} and obtain

$$BV_{CEO} = BV_{CBO}\sqrt[n]{1 - \alpha} = BV_{CBO}(h_{FE})^{-1/n} \tag{9-88}$$

With an n value between 2 and 4 in silicon and a large h_{FE}, the common-emitter breakdown voltage BV_{CEO} could be much lower than the common-base breakdown voltage.

The breakdown voltage between the collector and emitter can be made

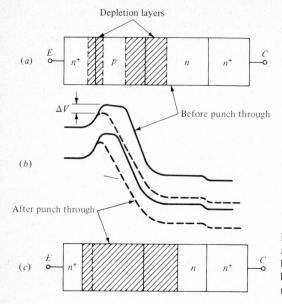

Figure 9-20 Punch-through in an n-p-n transistor: (a) space-charge regions before punch-through, (b) energy-band diagrams, and (c) space-charge regions after punch-through.

greater than BV_{CEO} by returning the base to the emitter through a base resistor R_B. If R_B is very large, the base is essentially an open circuit and the breakdown voltage approaches BV_{CEO}. If R_B is small and close to zero, the base is essentially short-circuited to emitter and the breakdown voltage approaches BV_{CBO}. For a finite R_B, the breakdown voltage is somewhere in between. The common-emitter breakdown characteristics are also shown in Fig. 9-19.

Punch-through Breakdown

A transistor is *punched through* if the space-charge region of the collector junction reaches the emitter junction before the avalanche can take place. The space charge and energy-band distributions of a n^+-p-n transistor are shown in Fig. 9-20. Under this condition, the emitter and collector regions are joined as a continuous space-charge region so that the potential barrier at the emitter is lowered by ΔV by the punch-through collector voltage. As a result, a large emitter current can flow in the structure and the breakdown occurs. Usually, the V-I curve for the punch-through breakdown is not as abrupt as the avalanche breakdown. The punch-through voltage is obtained by using the plot in Fig. 4-5 with x_B as the space-charge-layer width.

9-11 The p-n-p-n DIODES

A p-n-p-n device consists of four layers of semiconductor doped alternately with p- and n-type impurities. A p-n-p-n device having external connections only to the end regions is called a *p-n-p-n diode*. Figure 9-21 shows such a diode together with its voltage-current characteristic. The terminal connected to the p_1 region is called the *anode*, and the terminal connected to the n_2 region is called the *cathode*. A typical *V-I* characteristic for a silicon *p-n-p-n* diode is shown in Fig. 9-21*b*. When an external voltage is applied to make the anode positive with respect to the cathode, the *V-I* characteristic exhibits four distinct regions:

a-b At low voltages, junctions J_1 and J_3 are forward-biased and junction J_2 is reverse-biased. The external impressed voltage appears principally across the reverse-biased junction. The device behaves essentially like a reverse-biased *p-n* junction. Therefore, this is called the OFF or *high-impedance* region.

b-c As the voltage *V* is increased, the current increases slowly up to voltage called the *breakover voltage* V_{BO}, where the current increases abruptly.

c-d The device then traverses a region of differential negative resistance; i.e., the current increases as the voltage decreases sharply.

d-e The ON, or *low-impedance*, region. In this region J_2 is forward-biased, so the voltage across the device is essentially that of a single forward-biased *p-n* junction, i.e., about 0.7 V. If the current flowing through the diode is reduced, the diode will remain in its ON state until $I = I_H$. This current and the corresponding voltage V_H are called the *holding current* and *voltage*, respectively. When the current is reduced below this value, the diode switches back to its high-impedance state.

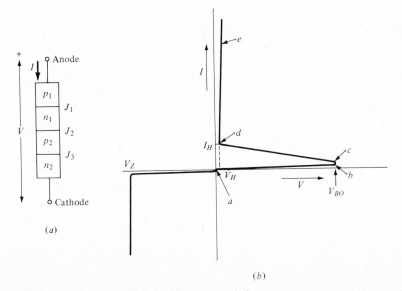

(*a*)

(*b*)

Figure 9-21 A *p-n-p-n* diode and its characteristic.

If the diode is biased negatively, the current remains very small because junctions J_1 and J_3 are reverse-biased. When the device is reverse-biased to the point that both junctions J_1 and J_3 break down, it behaves like an avalanche or Zener diode with $V_Z = V_{Z_1} + V_{Z_3}$. This region is of no particular interest.

In Fig. 9-22a, the p-n-p-n diode has been split into two parts to demonstrate that the device can be considered as two transistors connected back to back. One transistor is a p-n-p type, whereas the second is an n-p-n type. The n_1 region is the base of the p-n-p transistor and the collector of the n-p-n transistor. The p_2 region is the base of the n-p-n transistor and the collector of the p-n-p transistor. The junction J_2 is the collector junction shared by both transistors. In Fig. 9-22b the arrangement in Fig. 9-22a has been redrawn using transistor-circuit symbols, and a voltage source has been applied through a resistor across the switch, giving rise to a current I. Collector currents I_{C1} and I_{C2} for transistors Q_1 and Q_2 are indicated. In the active region the collector current of a transistor is given by Eq. (9-9). When we apply Eq. (9-9) to Q_1 and Q_2,

$$I_{C1} = -\alpha_1 I + I_{CO1} \tag{9-89}$$

$$I_{C2} = \alpha_2 I + I_{CO2} \tag{9-90}$$

Setting equal to zero the sum of the currents into transistor Q_1, we have

$$I + I_{C1} - I_{C2} = 0 \tag{9-91}$$

Combining Eqs. (9-89) to (9-91), we find

$$I = \frac{I_{CO2} - I_{CO1}}{1 - \alpha_1 - \alpha_2} = \frac{I_{CO}}{1 - \alpha_1 - \alpha_2} \tag{9-92}$$

where $I_{CO} = I_{CO2} - I_{CO1}$ is the total reverse saturation current of junction J_2.

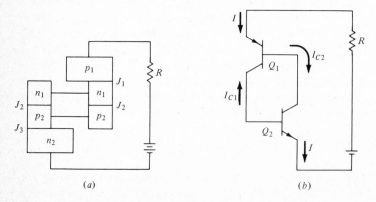

(a)   (b)

Figure 9-22 Equivalent circuit of a p-n-p-n diode.

We observe that as the sum $\alpha_1 + \alpha_2$ approaches unity, the current I increases without limit; i.e., the device breaks over.

In a transistor, the magnitude of α increases with collector voltage and with collector current at low current levels, the latter being shown in Fig. 9-6. The effect of collector voltage on α is particularly pronounced as the voltage is increased to the avalanche voltage.

Accordingly, as the voltage across the four-layer device is increased, the collector current and the magnitude of alphas in the two equivalent transistors increase. When the sum of α_1 and α_2 approaches unity, the current I increases sharply. The increase of current I, in turn, further increases the alphas. When the sum of the avalanche-enhanced alphas equals unity, $\alpha_1 + \alpha_2 = 1$, breakover takes place. At this point, the current is large, and α_1 and α_2 might be expected individually to attain values in the neighborhood of unity. If such were the case, Eq. (9-92) indicates that the current might be expected to reverse. What provides stability to the ON state of the switch is that in the ON state the center junction becomes forward-biased. Now both transistors are in saturation, and the current gain α is again small. Thus, stability is attained by virtue of the fact that the transistors enter saturation to the extent necessary to maintain the condition $\alpha_1 + \alpha_2 = 1$. The current I increases to the point at which it is limited primarily by the external circuitry. This region is the ON region. In the ON region, all three junctions in the diode are forward-biased, and normal transistor action is no longer effective. The voltage across the device is very nearly equal to the algebraic sum of these three saturation junction voltages. The magnitude of the ON voltage is of the order of 1.0 V because the voltage drop across the center junction J_2 is in the direction opposite that of the voltages across junctions J_1 and J_3.

In order to keep the diode in its low-impedance state, the condition $\alpha_1 + \alpha_2 = 1$ must be satisfied. The holding current corresponds to the minimum current at which $\alpha_1 + \alpha_2 = 1$ is satisfied. Further reduction of current violates the foregoing condition, and the device switches back to its high-impedance region.

In practical operation, when a positive voltage is applied to the anode to turn the device from the OFF to ON state, the junction capacitance across the junction J_2 is charged. The charging current flows through the emitter junctions of the two transistors. If the rate of change of the applied voltage with time is high, the charging current may be large enough to raise the alphas of the two transistors sufficiently to turn on the diode. This is called the *rate effect*. It may reduce the breakover voltage to half or less of its static value.

9-12 SILICON-CONTROLLED RECTIFIERS

In the p-n-p-n-diode operation, it is necessary to increase the external applied voltage so that the junction J_2 is in the avalanche multiplication region. The breakover voltage of the diode is fixed in fabrication. The shape of the voltage-current characteristic can be controlled if a third terminal, called the *gate*, is provided. The three-terminal device is commonly known as a *silicon-controlled rectifier* (SCR) or *thyristor*, and one of its possible structures is shown in Fig. 9-23. Using the standard planar technology, the p_1 and p_2 regions are diffused simultaneously, and the subsequent n_2 diffusion completes the four-layer device. The p_1-n_1-p_2 transistor is known as the *lateral transistor*, and its operation will be described in the next chapter. The current flowing through the gate terminal can now increase α_2 independently of V and I. The effect of gate current on the V-I characteristic is shown in Fig. 9-24. A family of V-I characteristics is obtained with I_G maintained constant at various values.

The most obvious effects of increasing I_G are to increase the OFF current and to decrease both the breakover voltage and the holding current. These changes can be explained qualitatively in terms of the two-transistor equivalent circuit. In the OFF state the device behaves essentially like a normal n-p-n transistor with the p-n-p transistor acting like an emitter-follower with a very small $h_{fe} \approx 0.1$. Increasing the gate current increases the collector current of the n-p-n transistor in the usual way. The increase of the collector current, in turn, increases the anode current. This larger current causes an increase in the alphas (Fig. 9-6). The condition $\alpha_1 + \alpha_2 = 1$ is therefore reached at lower values of the avalanche multiplication factor, and the breakover voltage is reduced. In the ON state, the flow of gate current

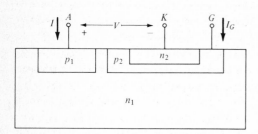

Figure 9-23 A low-power SCR structure.

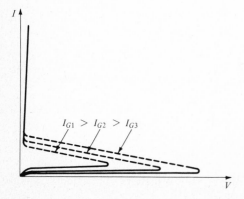

Figure 9-24 Current-voltage characteristics of a silicon-controlled rectifier.

again increases the alphas. Thus, the current I can fall to lower values at which the holding-point condition is still satisfied.

The voltage at which the device goes from an OFF to an ON state is controlled by a small gate signal. In high-power SCRs once the device is in the ON state, the gate circuit has little effect on the operation. However, in low-power SCRs the gate circuit can be used to turn the device both on and off. A negative current is usually required to turn the device off.

SCRs are mostly large-area devices because they need to handle a large amount of current. Therefore, lateral gate-current flow gives rise to a substantial potential drop, and the current-crowding effect (Sec. 9-6) tends to turn on the periphery of the device first. The turn-on condition is then propagated to cover the whole structure. During the turn-on transient, all the anode current passes through the small peripheral area momentarily, and the high-current density can burn out the SCR. For this reason, the interdigitated structure has been used to reduce the lateral effect. A small inductor connected in series with the anode can also be used to reduce the current due to the turn-on transient.

9-13 BIDIRECTIONAL p-n-p-n SWITCHES [12]

The undesirable lateral effect mentioned in the preceding section can be turned into an advantage if appropriate designs are employed. Let us consider the structure shown in Fig. 9-25, where the cathode (also called emitter) is shorted to the p_2 region. As a result of the lateral voltage drop, the central portion of the gate-to-cathode junction has the highest forward bias and thus the highest current density. At the same time, a portion of the anode current goes directly to the cathode, and this current component does not contribute to the current gain of the n_1-p_2-n_2 transistor. By adjusting the geometrical arrangement of the shorted and open p_2 regions, the variation of the overall current gain with the cathode current can be controlled. In addition, the rate effect can be minimized because of the shunting effect of the shorted cathode.

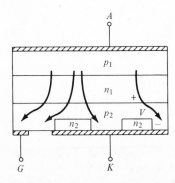

Figure 9-25 Shorted-emitter (cathode) p-n-p-n structure.

By making use of the concept of a shorted cathode, it is possible to fabricate a bidirectional diode switch, known as *diac*, which is shown in Fig. 9-26. Note that both the cathode and anode are short-circuited and the device is symmetrical. Applying a positive voltage at terminal A with terminal K grounded, we find that the junction J_3 is reverse-biased so that the n_3 region does not contribute to the functioning of the device. Therefore, the p_2-n_2-p_1-n_1 layers constitute a four-layer diode yielding the forward portion of the I-V characteristic shown in Fig. 9-26b. If a negative voltage is applied to the terminal A, current will conduct in the opposite direction and J_4 is reverse-biased. Therefore, the n_3-p_2-n_2-p_1 layers form the reverse p-n-p-n device leading to the negative I-V characteristics.

Using the same principle, we can design a bidirectional triode switch, known as a *triac*, shown in Fig. 9-27. In this structure, the gate current is employed to effect the turning on of the device.

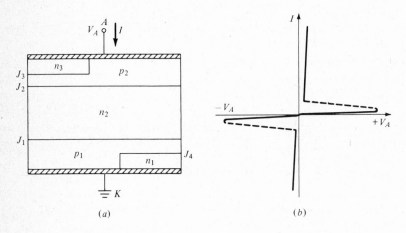

(a)

(b)

Figure 9-26 Bidirectional diode switch: (*a*) structure and (*b*) *V-I* curve.

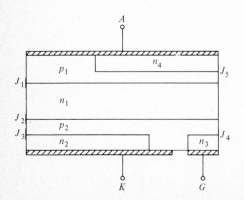

Figure 9-27 Bidirectional triode switch (triac) structure.

PROBLEMS

9-1 (*a*) Sketch the energy-band diagram for a *p-n-p* transistor at equilibrium and under the normal active mode of operation.

(*b*) Sketch a schematic diagram to represent the transistor and indicate all current components.

(*c*) Rewrite Eqs. (9-1) to (9-3) for this transistor.

9-2 Carry out the derivation of Eqs. (9-19) and (9-21). Plot the carrier distribution for $x_B/L_n = 0.5$ using both equations. What conclusion can you draw from your plot?

9-3 Consider a silicon *n-p-n* transistor with these parameters: $x_B = 2\,\mu\text{m}$, $N_a = 5 \times 10^{16}\,\text{cm}^{-3}$ in a uniformly doped base, $\tau_n = 1\,\mu\text{s}$, and $A = 0.01\,\text{cm}^2$. If the collector is reverse-biased and $I_{nE} = 1\,\text{mA}$, calculate the excess-electron density at the base side of the emitter junction, the emitter-junction voltage, and the base transport factor.

9-4 In the transistor of Prob. 9-3, assume that the emitter is doped with $10^{18}\,\text{cm}^{-3}$, $x_E = 0.5\,\mu\text{m}$, $\tau_{pE} = 10\,\text{ns}$, and $\tau_0 = 0.1\,\mu\text{s}$ in the emitter space-charge region. Calculate the emitter efficiency and h_{FE} at $I_nE = 1\,\text{mA}$.

9-5 An *n-p-n* transistor has the following specifications: emitter area = 1 square mil, base area = 10 square mils, emitter width = 2 μm, base width = 1 μm, emitter sheet resistivity = 2 $\Omega/\square$, base sheet resistivity = 200 $\Omega/\square$, collector resistivity = 0.3 Ω-cm, hole lifetime in emitter = 1 ns, electron lifetime in base = 100 ns. It is assumed that the emitter recombination current is constant and equal to 1 μA. Step junctions and uniform doping are also assumed. Calculate h_{FE} for $I_E = 10\,\mu\text{A}$, 100 μA, 1 mA, 10 mA, 100 mA, and 1 A. Plot on semilog axes. What is the controlling factor in the midcurrent range?

9-6 (*a*) Prove that Eqs. (9-35) and (9-37) remain valid for any value of x_B/L_n provided that the coefficients are changed to

$$a_{11} = -qAn_i^2\left[\frac{D_n}{N_aL_n}\left(\coth\frac{x_B}{L_n}\right) + \frac{D_p}{N_{dE}x_E}\right]$$

$$a_{12} = a_{21} = \frac{qAD_nn_i^2}{N_aL_n}\operatorname{csch}\frac{x_B}{L_n}$$

$$a_{22} = -qAn_i^2\left[\frac{D_n}{N_aL_n}\left(\coth\frac{x_B}{L_n}\right) + \frac{D_p}{N_{dc}L_{pc}}\right]$$

(*b*) Show that if $x_B/L_n \ll 1$, the expressions in (*a*) reduce to Eqs. (9-35) and (9-37).

9-7 (*a*) Ignoring the space-charge recombination currents, show that the exact expression for the common-emitter output characteristics of a transistor is

$$-V_{CE} = V_T\ln\frac{-I_{CO} + \alpha_NI_B - I_C(1-\alpha_N)}{-I_{EO} + I_B + I_C(1-\alpha_I)} + V_T\ln\frac{\alpha_I}{\alpha_N}$$

Note: First solve explicitly for the junction voltages in terms of the currents.

(b) If $I_B \gg I_{EO}$ and $I_B \gg I_{CO}/\alpha_N$, show that the foregoing equation reduces to

$$V_{CE} = V_T \ln \frac{1/\alpha_I + I_C/I_B h_{FEI}}{1 - I_C/I_B h_{FEN}}$$

9-8 Show that the emitter voltage-ampere characteristic of a transistor in the active region is given by

$$I_E \approx \frac{I_{EO}}{1 - \alpha_N \alpha_I} e^{V_E/V_T} + \frac{qAn_i W_E}{2\tau_o} e^{V_E/2V_T}$$

9-9 The Gummel number can be calculated from the collector-current-vs.-V_E plot if we assume $I_C = I_n$ in Eq. (9-46). Find the Gummel number for the transistor in Fig. 9-5. Use $D_n = 35\ \text{cm}^2/\text{s}$, $A = 0.1\ \text{cm}^2$, and $n_i = 1.5 \times 10^{10}\ \text{cm}^{-3}$.

9-10 (a) Show that Eq. (9-48) reduces to Eq. (9-26) for a uniformly doped base.

(b) Derive an expression of the base transport factor for a base with exponential distribution; that is, $N_a = N_o e^{-ax/x_B}$.

9-11 Derive the electron distribution in the base for the transistor of Prob. 9-10b under the normal active mode of operation. Plot the electron concentration for $a = 1$ and 10. Explain the physical meaning of the difference between these two cases.

9-12 The effect of dc base spreading resistance on the collector current can be expressed by $I_C = I_o \exp[(V_E - I_B r_{bb'})/V_T]$. Estimate $r_{bb'}$ by using this equation and the data shown in Fig. 9-5.

9-13 Obtain an approximate expression for the base resistance of the stripe-transistor structure shown in Fig. P9-13.

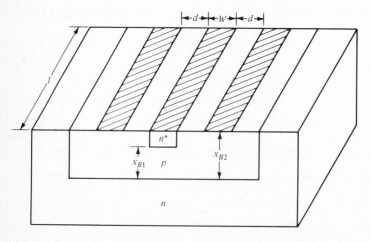

Figure P9-13.

[Margin handwritten notes:]

I_c at $v_E = 0$

$I_m \approx I_c$

Eqn (9-46) Find I_m

$G = \int_0^{x_B} N_a\, dx$

$= q A P_n T_i^2$

$\frac{I_n}{I_{v_E = 0}}$

(b)

$\int_x^{x_B} N_a\, dx'$

$= N_o \int_x^{x_B} e^{-ay/x_B}\, dy$

$B = 1 - \frac{x_B^2}{a^2 T_n} \left[1 \cdots \right.$

$\frac{x_B}{a}\left(1 - e^{-a}\right)\Big]$

9-14 (*a*) Derive an expression of the base transit time for the transistor with uniformly doped base. Assume $x_B/L_n \ll 1$.

(*b*) Repeat (*a*) for the base doping distribution $N_a = N_o \exp(-ax/x_B)$.

9-15 Consider the transistor with an impurity profile shown in Fig. 9-9. Let the emitter area and base area be the same (10 square mils) and $r_{sc} = 0$. The emitter current is 2 mA, and the collector reverse bias is 10 V. Calculate the cutoff frequency ω_α at 300 K. *Note:* Use $\tau_B = x_B^2/4D_n$ for a double-diffused transistor.

9-16 If the actual *CB* current gain is given by $\alpha = a_o e^{jm\omega/\omega_\alpha}/(1 + j\omega/\omega_\alpha)$, show that $\omega_T \approx \omega_\alpha/(1 + \alpha_o m)$ where ω_T is the frequency at which the common-emitter current gain is unity.

9-17 (*a*) Obtain an expression of I_o/I_i in Fig. 9-15 with output short-circuited.

(*b*) Find ω_β which corresponds to the 3-dB drop in the magnitude of I_o/I_i.

(*c*) Derive Eq. (9-71).

9-18 Estimate the hybrid-pi parameters for the transistor of Fig. 9-9 at $I_c = 2$ mA, $V_c = 10$ V and $h_{FE} = 50$. Let $A = 10$ square mils and assume abrupt junctions.

9-19 The silicon transistor shown in the circuit (Fig. P9-19) has $h_{FE} = 50$ and $\tau_s = 20$ ns. Estimate the storage time. Assume $V_{BE}(\text{sat}) = 0.8$ V.

9-20 Find BC_{CBO} and BV_{CEO} for the transistor of Prob. 9-5. *Note:* Use curves in Chap. 4 to obtain the avalanche voltage and compare it with the punch-through voltage.

9-21 Show that the punch-through voltage of a planar epitaxial double-diffused transistor is given by

$$BV = \frac{qG}{K_s \epsilon_o}\left(x_B + \frac{G}{2N_{dC}}\right)$$

where G is the Gummel number.

9-22 Derive an expression of the anode current I_A as a function of the gate current by using the two-transistor analog in an SCR.

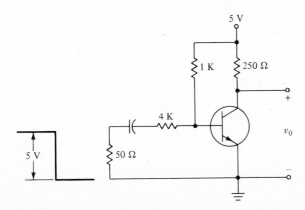

Figure P9-19.

9-23 A small-area SCR can be turned off by a negative gate current. The turnoff gain is defined as I_A/I_G, where I_A is the anode ON current, and I_G is the minimum gate current that turns off the device.

(a) Derive an expression for the turnoff gain.

(b) Specify the alphas that give rise to a high turnoff gain.

REFERENCES

1. R. M. Warner and J. N. Fordemwalt, "Integrated Circuits," p. 105, McGraw-Hill, New York, 1965.
2. J. J. Ebers and J. L. Moll, Large-Signal Behavior of Junction Transistors, *Proc. IRE*, **42**: 1761 (1954).
3. H. K. Gummel, Measurement of the Number of Impurities in the Base Layer of a Transistor, *Proc. IRE*, **49**: 834 (1961).
4. R. M. Burger and R. P. Donovan (eds.), "Fundamentals of Silicon Integrated Device Technology," vol. II, p. 103, Prentice-Hall, Englewood Cliffs, N.J., 1968.
5. R. M. Warner and J. N. Fordemwalt, pp. 108–109 in Ref. 1.
6. **J. L. Moll and I. M. Ross, The Dependence of Transistor Parameters on the Distribution of Base Layer Resistivity,** *Proc. IRE*, **44**: 72 (1956).
7. **D. E. Thomas and J. L. Moll, Junction Transistor Short-Circuit Current Gain and Phase Determination,** *Proc. IRE*, **46**: 1177 (1958).
8. C. T. Kirk, A Theory of Transistor Cutoff Frequency (f_T) Fall-off at High Current Density, *IEEE Trans. Electron. Devices*, **ED-9**: 164 (1962).
9. H. C. Poon, H. K. Gummel, and D. L. Scharfetter, High Injection in Epitaxial Transistors, *IEEE Trans. Electron. Devices*, **ED-16**: 455 (1969).
10. J. L. Moll, Large-Signal Transient Response of Junction Transistors, *Proc. IRE*, **42**: 1772 (1954).
11. R. Beaufoy and J. J. Sparks, The Junction Transistor as a Charge-controlled Device, *ATE J. (GB)*, **13**: 310 (October 1957).
12. F. E. Gentry, R. I. Scauce, and J. K. Flowers, Bidirectional Triode p-n-p-n Switches, *Proc. IEEE*, **53**: 355 (1965).

ADDITIONAL READINGS

Gentry, F. E., et al.: "Semiconductor Controlled Rectifiers," Prentice-Hall, Englewood Cliffs, N.J., 1964.
Gray, P. E., D. DeWitt, A. R. Boothroyd, and J. F. Gibbons: "Physical Electronics and Circuit Models of Transistors," SEEC Series, vol. II, Wiley, New York, 1964.
Lindmayer, J., and C. Y. Wrigley: "Fundamentals of Semiconductor Devices," Van Nostrand, Princeton, N.J., 1965.
Sze, S. M.: "Physics of Semiconductor Devices," chap. 6, Wiley, New York, 1969.

TEN

INTEGRATED DEVICES

Modern electronic circuits and systems are built on the foundation of discrete semiconductor devices and integrated circuits. More than any other electronic component, the integrated circuit (IC) is responsible for such low-cost electronic products as the hand-held calculator, microprocessor, and digital wristwatch. An IC consists of both active and passive elements formed on a single-crystal silicon substrate. Each circuit element is electrically isolated from the others, and interconnections are provided by a metallization pattern. This kind of IC is known as *monolithic* (meaning "single stone") IC. Only a few years ago, there was one standard fabrication process for bipolar circuits and another for MOS circuits. Now there are a variety of processes for both types.

This chapter begins with a description of the isolation technology for bipolar ICs. New approaches to isolating individual components are presented, and structural designs for overcoming or utilizing the limitations imposed by the isolation technology are then given. These structures include the multiple-emitter transistor, lateral *p-n-p* transistor, and integrated injection logic. The last device eliminates the isolation of components in the same logic gate. As for the MOS IC, there is no need for isolation because MOS transistors are self-isolated, and it is sufficient to have a metallization pattern joining the devices together to perform the desired electronic function. Our study of MOS circuits begins with the inverter using only *n*-channel MOS transistors. The complementary MOS inverter (CMOS), using an *n*-channel and *p*-channel transistor pair, is then described. This is followed by a discussion of the double-diffused MOS transistor (D-MOS) and the V-groove MOS transistor (V-MOS). In the last section, two types of nonvolatile MOS memory devices, the MNOS and FAMOS, are discussed.

10-1 ISOLATION TECHNOLOGY FOR BIPOLAR INTEGRATED DEVICES

In an IC, devices on the same substrate are usually isolated from one another so that there is no current conduction between them. Such an isolation may be provided by (1) a reverse-biased p-n junction or (2) isolating dielectric materials. There are a number of methods reported in the literature, all of which use either the junction or dielectric technique or a combination of both.

Junction Isolation

Basically, junction isolation uses an impurity diffusion to produce n-type islands surrounded by p-type materials. The standard industrial technology is the *epitaxial-diffused process* (EDP). Fabrication using EDP begins with a p-type silicon substrate upon which an n-type epitaxial layer, typically 3 to 15 μm thick, is grown (Fig. 10-1a). By means of the oxidation and photoresist steps described in Sec. 3-2, areas of n-type islands are defined, and acceptor impurity atoms are diffused into regions between the islands. The acceptor concentration in the newly diffused regions, known as the *isolation regions*, must be higher than the donor concentration in the epitaxial layer. In addition, acceptors must diffuse through the epitaxial layer so that the isolation regions join with the p-type substrate. After the isolation diffusion, an n island is electrically isolated, as shown in Fig. 10-1b and is ready to have components built in it. For example, let us consider a transistor fabricated in the island shown. The n-type island is used as the collector region, and two diffusion steps, a p diffusion followed by an n^+ diffusion, are made to form respectively the base and emitter regions. During the emitter diffusion, additional donor impurities are diffused directly into the n collector to facilitate ohmic contacts (Sec. 5-6) to the collector. The final structure after the deposition of metal contacts is shown in Fig. 10-1c.

A major disadvantage of the integrated transistor shown in Fig. 10-1c is its high collector series resistance since the collector current is horizontally routed through the lightly doped collector. This difficulty is overcome by making an n^+ diffusion before growing the epitaxial layer, as illustrated in Fig. 10-1d. The n^+ layer, known as the *buried layer*, provides a low-resistance path from the collector contact to the active portion of the transistor (Fig. 10-1e). Presently, the buried layer is used in all bipolar ICs fabricated by the EDP process. The thickness of the epitaxial layer is usually less than 10 μm.

During the isolation diffusion, acceptor atoms diffuse in both the vertical and lateral directions, the lateral diffusion being about 75 to 80 percent of the

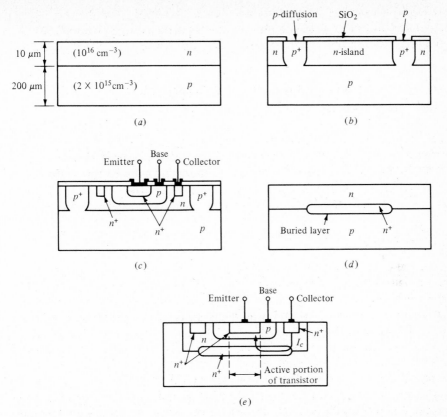

Figure 10-1 Fabrication of IC using epitaxial-diffused process: (*a*) *p* substrate with *n* epitaxial layer, (*b*) *p*-type isolation diffusion, (*c*) completed transistor, (*d*) buried n^+ layer, and (*e*) integrated transistor with buried layer.

vertical penetration. For this reason, the width of the isolation region is approximately twice the epitaxial-layer thickness in order to have the acceptors penetrate through the epitaxial layer. Since the isolation regions do not play an active role electronically, they should be minimized.

Processing complexity can be reduced by using the collector-diffusion-isolation (CDI) structure shown in Fig. 10-2. Fabrication starts with an n^+

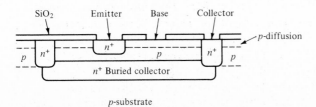

Figure 10-2 Collector-diffusion isolation. *(After Murphy et al. [1].)*

diffusion to form a buried layer in a p-type substrate. A p-type layer, rather than an n-type layer, is epitaxially grown over the surface and is followed by a p diffusion over the entire wafer. The p diffusion is needed to avoid an n-type surface inversion layer induced by positive charge in the oxide (Chap. 8) as well as providing a better ohmic contact to the base (Sec. 5-6). Subsequently, n^+-type collector contacts are selectively diffused through the p-type epitaxial layer to join with the buried layer, which serves as the collector. This diffusion is carried out around the periphery of every buried collector so that the p-type epitaxial layer above the n^+ buried collector is completely isolated from the substrate. An n^+ emitter diffusion is now performed selectively, as shown, and contact windows are etched in the oxide to provide metallization connections. In using the CDI, the p-type epitaxial layer is taken as the base region to eliminate the base-diffusion step employed in the EDP, thus simplifying the processing steps. Further simplification can be obtained using the *base-diffusion-isolation* structure shown in Fig. 10-3. Isolation is achieved by applying a large negative voltage to the p^+ isolation stripe so that the depletion layer *reaches through* the substrate. This depletion layer becomes part of the surrounding isolation. The processing steps necessary to form the BDI device are left as a problem. At the present, the CDI and BDI processes are seldom used because the resulting ICs are inferior to those fabricated by the EDP.

Dielectric Isolation

An alternative approach in bipolar-integrated-device fabrication is dielectric isolation. Usually, this process begins with an n^+ diffusion on the entire surface of an n substrate. It is followed by an etching step to obtain channels, as shown in Fig. 10-4a. An oxide layer is then grown to cover the wafer, and polycrystalline silicon is deposited (Fig. 10-4b). The structure is now turned over and the n-silicon substrate is lapped off to reach the oxide on the etched channel, as depicted in Fig. 10-4c. Thus, isolated islands are obtained. Additional diffusions identical to those used in the EDP may be made on the islands. The advantages of this method include low parasitic capacitances and elimination of substrate bias. The disadvantages are extra processing steps and the difficulty of lapping the silicon precisely parallel to the etched

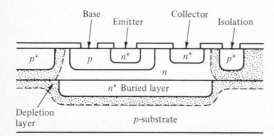

Depletion layer

p-substrate

Figure 10-3 Base-diffusion isolation. *(After Murphy et al. [2].)*

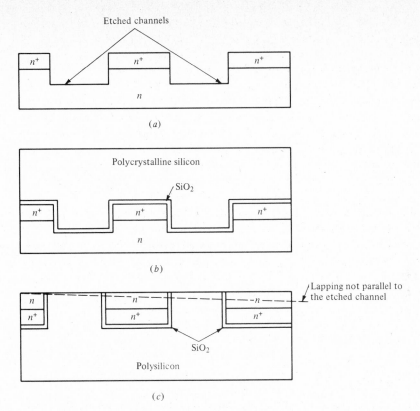

Figure 10-4 Dielectric isolation: (*a*) etching of channels, (*b*) deposition of polysilicon, and (*c*) lapping off of substrate to form islands.

channels. The dotted line in Fig. 10-4*c* indicates what could happen if the lapping plane formed a small angle to the etched channels. Because of its high cost, dielectric isolation is not widely used at present.

Junction-Dielectric Techniques

A combination of junction and dielectric isolation is frequently used to form isolation islands. Among the available processes, the isoplanar and V-groove methods have demonstrated better performance than the EDP and are described here briefly.

The isoplanar process [3] The isoplanar method, developed by Fairchild Semiconductor, begins with the silicon wafer shown in Fig. 10-1*d*, upon which a silicon nitride layer is deposited. After photoresist, an etching step removes the *n*-type epitaxial layer in areas between isolated islands. Because the silicon nitride inhibits oxide growth, a subsequent thermal oxidation fills only

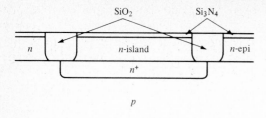

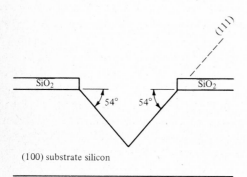

Figure 10-5 Isoplanar isolation.

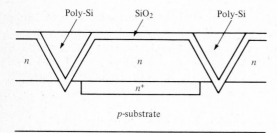

Figure 10-6 Preferential V-groove etching of (100) substrate.

Figure 10-7 The V-groove process.

the etched areas, as indicated in Fig. 10-5. Thus, the isolated island has a p-n junction isolation on the bottom and silicon dioxide isolation on the sides. The disadvantages of this process include a thin epitaxial layer and long oxidation time.

V-groove process [4] When silicon is chemically etched, the etching rate along the (111) crystal plane is about 30 times faster than that along the (100) plane. If a wafer having a (100) surface orientation is used, the etched silicon surface develops a V-shaped groove, making a 54° angle with the surface, as shown in Fig. 10-6. The etching process stops when the two (111) planes intersect at a depth d below the surface. The depth of the groove is determined by the width of the opening at the surface because of the fixed angle in the preferential etching. If we start with the silicon wafer of Fig. 10-1d, the depth of isolation grooves should be slightly greater than the thickness of the epitaxial layer. After the V-groove etching, an oxide is grown on the entire surface, and polycrystalline silicon is deposited to fill the groove. The structure of an isolated island is illustrated in Fig. 10-7, in which additional steps can be made to produce integrated devices.

10-2 THE INTEGRATED BIPOLAR TRANSISTOR

An integrated transistor fabricated by the EDP is redrawn in Fig. 10-8, which includes the top view of the transistor to show its standard layout. The EDP without a buried layer needs five masks for the following functions: isolation, base diffusion, emitter and n^+ diffusion, contact-window opening, and metallization pattern etching. The dotted line in Fig. 10-8b indicates the lateral diffusion of the isolation step so that the actual isolation island is smaller than the isolation oxide mask. Lateral diffusions of other steps are not significant and are neglected here. Typically, the width of aluminum stripes is 5 to 8 μm, and the clearance between lines is about the same. In general, the theory developed in Chap. 9 is applicable to the integrated transistor, whose impurity profile is essentially the same as the double-diffused transistor. The differences are mostly parasitic effects introduced by the isolation.

When the substrate is included, the integrated transistor has a four-layer structure. The n-p-n section is the desired transistor, and the p-n-p section is a parasitic element. It was shown in Sec. 9-11 that the p-n-p-n structure has a built-in positive feedback action such that it latches onto an ON position (all junctions become forward-biased) if the emitter and substrate-collector junctions are allowed to be forward-biased simultaneously. Therefore, a reverse

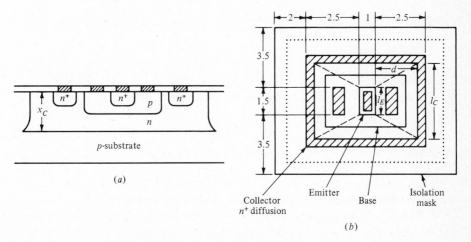

Figure 10-8 Layout of the integrated transistor. All dimensions are in mils, and all clearances are 0.5 mil. Dashed lines denote boundaries of trapezoids for r_{sc} calculation.

bias must be applied between the collector and substrate. This is accomplished by connecting the substrate to the most negative dc voltage in the electronic circuit. The second parasitic element is the isolation-junction capacitance partly due to the sidewall junction and partly the bottom junction. In a practical integrated transistor, the bottom junction is a step junction, and the capacitance is determined by the substrate doping. The sidewall junction can be approximated by a linearly graded junction and can be obtained by using Eqs. (4-87) and (4-24).

The requirement for making all contacts on the surface of the integrated transistor forces the collector current to traverse the lightly doped collector region laterally. As a result, the collector series resistance r_{cs} in Fig. 10-1c can be quite large. A large r_{cs} decreases the cutoff frequency and introduces an ohmic drop which increases the saturation voltage. Since both effects degrade the transistor performance, it is necessary to estimate the value of r_{cs}. One of the techniques used to estimate the collector resistance is known as the *trapezoid method*. The collector region is divided into four trapezoids, shown by the dashed lines in Fig. 10-8b. The resistance of each trapezoid is approximated by

$$r = \frac{\rho_c d}{A_{\text{eff}}} \qquad (10\text{-}1)$$

where ρ_c = collector resistivity

d = distance between emitter-base junction and collector contact

A_{eff} = cross-sectional area at middle of trapezoid

Thus,

$$A_{\text{eff}} = \frac{l_E + l_C}{2} x_C \qquad (10\text{-}2)$$

where l_E = top length of trapezoid

l_C = bottom length of trapezoid

x_C = thickness of collector layer

The collector resistance is now given by the resistance of the trapezoids in parallel. The collector resistance can be reduced significantly if a buried layer is incorporated. The lateral collector current now flows through the heavily doped buried layer, which has negligible resistance. Thus r_{cs} is given by the sum of series resistances in the vertical direction under the emitter and collector contacts:

$$r_{cs} = \rho_c \left(\frac{x_C}{A_E} + \frac{x_C}{A_{CC}} \right) \qquad (10\text{-}3)$$

where A_E and A_{CC} are the emitter and collector-contact areas, respectively. The buried layer is used in practically all industrial bipolar processes in which the effect of r_{cs} becomes negligible.

The parasitic capacitance can be reduced by using the isoplanar or V-groove isolation method, and it is essentially eliminated in a dielectric isolated transistor. An additional benefit of dielectric isolation is the elimination of the parasitic p-n-p transistor. Improvement of the transistor switching speed can be realized by making use of a Schottky-diode clamp, as described in Chap. 5.

10-3 MULTIPLE-TRANSISTOR STRUCTURES

In the last section, we considered the integrated transistor, which occupies an isolation island all by itself. Using such transistors, we can build integrated circuits by making use of a metallization and etching step to provide interconnections. However, it is possible to combine two or more transistors into one isolation island in circuits in which collectors are connected electrically. For example, two transistors with a common collector are shown in Fig. 10-9 along with the equivalent circuit. Note that the base and emitter diffusions are made in two regions in the same isolation island. When both the collector and base regions are shared by a number of transistors, the *multiple-emitter transistor* is obtained, as depicted in Fig. 10-10.

By combining the collector region or the collector and base regions we achieve a reduction of the total isolation area, resulting in smaller isolation

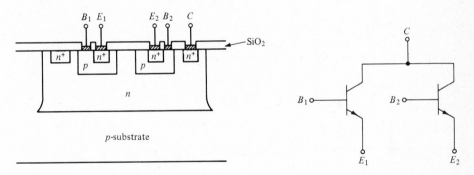

Figure 10-9 Integrated common-collector transistors and equivalent circuit.

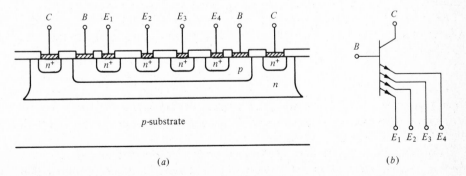

Figure 10-10 Multiple-emitter transistor: (*a*) cross section and (*b*) equivalent circuit.

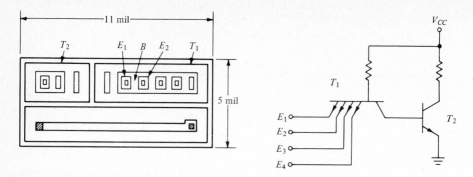

Figure 10-11 Transistor-transistor logic IC implementation and its equivalent circuit.

capacitance, simpler diffusion masks, and fewer interconnections. This design technique lowers the cost and improves the reliability of integrated circuits and should be employed whenever possible. At present, most commercial digital bipolar ICs are of the type known as *transistor-transistor logic* (TTL), which utilizes the multiple-emitter transistor as its basic building block. A simple TTL circuit is shown in Fig. 10-11 along with its IC implementation by means of the EDP. The integrated circuit shown is the top view, excluding the interconnection pattern. Note that the base diffusion is used to define the resistor stripe since the base-sheet resistivity is relatively high (typically $200 \, \Omega/\square$). If there is more than one resistor, all resistors may be formed in the same isolation island and isolation between the resistors is realized by reverse biasing the base-collector junction of this isolation island. The emitters are usually considered to be independent although interaction does take place when they are physically close. For example, if E_1 is forward-biased and E_2 is reverse-biased, E_1-B-E_2 forms an *n-p-n* transistor with E_2 as the collector. It results in current flow in E_2, which destroys the logic function of E_2. For this reason, the gain of the E_1-B-E_2 transistor should be made very small.

10-4 THE LATERAL *p-n-p* TRANSISTOR [6, 7]

In an integrated circuit, the basic structure is the double-diffused *n-p-n* transistor described in the previous sections. Occasionally, it is desirable to have both *n-p-n* and *p-n-p* transistors fabricated on the same chip. This complementary circuit can be realized by various techniques. The most common method utilizes the lateral *p-n-p* structure shown in Fig. 10-12. The *p*-type emitter and collector regions are formed simultaneously during the base diffusion of the *n-p-n* transistor and the epitaxial *n* layer provides the base region. Since the current flow is along the lateral direction, it is known as the *lateral transistor*. The advantage of the lateral structure is that it does not require any processing steps beyond those necessary for the fabrication

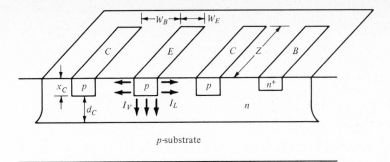

Figure 10-12 The lateral p-n-p transistor.

of the n-p-n transistor. However, the performance of the p-n-p device is significantly inferior to that of the vertical n-p-n transistor.

Let us examine the idealized lateral p-n-p transistor illustrated in Fig. 10-12. With the emitter junction forward-biased and the collector junction reverse-biased, the injected current is divided into vertical and lateral components I_V and I_L. By assuming that $W_B \ll L_p \ll d_C$, where L_p is the hole diffusion length in the n-type base, and making use of

$$I_p = -qAD_p \frac{dp}{dx}\bigg|_{x=0} \tag{10-4}$$

we derive (Prob. 10-7)

$$I_L = -2qZx_C D_p \frac{P_o}{W_B} \tag{10-5}$$

$$I_V = -qZW_E D_p \frac{P_o}{L_p} \tag{10-6}$$

where P_o is the injected-hole density at the edge of the depletion layer of the emitter junction. Since the vertically injected carriers recombine in the base, they are not collected by the collector and the current I_V constitutes the base current. Therefore, the common-emitter current gain is given by

$$h_{FE} \equiv \frac{I_C}{I_B} = \frac{I_L}{I_V} = \frac{2L_p x_C}{W_E W_B} \tag{10-7}$$

In a practical device, the current gain is between 1 and 5. In addition to reducing the current gain, the vertical injection current also degrades the frequency response of the transistor. An equivalent circuit of the lateral structure is shown in Fig. 10-13, where a shunting diode represents the effect of vertical hole injection. This shunting diode acts as a large capacitance and reduces the cutoff frequency of the composite structure. The gain-frequency characteristic shows a 3-dB/octave slope with a typical 3-dB frequency less than 500 kHz for $W_B \approx 5 \mu$m.

The vertical carrier injection can be reduced significantly if the lateral

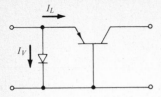

Figure 10-13 Equivalent circuit of the lateral transistor.

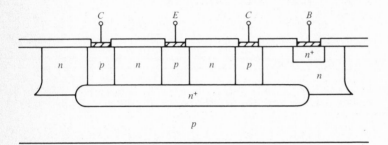

Figure 10-14 An improved lateral-transistor structure.

transistor is fabricated on a region with an n^+ buried layer. The resultant structure is shown in Fig. 10-14. Since the doping level of n^+ is higher than that of the p emitter, hole injection into the n^+ layer becomes very small. However, there will be electron injection from the n^+ layer into the p emitter, and the injection efficiency is reduced. The overall gain of this modified lateral transistor is improved to about 50, and the frequency response follows a 6-dB/octave decrease, as in the vertical double-diffused n-p-n transistor.

10-5 INTEGRATED INJECTION LOGIC [7, 8]

Integrated injection logic (I^2L), also known as *merged-transistor logic*, is a new approach to bipolar large-scale-integration chip design. The I^2L gate is laid out in such a fashion that the individual transistor-isolation step is eliminated and the large-area passive resistors are replaced by small-area lateral transistors in the form of current sources. The device's self-isolation and elimination of resistors are the key to high functional density and low power consumption.

To understand the operation of an I^2L, let us first examine the logic circuit shown in Fig. 10-15a. This is basically a direct-coupled transistor logic (DCTL) with the resistors replaced by constant current sources. The current sources may be provided by lateral p-n-p transistors, as shown in Fig. 10-15b. Implementation of the logic gate in integrated form is illustrated in Fig. 10-16. Fabrication of this device begins with an n^+ substrate on which an n-type epitaxial layer is grown. Subsequently, two diffusions are made to obtain the

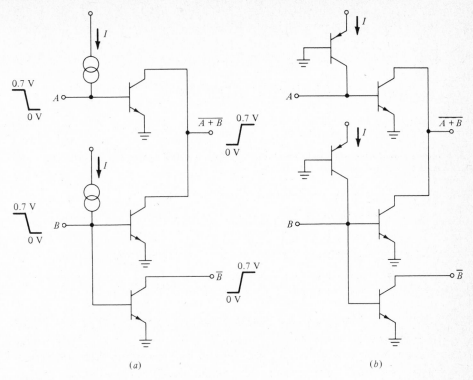

Figure 10-15 A direct-coupled transistor logic gate with (*a*) base-current sources and (*b*) *p-n-p* transistor base drive.

structure shown. The *n*-on-n^+ layer serves as the emitter and common ground plane. The *p*-diffused region serves as both the base of the vertical *n-p-n* transistor and the collector of the lateral *p-n-p* transistor. Thus, the lateral *p-n-p* transistor is integrated into the vertical *n-p-n* transistor and does not exist as a separate component. The n^+-diffused regions are the multiple collectors. In terms of the standard double-diffused transistor, the vertical *n-p-n* transistor here is operated upside down. Special care is necessary to obtain high current gain for the upside-down transistor.

The previous paragraph describes the I²L circuit in its nonisolated form. It is possible to isolate the I²L gate using junction or dielectric isolation. The isolated integrated injection logic allows all other standard bipolar and MOS circuits to be combined directly with the I²L gates on the same chip.

10-6 MOS INVERTERS

The most common applications of MOS transistors are in integrated digital logic gates and memory arrays, in which the MOS inverter is the basic circuit.

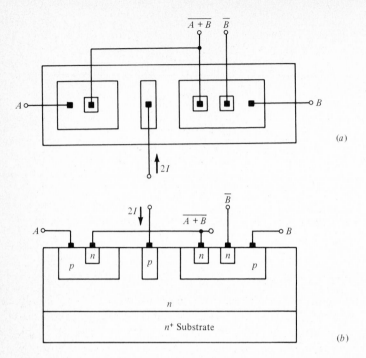

Figure 10-16 Implementation of the circuit shown in Fig. 10-15b in integrated form: the I^{2}L gate. (a) Top view and (b) cross-sectional view.

We consider several types of inverters used in IC designs. Each inverter is characterized by the arrangement of its load. By examining the load line drawn in the output characteristics, we can understand the basic operation of each circuit. Design consideration is given from two points of view: (1) circuit performance including power dissipation, maximum voltage swing, and speed and (2) fabrication factors, including isolation, device area, and cost.

Linear Resistive Load

An MOS inverter circuit with a linear resistive load is shown in Fig. 10-17, and the load line is plotted in Fig. 10-17b. This circuit has a large output-voltage swing, and its switching speed is limited by the product $R_L C_L$. Because C_L is fixed by parasitic capacitances, high speed is achieved with a smaller R_L, which leads to high power dissipation and low voltage swing (see Fig. 10-17b). The more detrimental factor of the circuit is the large area required for a linear diffused resistor, which also needs electrical isolation. At present, the MOS inverter with a linear resistor is not competitive with other inverter circuits described below. However, it is conceivable that in the future we can use an ion-implanted region or doped polycrystalline silicon embedded in the oxide as the load resistance. The advantages of a resistive load include low temperature dependence and reasonable power-speed product.

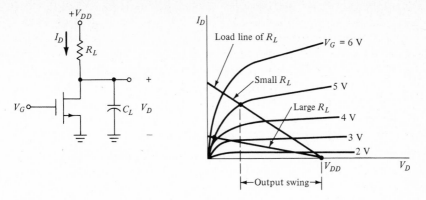

Figure 10-17 An MOS inverter with linear resistive load: (*a*) circuit and (*b*) output characteristics and load line.

Saturated MOS Load

Most MOS logic circuits use an MOS transistor load, as shown in Fig. 10-18*a*. The MOS load is used in ICs because it takes up a much smaller area than a resistive load. The operation of the circuit can be understood through the current-voltage characteristics shown in Fig. 10-18*b*, where the curve under the condition $V_G = V_D$ is plotted to represent the load characteristics with the gate tied to the drain. If the inverter device and the load device are identical, the load line is obtained by using the $V_G = V_D$ curve and plotting it as shown by the dotted line. This arrangement, however, produces an output swing significantly less than V_{DD} and is not desirable. The output swing can be increased if the transconductance of the load device is reduced to one-tenth of the inverter

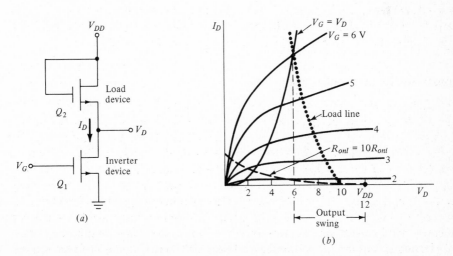

Figure 10-18 An MOS inverter with a saturated MOS load: (*a*) circuit and (*b*) *I-V* characteristics with load curves.

device. This change leads to an ON resistance of the load device that is 10 times that of the inverter transistor. In practical design, we make use of Eq. (8-42) to obtain

$$R_{\text{on},l} = \frac{L_l}{\mu_n C_o Z_l (V_{Gl} - V_{\text{TH}})} \tag{10-8}$$

for the load device and

$$R_{\text{on},i} = \frac{L_i}{\mu_n C_o Z_i (V_{Gi} - V_{\text{TH}})} \tag{10-9}$$

for the inverter device. Since $V_{Gl} = V_{DS} \approx V_{DD}$ and $V_{Gi} \approx V_{DD}$, the resistance ratio becomes

$$\frac{R_{\text{on},l}}{R_{\text{on},i}} = \frac{L_l / Z_l}{L_i / Z_i} \tag{10-10}$$

For example, a resistance ratio of 10 can be obtained by setting

$$\frac{L_l}{Z_l} = 3 \quad \text{and} \quad \frac{L_i}{Z_i} = \frac{1}{3}$$

As we can see, the length-to-width ratio of the MOS transistor is the most important design parameter in a saturated MOS load inverter. The layout diagram shown in Fig. 10-19a illustrates the required area for the complete inverter. The new load line is sketched as the dashed curve in Fig. 10-18b. The output swing is now substantially improved, and the transfer characteristic is displayed in Fig. 10-19b. Note in Fig. 10-18b that the load resistance is not linear and has a high value at V_{DD} and a low value at $V_D \rightarrow 0$ V. With the same parasitic capacitance, we find that the charging time through the load device is much slower than the discharging time through the inverter device.

Nonsaturated MOS Load

If an additional power supply is available, we can bias the gate of the load device instead of having it shorted to the drain. This bias arrangement is shown in Fig. 10-20a, in which the load device changes from a highly nonlinear resistor to a linear one and approaches a linear resistor for high gate bias voltage. The load line is shown in Fig. 10-20b. The charging time through the load device is significantly reduced because of the lower load impedance. In addition, the voltage swing is increased by V_{TH} compared with the saturated load.

Depletion MOS Load

When the gate is short-circuited to the source, a depletion MOS transistor has a current-voltage characteristic shown in Fig. 10-21. With the inverter circuit shown in Fig. 10-22a, the output characteristics and load line are shown in Fig.

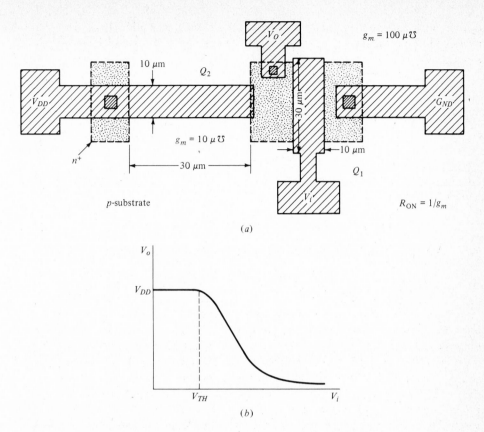

(a)

(b)

Figure 10-19 (a) Layout diagram and (b) transfer characteristic of an MOS inverter with saturated MOS load.

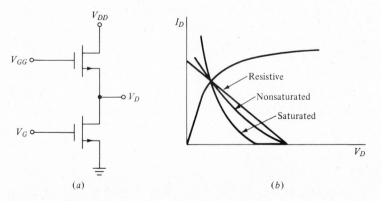

(a)　　　　　　　(b)

Figure 10-20 (a) MOS-inverter circuit with nonsaturated load and (b) its load curve in comparison with the resistive and saturated loads. *(After Hung [9].)*

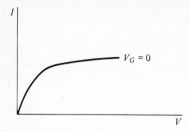

Figure 10-21 I-V characteristic of the depletion load.

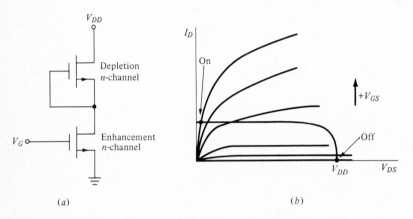

(a)

(b)

Figure 10-22 The depletion-load MOS inverter with load curve.

10-22b. An important feature of the depletion-load circuit is its essentially symmetrical charging and discharging time constants. The switching speed is about 3 times faster than the all-n-channel MOS inverter. Implementation of this inverter using D-MOS technology (described later) eliminates the need of the separate isolation step required in earlier structures. Power dissipation of this inverter is quite low.

10-7 COMPLEMENTARY MOS (CMOS) INVERTER [10]

The standard dc power dissipation of an MOS inverter can be reduced to very small (10-nW) levels by using a complementary p-channel and n-channel pair connected as shown in Fig. 10-23a. Transistor Q_1 is an n-channel device, and Q_2 is a p-channel device. When the input voltage is high (V_{DD}), Q_1 is turned on and Q_2 is turned off. When the input voltage is low (0 V), Q_1 is turned off and Q_2 is turned on. The operating points are shown by the curves in Fig. 10-23b. Note that under either input condition, very little current is drawn in the steady state. This current is the leakage current of the OFF device. By examining the load lines, it is seen that the turn-on and turn-off time constants are about the same. The total power dissipation of a CMOS inverter is the

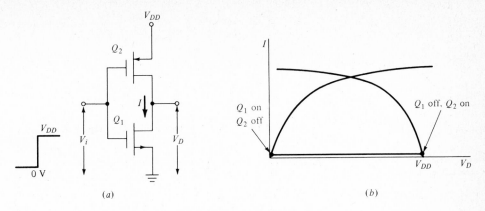

(a)

(b)

Figure 10-23 The complementary MOS inverter and its load curves.

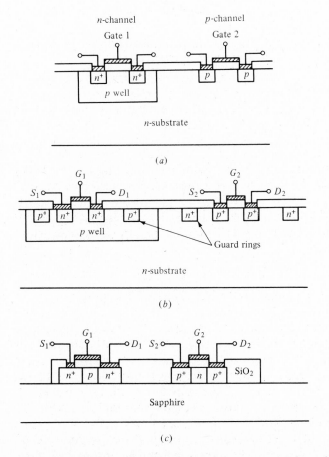

(a)

(b)

(c)

Figure 10-24 CMOS structures: (a) CMOS without guard rings, (b) CMOS with guard rings, and (c) silicon-on-sapphire (SOS) CMOS.

sum of the standby power and the transient power loss during switching. The latter is given by fCV^2, where f is the switching frequency, C is the total output capacitance, and V is the output voltage.

A simple IC implementation of the CMOS is shown in Fig. 10-24a, where a p diffusion is made in the n substrate to form a p-type well in which the n-channel device is fabricated. Subsequent selective n^+ and p^+ diffusions produce the n- and p-channel transistors. This structure unfortunately leads to difficulty when very high positive or negative voltage is applied at the source or drain. Such voltage could induce a negative inversion channel joining the n^+ region to the n substrate or p^+ region to the p-well region. In practical circuits, guard rings are diffused to achieve isolation, as shown in Fig. 10-24b. Isolation can also be achieved by the silicon-on-sapphire technology (Fig. 10-24c).

10-8 D-MOS AND V-MOS TRANSISTORS [11]

Presently, new processes are being explored by investigators to make MOS circuits competitive with bipolar I^{2}L circuits for large-scale-integration applications. Among recent designs, the most promising structures are the *double-diffused* MOS (D-MOS) and *V-groove or vertical* MOS (V-MOS) transistors. In a standard MOS transistor, present photolithographic procedures must be stretched to their utmost to yield a short channel length (4 or 5 μm). In the D-MOS or V-MOS process, a channel length of 1 μm is easily obtainable because the channel dimension is defined by solid-state diffusion. As a result, the D-MOS and V-MOS devices have very small size and high performance.

The cross section of a standard n-channel double-diffused MOS transistor is shown in Fig. 10-25. The substrate is a π-type (lightly doped p-type) material on which a p diffusion is made in selected regions. This p diffusion is the same as the base diffusion in a double-diffused bipolar transistor, and it is used as the channel for the MOS device. Subsequently, an n^+ diffusion is made through the same mask window to form the source and drain regions as shown. The lateral dimension of the p region is the channel width. Assuming the diffusion depth in the lateral direction to be the same as in the vertical direction, we can control the channel length to less than 1 μm, just like the base width of a bipolar transistor. The π layer yields a high breakdown voltage and low feedback capacitance and improves the device performance. However, the D-MOS is not a symmetrical device in that the exchange of the

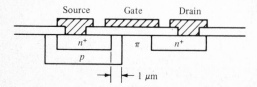

Figure 10-25 Double-diffused MOS transistor. *(After Altman [11].)*

source and drain terminals leads to different current-voltage characteristics. This is because the channel is formed by diffusion, so that its impurity concentration is higher at the source end than that at the drain end. A D-MOS inverter with a depletion MOS load is shown in Fig. 10-26. The n^+ diffusion for the source and drain regions of the depletion-load device can be made at the same time as the source region of the D-MOS. Thus, no additional processing step is needed. The π region under the gate of the load device is inverted to n because of positive charge in the oxide. This D-MOS inverter design results in a compact structure that boosts performance significantly over the all-n-channel designs. Fully loaded gates have a power dissipation of 0.7 mW and a delay within 5 ns. The only drawback is that the lateral diffusion just under the surface cannot be estimated precisely in practice and the channel length is not accurately known.

Figure 10-27 shows the cross section of a V-MOS structure. Fabrication begins with an n^+ substrate on which a thin layer (1 μm) of p-type and π-type materials is epitaxially grown sequentially. The dimension of the p layer corresponds to the length of the channel and can be more easily controlled than in the D-MOS process. An n^+ diffusion is made to form the drain region. Then a V-shaped groove is etched through the channel and source regions using a preferential etchant. As described in Sec. 10-1, the dimensions are determined only by the width of the oxide window. The process is completed by growing silicon dioxide over the V-groove gate region and then applying metallization. Because the n^+ substrate is used as the common source for all V-MOS devices, it imposes a certain limitation on circuit applications. For

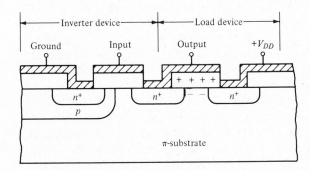

Figure 10-26 D-MOS inverter with a depletion MOS load. (*After Altman [11].*)

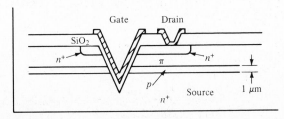

Figure 10-27 Cross section of the V-MOS transistor structure. (*After Altman [11].*)

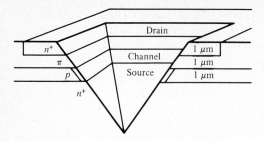

Figure 10-28 Three-dimensional view of a V-MOS structure. *(After Altman [11].)*

example, we cannot use a V-MOS device as the load device in an inverter or as a memory element in which the source terminals are not common for all transistors. Nevertheless, V-MOS devices can be combined with standard n-channel and ion-implanted load resistors on the same chip. Access to the controlled gate is through the V groove, as shown in the three-dimensional schematic diagram in Fig. 10-28. Because of the four sides of the groove, a V-groove opening of 10 by 10 μm^2 can yield a channel width of 25 μm. For this reason, V-groove structure is also useful for power MOS transistor applications.

10-9 NONVOLATILE MIS MEMORY DEVICES: MNOS AND FAMOS STRUCTURES [12–14]

Before discussing nonvolatile devices, let us first explain two semiconductor memory systems: the read-only memory (ROM) and read-write random-access memory (RAM). In an ROM, "programmed" information is stored in the memory, and only the *read* operation is performed. Examples of ROM applications include look-up tables and code-conversion systems. The RAM is an array of memory cells that stores information in binary form in which information can be randomly written into, or read out of, each cell as required. In other words, the RAM is a *read-write memory.* Most semiconductor memory systems utilize the bistable flip-flop or capacitive-charge-storage circuit as their basic memory cell in that the stored information is lost when the power supply is interrupted. It is often desirable to have nonvolatile semiconductor memories so that they can stand power failure or be stored and shipped without being energized all the time. In this section, we introduce nonvolatile memory devices which make use of the MIS structure. Non-volatility of an MIS device is usually achieved by storing charge in the gate structure. The stored charge alters the threshold voltage of the MIS transistor so that two stable states can be detected electrically. Presently, nonvolatile memory devices are used in ROMs only because the writing speed is too slow. Undoubtedly, as the switching speed is improved, nonvolatile devices will also be used in other forms of memory systems.

Among the nonvolatile MIS devices, we limit our discussion to the metal-nitride-oxide-silicon (MNOS) and floating-gate avalanche-injection MOS (FAMOS) structures. Used as a ROM, the FAMOS is electrically programmable but it can be erased only with ultraviolet light. It can be reprogrammed and is known as the *electrically programmable ROM* (EPROM). An MNOS device can be electrically programmed and erased so that it is capable of functioning as a nonvolatile read-write memory, but the speed in writing is rather slow (in microseconds).

MNOS Transistor

A cross section of a p-channel MNOS transistor is shown in Fig. 10-29. It is a conventional MOSFET in which the oxide is replaced by a double layer of nitride and oxide. The oxide layer is very thin, typically 50 Å thick, and the nitride layer is about 500 Å thick.

For low values of negative voltage applied to the gate, the MNOS transistor behaves like a conventional p-channel MOSFET. Upon application of a sufficiently high positive charging voltage V_C to the gate, as shown in Fig. 10-30a, electrons will tunnel from the silicon conduction band to reach traps in the nitride-oxide interface and nitride layer resulting in negative charge accumulation there. The stored charge as a function of time is given by

$$\frac{dQ}{dt} = J_o - J_n \tag{10-11}$$

where J is the current density and the subscripts denote the corresponding layers. The voltages across the oxide and nitride layers are, respectively, V_o and V_n, and they establish an electric field $\mathscr{E}_o$ in the oxide and $\mathscr{E}_n$ in the nitride. The relationship between the potential drops across the structure is

$$V_C = V_o + V_n = \mathscr{E}_o x_o + \mathscr{E}_n x_n \tag{10-12}$$

If the charge Q can be considered to have been stored in the interface only, the continuity of the electric flux leads to

$$K_o \epsilon_o \mathscr{E}_o = K_n \epsilon_n \mathscr{E}_n + Q \tag{10-13}$$

The stored charge as a function of time can be found by solving Eqs. (10-11)

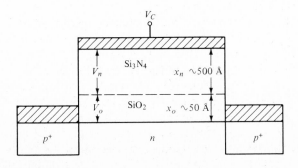

Figure 10-29 The MNOS transistor.

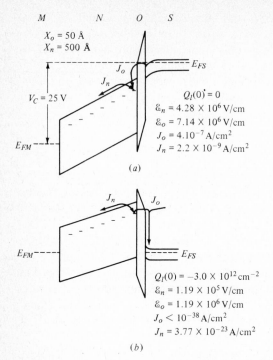

$X_o = 50$ Å
$X_n = 500$ Å

$V_C = 25$ V

E_{FM}

$Q_l(0) = 0$
$\mathscr{E}_n = 4.28 \times 10^6$ V/cm
$\mathscr{E}_o = 7.14 \times 10^6$ V/cm
$J_o = 4.10^{-7}$ A/cm^2
$J_n = 2.2 \times 10^{-9}$ A/cm^2

(a)

E_{FM}

E_{FS}

$Q_l(0) = -3.0 \times 10^{12}$ cm^{-2}
$\mathscr{E}_n = 1.19 \times 10^5$ V/cm
$\mathscr{E}_o = 1.19 \times 10^6$ V/cm
$J_o < 10^{-38}$ A/cm^2
$J_n = 3.77 \times 10^{-23}$ A/cm^2

(b)

Figure 10-30 Energy-band diagram for an MNOS transistor: (a) in the charging mode under an applied positive voltage $V_C = +25$ V and (b) at storage after voltage removal. (*After D. Frohman Bentchkowsky [12].*)

to (10-13) provided that the currents as functions of the electric field are known. Since the derivation of the current-field relationships is beyond the scope of this text, we shall discuss one simple case where these equations are not needed.

Since the majority of presently available MNOS devices have an oxide conduction current considerably larger than the nitride conduction current, we assume that $J_o \gg J_n$. As a result, electrons tunneling through the oxide will accumulate in the interface traps until the electric field in the oxide is reduced to zero. Therefore, Eq. (10-12) reduces to

$$V_C = \mathscr{E}_n x_n \qquad (10\text{-}14)$$

and the stored charge is given by

$$Q = -K_n \epsilon_o \mathscr{E}_n = -\frac{K_n \epsilon_o}{x_n} V_C = -C_n V_C \qquad (10\text{-}15)$$

where $C_n = K_n \epsilon_o / x_n$. With a stored charge Q at the interface, the threshold voltage is shifted by

$$\Delta V_{\text{TH}} = -\frac{Q}{C_n} = V_C \qquad (10\text{-}16)$$

and the corresponding turn-on characteristics are shown in Fig. 10-31.

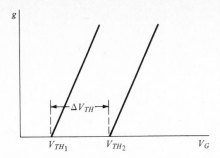

Figure 10-31 The shift of V_{TH} after charging gate.

On the other hand, if the interface charge does not reduce $\mathscr{E}_o$ by a significant amount, and if the oxide current as a function of the electric field is known, we can write the stored charge as

$$Q = \int J_o \, dt \approx J_o \, \Delta t \qquad (10\text{-}17)$$

Thus, the corresponding change in threshold voltage is given by

$$\Delta V_{\text{TH}} = -\frac{J_o \, \Delta t}{C_n} \qquad (10\text{-}18)$$

Further understanding of this charge storage effect can be gained by working out Prob. 10-14. Equation (10-18) is a reasonable approximation if ΔV_{TH} is small compared with V_C.

To switch to the low-threshold voltage, a large negative voltage pulse is applied to the gate, and electrons are driven out of the interface traps and returned to the silicon substrate. It is important to note that while the tunneling current through the oxide is high at high electric fields present during write and erase operations, there is essentially no current conduction through the oxide at low electric fields during the read or storage period. In other words, the charge stored in the nitride traps is considered permanent.

Besides calculating the change in the threshold voltage one would like to be able to estimate the switching speed and charge-retention time. Unfortunately, it is difficult to obtain a reasonable estimate because of the lack of thorough knowledge of the electrical properties of nitride traps, e.g., energy levels, densities, and capture cross sections. It has been found experimentally that the switching speed goes up as the oxide thickness is reduced or the nitride-trap density is increased or both. But as the switching speed is increased, the ability to keep the stored charge, known as the *retentivity*, of the device is decreased. This tradeoff between retentivity and switching speed generally holds for all nonvolatile semiconductor memory devices. With MNOS structures, a switching speed on the order of nanoseconds can be achieved if a short retention time is tolerable. On the other hand, devices having a retention time estimated at tens of years can be fabricated if a slow switching speed is acceptable.

FAMOS Transistor

A floating-gate avalanche-injection MOS transistor is shown in Fig. 10-32a. Like the MNOS devices, the floating-gate structure is also a charge-storage transistor, but the charge is stored in a polycrystalline-silicon floating gate rather than the nitride traps. The oxide is very thick, so that injection of charge by tunneling is not possible. To charge the floating gate of the FAMOS device, the drain junction is reverse-biased to avalanche breakdown. As a result, electrons are accelerated in the depletion region, and some electrons are injected from the silicon substrate into the silicon dioxide. These electrons are attracted to the floating gate because of the field set up by the voltage-divider action shown in Fig. 10-32b. The negative charge accumulated at the floating gate creates a p channel and gives rise to a low threshold voltage. Under steady-state conditions, the electrons are trapped in the floating gate, as depicted in Fig. 10-33. Because the electrons are trapped in a deep potential

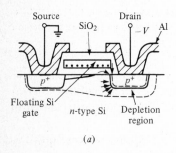

(a)

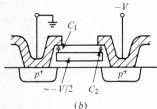

(b)

Figure 10-32 The FAMOS device: (a) avalanche injection of electrons into the floating gate and (b) potential of the floating gate due to capacitive voltage-divider action. *(After Chang [13].)*

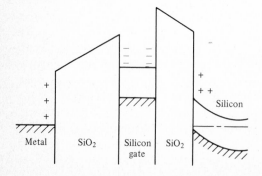

Figure 10-33 Energy-band diagram of the FAMOS device with charge stored in the silicon gate. *(After Card and Worrall [14].)*

well, FAMOS devices have a very long retention time. Discharge of the floating gate requires an ultraviolet-light source with enough energy to excite the trapped electrons into the conduction band of the SiO_2-gate dielectric. Therefore, erasing is not convenient, and the FAMOS is used only in read-only memory systems.

PROBLEMS

10-1 Describe the processing steps in fabricating a bipolar transistor by base-diffusion isolation.

10-2 Tabulate the number of mask, oxidation, diffusion, and epitaxy steps required for EDP, CDI, and BDI.

10-3 Consider the V-grooved isolation. The thickness d of the epitaxial layer is between 5 and 20 μm. Find the minimum width of the initial oxide opening W necessary for the preferential etching and plot W vs. d.

10-4 Using the transistor of Fig. 10-8 with a buried layer, draw a monolithic layout of the circuit shown. Draw a complete set of (individual) masks. Darken areas which are not to pass light. (Tracing from your layout suggested.) Use rectangular coordinate paper with 20 lines to the inch, and a scale of $\frac{1}{20}$ in per 0.1 mil. You may use any necessary chip size, but about 20 square mils is reasonable. In layout use green for isolation-diffusion-mask boundaries, red for base, blue for emitter, and black for contact windows and metal. (See Fig. P10-4.)

10-5 For the integrated transistor of Fig. 10-8 with an impurity profile shown in Fig. 9-9 (but changing it to a p substrate of $N_a = 2 \times 10^{15}\,cm^{-3}$ and an n-type collector of $N_d = 10^{16}\,cm^{-3}$), estimate the collector-base and collector-substrate capacitances at 5 V. Also find the collector series resistance and breakdown voltages of the collector, emitter, and substrate junctions.

10-6 Calculate the cutoff frequency of the transistor of Prob. 10-5 at $I_E = 1$ mA.

10-7 (a) In integrated circuits, a transistor may be connected as a diode.

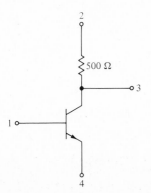

Figure P10-4.

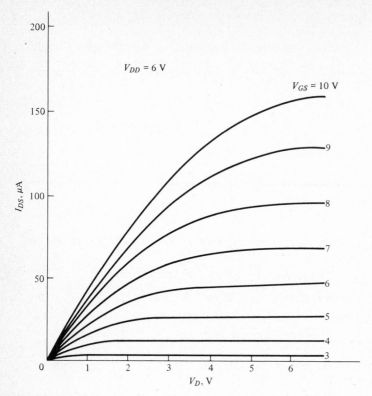

Figure P10-11.

Sketch five possible diode configurations along with the minority-carrier distribution in the base for $V \gg V_T$.

(*b*) Use the doping profile and carrier distribution to compare the diode performance in speed, conductance, and breakdown voltage if the impurity profile of the double-diffused transistor is assumed.

10-8 Derive Eqs. (10-5) and (10-6).

10-9 (*a*) The following parameters are given for a lateral transistor: the epitaxial layer is $20\ \mu$m, the junction depth of the p diffusion is $3\ \mu$m, $W_B = 3\ \mu$m, $W_E = 1.5$ mils, and $L_p = 15\ \mu$m. Calculate h_{FE}.

(*b*) Repeat (*a*) for $W_B = 10\ \mu$m.

10-10 Draw the first four masks for the standard p-channel MOS transistor. Specify the device dimensions.

10-11 Plot the transfer characteristics of an n-MOS inverter with an MOS load where the load device is biased with (*a*) $V_{GS} = V_{DD}$ and (*b*) $V_{GS} - V_{DD} = V_{TH} = 4$ V. The V-I characteristic of the inverter device is shown in Fig. P10-11. Assume that the transconductance of the load device is one-tenth that of the inverter device.

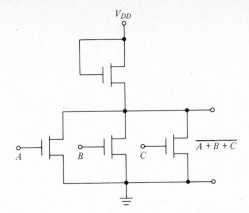

Figure P10-12.

10-12 Draw a layout of the three-input monolithic MOS NOR/NAND gate shown in Fig. P10-12.

10-13 Draw a set of masks for the D-MOS inverter of Fig. 10-26.

10-14 Assume that the insulating layer above the floating gate in an MNOS memory transistor has $K_n = 40$ and $x_n = 1000$ Å and the lower insulator has $K_o = 4$ and $x_o = 100$ Å. If the conductive properties of both insulators are the same and given by $J = \sigma\mathscr{E}$, where $\sigma = 10^{-7}$ mho-cm, find the approximate threshold-voltage shift of the transistor induced by a voltage of 10 V applied to the external gate for 1 μs.

10-15 Explain the conditions (physical parameters of insulators) under which an ideal nonvolatile MIS memory transistor would be achieved.

REFERENCES

1. B. T. Murphy, V. J. Glinski, P. A. Gray, and R. A. Pedersen, Collector Diffusion Isolated Integrated Circuits, *Proc. IEEE*, **57**: 1523 (1969).
2. B. T. Murphy, S. M. Neville, and R. A. Pedersen, Simplified Bipolar Technology and Its Applications to Systems, *IEEE J. Solid-State Circuits*, **SC-5**: 7 (1970).
3. D. L. Peltzer and W. H. Herndon, Isolation Method Shrinks Bipolar Cells for Fast, Dense Memories, *Electronics*, Mar. 1, 1971, p. 52.
4. J. Mudge and K. Taft, V-ATE Memory Scores a New High in Combining Speed and Bit Density, *Electronics*, July 17, 1972, p. 65.
5. H. C. Lin et al., Lateral Complementary Transistor Structure for the Simultaneous Fabrication of Functional Blocks, *Proc. IEEE*, **52**: 1491 (1964).
6. J. Lindmayer and W. Schneider, Theory of the Lateral Transistor, *Solid-State Electron.*, **10**: 225 (1967).
7. H. H. Berger and S. K. Wiedmann, The Bipolar LSI Breakthrough, I: Rethinking the Problem, *Electronics*, Sept. 4, 1975, p. 89.
8. R. H. Horton, J. Englade, and G. McGee, I^2L Takes Bipolar Integration: A Significant Step Forward, *Electronics*, Feb. 6, 1975, p. 83.
9. R. Hung, High Performance Ion-implanted MOSFET Technology, *Int. Conf. Integrated Circuits, Paris, Dec. 3–6, 1974*, pp. 35–42.

10. L. Altman, Special Report: C-MOS Enlarges Its Territory, *Electronics*, May 15, 1975, p. 77.
11. L. Altman, MOS Advances in Designs and New Processes Yield Surprising Performance, *Electronics*, Apr. 1, 1976, p. 73.
    T. J. Rodgers and J. D. Meindl, VMOS: High Speed TTL Compatible MOS Logic, *IEEE J. Solid-State Circuits*, **SC-9**: 239 (1974).
12. D. Frohman Bentchkowsky, The Metal-Nitride-Oxide-Silicon (MNOS) Transistor: Characteristics and Applications, *Proc. IEEE*, **58**: 1207 (1970).
13. J. J. Chang, Nonvolatile Semiconductor Memory Devices, *Proc. IEEE*, **64**: 1039 (1976).
14. H. C. Card and A. G. Worrall, Reversible Floating-Gate Memory, *J. Appl. Phys.*, **44**: 2326 (1973).

ADDITIONAL READINGS

Altman, L.: Five Technologies Squeezing More Performance from LSI Chips. *Electronics*, August 18, 1977, pp. 91–112. (This includes three up-to-date papers on integrated circuits.)
Hamilton, D. J., and W. G. Howard: "Basic Integrated Circuit Engineering," McGraw-Hill, New York, 1975.
Penney, W. M. (ed.): "MOS Integrated Circuits," Van Nostrand Reinhold, New York, 1972.
Warner, R. M. Jr., and J. N. Fordemwalt: "Integrated Circuits," McGraw-Hill, New York, 1965.

CHARGE-TRANSFER DEVICES

The search for simplicity in designing ICs has led to the invention of a new class of functional structures called *charge-transfer devices* (CTDs). In the family of the CTDs there are two types of devices which are different in operating principles, the charge-coupled device (CCD) and the bucket-brigade device (BBD). In this chapter, the physical theory of CTDs is presented, and practical device structures are described. Applications to digital shift registers and memory, analog signal processing, and imaging systems are considered. The emphasis will be on the CCD because it can be understood only on the device level.

11-1 THE CONCEPT OF CHARGE TRANSFER

The concept of charge transfer can be explained by using a chain of amplifiers with unity gain and infinite input impedance connected as shown in Fig. 11-1a. Upon the closing of the switch S_1, the input signal is stored in the form of a charge packet in the capacitor C_1. Now, let us open S_1 and then close S_2; the stored charge will be transferred to the capacitor C_2. Following the same procedure, the charge will eventually reach the output terminal. It is obvious that this system can be used as a digital shift register or an analog delay line. If we replace each amplifier-and-switch pair by an MOS transistor, we obtain the circuit shown in Fig. 11-1b. The transistors can be turned on and off sequentially by applying voltage pulses at the respective gate electrode, and

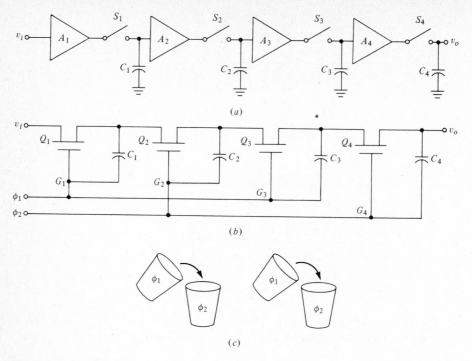

Figure 11-1 Charge-transfer systems using (*a*) operational amplifiers, (*b*) MOS transistors, and (*c*) water buckets.

the charge is stored and transferred just as in Fig. 11-1*a*. In practical systems, gates 1 and 3 are connected and pulsed together and gates 2 and 4 are similarly joined. The transfer of charge is analogous to filling and emptying water buckets, as illustrated in Fig. 11-1*c*. The circuit shown in Fig. 11-1*b* is known as the *bucket-brigade device*. It is a *two-phase* system because two separate clock pulses ϕ_1 and ϕ_2 are needed.

In the bucket-brigade device, the charge transfer is implemented on the *circuit* level by using either discrete or integrated components. Implementation of charge transfer on the *device* level is realized by the *charge-coupled device* (CCD). In the CCD, minority carriers are stored in potential wells created at the surface of a semiconductor. These carriers are transported along the surface by filling and emptying a series of potential wells sequentially. In its simplest form, the CCD is a string of closely spaced MOS capacitors, as illustrated in Fig. 11-2. If electrode 2 is biased at 10 V, more positively than its two adjacent electrodes (at 5 V), a potential well is set up, as depicted by the dotted line, and charge is stored under this electrode, as in Fig. 11-2*a*. Now let us bias electrode 3 with 15 V; a deeper potential well is established under electrode 3 (Fig. 11-2*b*). The stored charge seeks the lowest potential and therefore travels along the surface when the potential wells are

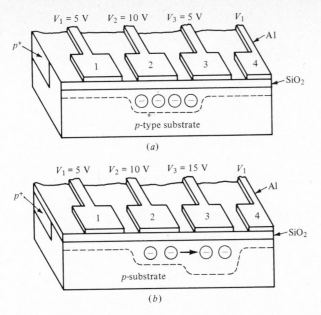

Figure 11-2 The basic operation of a three-phase CCD: (a) charge storage and (b) charge transfer. The p^+ diffusion is used for channel confinement; its function will be explained later.

moved. Note that we need three electrodes in this structure to facilitate charge storage and transfer in one direction only. The three electrodes will be referred to as one *stage* or *cell* of the device.

11-2 TRANSIENT CHARACTERISTICS OF THE MOS CAPACITOR

The charge-coupled device stores and transports minority carriers between potential wells created by voltage pulses on closely spaced MOS capacitors. Therefore, it is important to understand an MOS capacitor under pulsed conditions. Figure 11-3 shows the structure and energy-band diagrams of an MOS capacitor on a p-type substrate. Upon the application of a positive gate voltage, a depletion layer is formed under the gate. The applied voltage is large enough to induce an inversion layer. However, no inversion layer is formed at $t = 0+$ because no minority carriers are available. As a result, a *deep* depletion layer exists, as seen in the energy-band diagram in Fig. 11-3b. Under the condition of deep depletion, a substantial portion of the external voltage is across the depletion layer, and the surface potential is large. As time elapses, electron-hole pairs are generated in the depletion region. The electrons accumulate at the Si–SiO$_2$ interface, and they influence the charge distribution and energy-band diagram. With a positive gate voltage, holes are

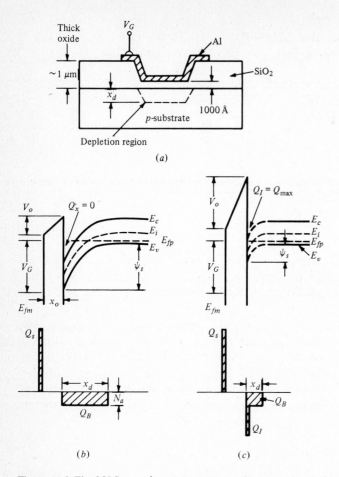

Figure 11-3 The MOS capacitor: (*a*) structure, (*b*) energy-band diagram and charge distribution under deep depletion at $t = 0+$, and (*c*) energy-band diagram and charge distribution at thermal equilibrium ($t = \infty$).

driven to the substrate to reduce the depletion-layer width, and at the same time electrons are attracted to the surface to form an inversion layer. When a sufficient number of electrons has been collected under the surface, a saturation condition is reached in which the electron diffusion current away from the surface is exactly balanced by the electron drift current toward the surface. This condition is illustrated by the energy-band diagram in Fig. 11-3*c*. If the accumulated electrons are less than the saturation value, the net flow will be toward the surface; but if the saturation value is exceeded, there will be a net flow into the undepleted bulk, where the minority carriers recombine. The time required to reach the saturation condition is known as the *thermal-relaxation time*. The thermal-relaxation time at room temperature has been measured from 1 s to several minutes, depending on the structure and fabrication processes of the CCD. Since a useful potential well does not exist in the saturation condition, the CCD is basically a *dynamic* device in which charges can be stored for a time *much shorter* than the thermal-relaxation time.

Let us now derive the surface potential ψ_s as a function of the applied voltage V_G at the gate electrode. The surface potential of an MOS capacitor is given by Eq. (8-15), repeated here,

$$V_G = -\frac{Q_s}{C_o} + \psi_s \tag{11-1}$$

where Q_s is the total charge per unit area in the semiconductor surface region and C_o is the oxide capacitance, given by

$$C_o = \frac{K_o\epsilon_o}{x_o} \tag{11-2}$$

where x_o is the oxide thickness. Equation (11-1) was derived under the assumption that the energy bands in the semiconductor are flat at $V_G = 0$. Therefore, it must be modified to include the flat-band voltage V_{FB}:

$$V_G - V_{FB} = -\frac{Q_s}{C_o} + \psi_s \tag{11-3}$$

For a time much smaller than the thermal-relaxation time, there is no inversion layer in a CCD. Therefore, the charge Q_s is the sum of the depletion-layer charge and an externally introduced *signal charge* Q_{sig}. The depletion-layer charge is given by Eq. (8-5), so that the total surface charge is

$$Q_s = -qN_ax_d - Q_{sig} \tag{11-4}$$

where x_d, the depth of the potential well, is given by Eq. (8-6):

$$x_d = \left(\frac{2K_s\epsilon_o}{qN_a}\psi_s\right)^{1/2} \tag{11-5}$$

Substituting Eqs. (11-4) and (11-5) into Eq. (11-3) yields

$$V_G - V_{FB} - \frac{Q_{sig}}{C_o} = \psi_s + \frac{1}{C_o}(2qK_s\epsilon_oN_a\psi_s)^{1/2} \tag{11-6}$$

Solving for the surface potential from the foregoing equation leads to

$$\psi_s = V - B\left[\left(1 + \frac{2V}{B}\right)^{1/2} - 1\right] \tag{11-7}$$

where
$$V = V_G - V_{FB} - \frac{Q_{sig}}{C_o} \tag{11-8}$$

and
$$B = \frac{qK_s\epsilon_oN_a}{C_o^2} \tag{11-9}$$

Equation (11-7) is very important in the design of a CCD because the gradient of this potential governs the motion of minority carriers. The value of ψ_s also specifies the depth of the potential well by means of Eq. (11-5). In Eqs. (11-7) to (11-9), we find that the surface potential is controlled by the substrate doping concentration N_a and the oxide thickness x_o, which determines C_o. If we set V to be constant, then ψ_s increases as N_a and x_o are reduced. Equation (11-7) is plotted as a function of V in Fig. 11-4 with N_a and x_o as parameters. Since Eq. (11-8) indicates that V decreases with an increase of Q_{sig}, the surface potential is also a function of the magnitude of the signal charge.

The capacitance C_{GS} between the gate electrode and substrate is the series combination of the oxide and depletion capacitances. By making use of Eqs. (8-13), (8-14), (11-2), and (11-5) we derive

$$C_{GS} = \frac{C_o}{1 + (2\psi_s/B)^{1/2}} \tag{11-10}$$

from which we can calculate ψ_s if C_{GS} is measured. The signal charge can then be calculated from Eqs. (11-7) and (11-8). Alternatively, the signal charge can be estimated from the consideration of charging up the oxide and depletion-layer capacitances. Usually, the depletion capacitance is much smaller than the oxide capacitance, and it can be neglected. For an oxide thickness of 1000 Å and electrode area of 10 by 20 μm, the oxide capacitance is calculated to be 0.068 pF [Eq. (11-2)]. Assuming that half the gate voltage

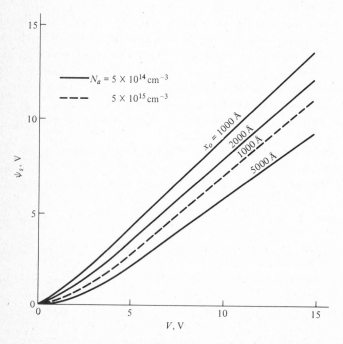

Figure 11-4 Surface potential as a function of voltage in Eq. (11-7). (*After Amelio et al. [1].*)

(10 V) is across the oxide capacitor, we find each charge packet to be 0.34 C. Since an electron has a charge of 1.6×10^{-19} C, there are 2×10^6 electrons in each packet, giving an electron density of 10^{12} cm^{-2}.

For time intervals that are short compared with the thermal-relaxation time, the MOS capacitor serves as a storage element for analog information represented by the amount of charge in the well.

11-3 ELECTRODE ARRANGEMENTS AND FABRICATION TECHNOLOGY

The transfer of charge can be implemented by different MOS structures and electrode arrangements. The approach in designing a CCD depends on consideration of electrical performance, fabrication difficulty, and cell size. One question is how many phases we should have in the system, as it is practical to build a two-, three-, or four-phase system. In this section, we present the basic principles of two- and three-phase CCDs and discuss some merits of each system along with representative fabrication technology.

The Three-Phase CCD

A three-phase CCD is a linear array of closely spaced MOS capacitors with three electrodes per stage or cell. As shown in Fig. 11-5a, every third electrode is connected to the same clock voltage so that three separate clock generators are required. The basic principle was discussed in connection with Fig. 11-2. In practice, the driving clock pulses display the special features shown in Fig. 11-5e. These waveforms are designed to achieve better efficiency in charge transfer, as explained in the following paragraph.

If a positive voltage applied to ϕ_1 is higher than that applied to ϕ_2 and ϕ_3, surface potential wells will be formed under the ϕ_1 electrodes. Charge packets, which have been introduced either optically or electrically, are accumulated in these wells at $t = t_1$. These charge packets may be of different magnitudes, as depicted in Fig. 11-5a. To facilitate charge transfer to the right, a positive voltage step is applied to ϕ_2 so that the potential wells under ϕ_1 and ϕ_2 electrodes are the same in depth. Thus, the stored charge packets spread out, as seen at $t = t_2$. Almost immediately after the application of the positive pulse at ϕ_2, the voltage at ϕ_1 starts to decrease linearly, so that the potential wells under ϕ_1 electrodes rise slowly rather than abruptly. The charge packets tend to spill over to the potential wells under gates 2 and 5, as shown at $t = t_3$. The slow rise of the potential wells under gates 1 and 4 provides a more favorable potential distribution for the complete transfer of charge. When we reach the condition at $t = t_4$, the charge has been transferred over to the wells under ϕ_2 electrodes. Note that the charge is prevented from moving to the left by the barriers under the ϕ_3 electrodes. Repeating the same procedure, we can

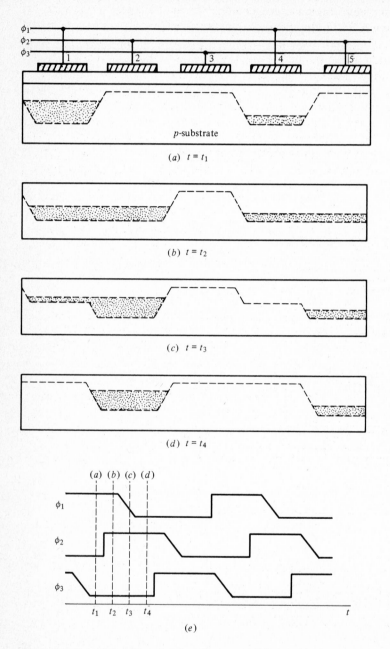

Figure 11-5 Potential wells and timing diagram at different time intervals during the transfer of charge in a three-phase CCD.

move the charge from ϕ_2 to ϕ_3 and then from ϕ_3 to ϕ_1. As a full cycle of clock voltages is completed, the charge packets advance one stage to the right.

In designing a CCD, the capacitors must be physically close together so that the depletion layers overlap strongly and the surface potential has a smooth transition at the boundaries between neighboring electrodes. The first realization of the CCD is by means of a *single metal gate*, shown in Fig. 11-5. Typically, the oxide thickness is between 1000 and 2000 Å, and the gap or spacing between aluminum gates is about 2.5 μm. This gap size is smaller than the minimum size of the standard photolithographic process, which is approximately 5 μm. Therefore, etching the gaps between metal electrodes is difficult, and slight flaws in masking or photoresist steps could short-circuit the electrodes or (at the other extreme) could cut down the charge-transfer

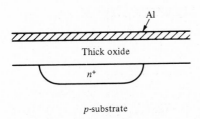

Figure 11-6 A diffused cross-under used as a conductor normal to the paper.

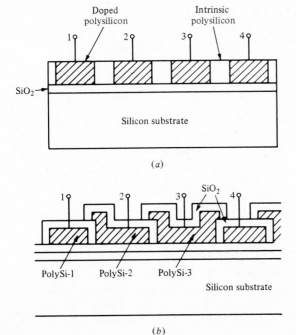

(a)

(b)

Figure 11-7 Three-phase CCD structures: (*a*) doped-polysilicon gate (*after Kim and Snow [2]*) and (*b*) triple polysilicon gate (*after Bertram et al. [3]*).

efficiency significantly in electrodes that are far apart. In addition, the channel oxide in the gaps is exposed, and electrostatic charge residing there leads to device instabilities. Furthermore, the need for three separate interconnections for three phases means that two conducting bus lines must cross over to address all electrodes. To avoid making electrical contact between conducting bus lines, a *cross-under* diffusion (Fig. 11-6) or two-level metallization is required in the area away from the active-device region. An alternative to the single-metal-gate structure is the *doped-polysilicon-gate* CCD, depicted in Fig. 11-7*a* where the bare gaps are covered with high-resistivity polysilicon to eliminate device instabilities. The difficulty with this method is that the doping of polysilicon is not precisely localized, resulting in large cell dimensions. Another sealed-channel structure uses a *triple-polysilicon gate*, as illustrated in Fig. 11-7*b*. Each polysilicon gate is covered with oxide and is isolated from others so that shorts between gates are unlikely. With this technology it is possible to fabricate a small CCD. The drawback is the complexity of the process.

The Two-Phase CCD

In a three-phase CCD, the potential well is symmetrical, so that charge can flow to the right or left. Directionality of the signal flow is provided by blocking the charge transfer in one direction with appropriate *external* gate

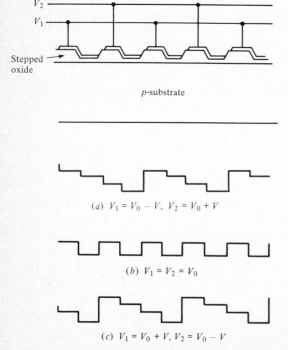

(a) $V_1 = V_0 - V$, $V_2 = V_0 + V$

(b) $V_1 = V_2 = V_0$

(c) $V_1 = V_0 + V$, $V_2 = V_0 - V$

Figure 11-8 Potential-well diagrams of a two-phase charge-coupled device.

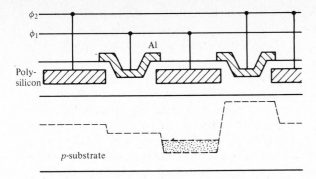

Figure 11-9 Two-phase poly-silicon-aluminum-gate structure. *(After Kosonocky and Carnes [4].)*

voltages. If the potential well is constructed to provide *built-in* directionality, we obtain a *two-phase* CCD system, as shown in Fig. 11-8. Note that the oxide thickness is stepped, so that a different potential appears beneath each individual electrode. To facilitate charge transfer, the potentials on adjacent electrodes are alternated between $V_o + V$ and $V_o - V$ to obtain unsymmetrical potential distribution. In both potential diagrams shown in Fig. 11-8a and c, the signal is always directed toward the right. Because of the stepped oxide, the two-phase CCD can be operated satisfactorily without overlapping clock-voltage pulses, as in a three-phase system. Interconnections to the electrodes can be made easily since no cross-under is needed.

A popular fabrication technology for building a two-phase CCD is the *polysilicon-aluminum gate* using the silicon-gate process described in Sec. 8-9. After the formation of the silicon gates, a thermal oxidation is performed to cover the entire wafer. Subsequently, aluminum gates are produced by metallization and etching in areas between the polysilicon gates leading to the structure shown in Fig. 11-9 [4, 5]. Commonly used layout rules are 0.1-mil gate overlap and 0.2-mil silicon gate. The main disadvantage of this structure is its large RC time-constant delay in charging the electrodes due to the high resistivity of the polysilicon gates.

11-4 TRANSFER EFFICIENCY

When a charge packet is traveling along the CCD, a small fraction of charge is left behind at each transfer. The fraction of charge transferred from one potential well to the next is known as the *transfer efficiency* η. The fraction left behind is known as *transfer inefficiency* ϵ so that $\eta + \epsilon = 1$. It is obvious that the longer we allow the transfer to take place, the more charge is moved over to the next potential well. Therefore, the fraction of charge left behind is a function of time. Experimentally, it has been observed that most of the charge appears to transfer rapidly, whereas a small fraction b of the total charge packet transfers more slowly with an exponential time constant τ.

Thus, the slower charge transfer limits the frequency response of the device, and the transfer efficiency conforms to

$$\eta = 1 - be^{-t/\tau} \tag{11-11}$$

Let us examine the carrier behavior during the process of transfer carefully. In the beginning, the charge packet is very dense and localized with large density gradient at the edges of the well. For a short time, a large fraction of the packet is transferred over due to the strong repulsive force between the electrons. As time evolves, this repulsive force decreases, and electrons are transported by means of thermal diffusion and/or drift caused by the fringing field. Usually, it is the last two mechanisms that are responsible for the loss of charge, although the first mechanism can also be used to improve the transfer efficiency.

For a small amount of signal charge, the transfer mode is governed by thermal diffusion. This mechanism leads to an exponential decay of charge under the transferring electrode with a time constant given by [6]

$$\tau = \frac{L^2}{2.5D} \tag{11-12}$$

where D is the carrier diffusivity and L is the center-to-center electrode spacing. By means of thermal diffusion alone in a p-channel CCD, 99.99 percent of the charge is removed each cycle at a clock frequency of

$$f_c = \frac{5.6 \times 10^7}{L^2} \quad \text{Hz} \tag{11-13}$$

where $D = 6.75 \text{ cm}^2/\text{s}$ for holes at the surface and L is in micrometers.

The charge-transfer process can be speeded up by the fringing field established between the electrodes in the direction of charge propagation along the channel. This field has its maxima at the boundaries between electrodes and minima at the centers of the transferring electrodes. The magnitude of the fringing field increases with gate voltage and oxide thickness and decreases with gate length and substrate doping density. Computer simulation has been performed for this process, and the results are plotted in Fig. 11-10 for transfer time at 99.99 percent of charge transfer vs. gate length with the substrate doping as a variable. The thermal-diffusion time is also plotted. Above the dashed line, it takes longer to remove the signal charge by means of the fringing field, so that the effect of thermal diffusion dominates. Below the dashed line, the charge is transferred by the fringing field. Thus, we obtain 99.99 percent charge transfer at a clock frequency of 10 MHz if the substrate doping is 10^{15} cm^{-3} and $L = 7 \, \mu\text{m}$.

The self-induced drift in the beginning of transfer is caused by the repulsive force between the carriers with the same sign of charge. For this reason, it is important only when the signal density is large (typically greater than 10^{10} cm^{-2}), and in most cases the transfer of the first 99 percent of charge obeys this mechanism. In some CCDs, the entire channel is filled with a large

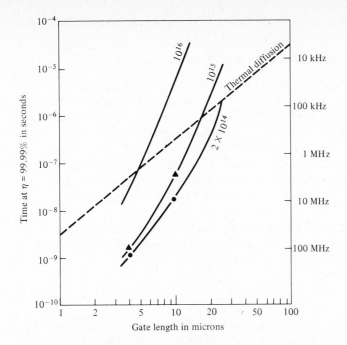

Figure 11-10 Time required to achieve 99.99 percent transfer efficiency as a function of gate length for various substrate doping densities; $\mu_p = 250 \text{ cm}^2/\text{V-s}$, $x_\sigma = 2000$ Å, $V_{\text{substrate}} = 7$ V, and $V_G = 10$ V(pulse). *(After Carnes et al. [7].)*

background charge, known as *fat zero*, to improve the transfer efficiency. Self-induced drift is significant under this mode of operation.

In practical devices, the above consideration overestimates the transfer efficiency because charges can be trapped in surface states. While the capture rate of carriers in these states is proportional to the free-carrier density, the empty rate depends only on the energy level of the surface states. Consequently, the filling rate can be much faster than the empty rate, and some carriers are trapped there to be released later as noise. This type of loss can be reduced by propagating a small background charge throughout the channel to fill these traps.

If a charge packet with an initial amplitude of A_o is traveling down a three-phase CCD shift register, the charge amplitude after m stages is

$$A_n = A_o(1 - 3m\epsilon) = A_o(1 - n\epsilon) \tag{11-14}$$

where $n = 3m$ is the number of transfers. For a sinusoidal input signal with a frequency f, the output amplitude has been derived and is given here without proof [8]:

$$\frac{A_n}{A_o} = \exp\left[-n\epsilon\left(1 - \cos\frac{2\pi f}{f_c}\right)\right] \tag{11-15}$$

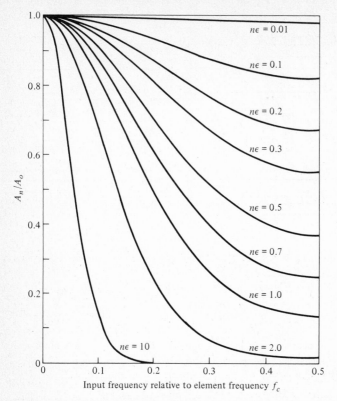

Figure 11.11 The frequency response of a CCD plotted for different values of the transfer inefficiency $n\epsilon$. *(After Joyce and Bertram [8].)*

where $f_c = 1/T$ is the operating clock frequency. This result is plotted in Fig. 11-11. We observe that the attenuation in amplitude is significant for a large $n\epsilon$ product and for $f/f_c > 0.1$. An additional phase delay with respect to the ideal case is

$$\Delta\phi = n\epsilon \, \sin \frac{2\pi f}{f_c} \qquad (11\text{-}16)$$

Equations (11-15) and (11-16) can be used to give an estimate of performance degradation in a particular application due to transfer inefficiency.

11-5 CHARGE INJECTION, DETECTION, AND REGENERATION

Charge packets are injected either electrically or optically. While optical injection is necessary in an image sensor, electrical injection is preferred in a shift register or delay line. Figure 11-12*a* illustrates the method using a *p-n*

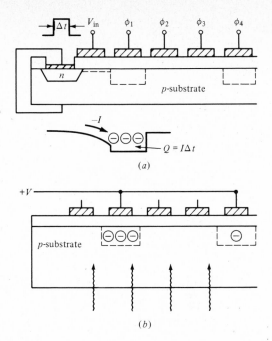

Figure 11-12 Charge injection with (a) a p-n junction and (b) a light source.

junction to introduce minority carriers. The n-type diffusion, also known as the *source*, is short-circuited to the substrate. When a positive pulse is applied to the input gate, it allows the electrons to flow from the source to the potential well under the ϕ_1 electrode. The current source keeps filling the first potential for the duration of the pulse Δt of the input. This is similar to the carrier flow from source to drain in the inversion layer of an MOS transistor. By biasing the source and input gate, the p-n junction injection can be made more efficient. A possible optical-injection technique is shown in Fig. 11-12b, where optically generated minority carriers are attracted by the electrodes and accumulate in the potential wells.

Methods of charge detection are illustrated in Fig. 11-13. A simple approach of sensing the charge in a CCD is by means of a p-n junction diode at the end of the line, as shown in Fig. 11-13a. The diode is reverse-biased, so that it acts as a drain. When the signal charge reaches the drain, a current spike is detected at the output as a capacitive charging current. This is known as the *current-sensing method*. The most popular scheme in charge detection is the *charge-sensing* method shown in Fig. 11-13b, in which the floating diffusion region is periodically reset to a reference potential V_D through the reset gate. When the signal charge of the CCD reaches the floating diffusion region, the voltage there becomes a function of the signal charge. The variation of voltage at the floating region is detected via an MOS transistor amplifier, as shown.

As a charge packet is traveling along a CCD, its magnitude is changed

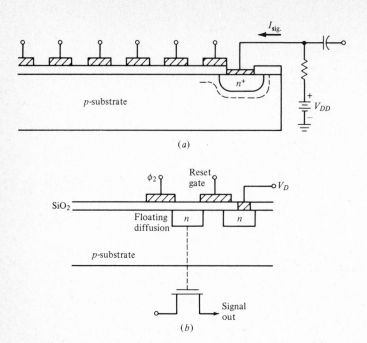

Figure 11-13 Charge-detection methods: (*a*) a reverse-biased *p-n*-junction collector and (*b*) a charge-sensing floating-gate amplifier.

because of both transfer inefficiency and thermal generation of carriers. If we want to store information longer than the thermal-relaxation time, the charge packet must be regenerated or refreshed periodically. In the case of binary bits, a circuit can be set up to measure the charge packets at the end of a transfer channel and to compare them with a given threshold value. For each packet that exceeds the threshold, a new full-sized packet is injected to replace the old one. A refreshing scheme is shown in Fig. 11-14 in its simplified form. Note that the two CCD transfer channels are oriented vertically rather than horizontally, as in previous CCD diagrams. A sensing *n* diffusion is fabricated under a transfer electrode in the first channel. Since the silicon surface is always depleted in charge-coupled operation, this diffusion region will adopt the surface potential beneath the electrode. The surface potential can now be measured by connecting the sensing diffusion to the gate of an MOSFET which serves as the input stage of the second transfer channel. The drain of the MOSFET is under the first transfer electrode, and the source is either grounded or biased positively. When a full charge packet (logic ONE) arrives at the sensing diffusion, the surface potential ψ_s is at its lowest value (Fig. 11-3*c*). Therefore, the gate voltage is below the threshold voltage of the transistor, and the drain registers a logic ZERO. On the other hand, the absence of a charge packet in the sensing diffusion produces the highest ψ_s (Fig. 11-3*b*), so that the MOSFET is turned on and a charge packet

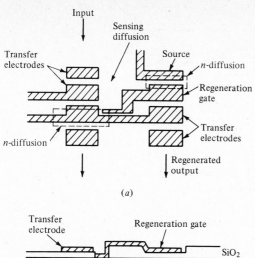

Figure 11-14 A charge-refreshing scheme: (*a*) top view and (*b*) cross-sectional view. *(After Tompsett [9].)*

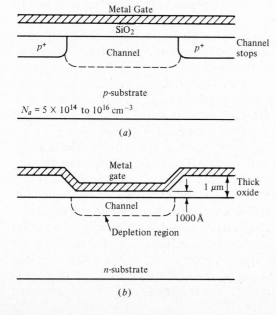

Figure 11-15 Channel-confinement techniques: (*a*) channel-stop diffusion and (*b*) step oxide structure. Note that the channel is oriented in the direction normal to the paper, i.e., a 90° turn from that of Fig. 11-2.

is injected from the source to the drain delivering a logic ONE. In either case an inverse bit is regenerated.

An efficient charge transfer requires the charge packets to be confined to a narrow channel to prevent charge leaking and to reduce charge trapping by surface states. This is accomplished either by a channel-stop diffusion or an oxide step, as shown in Fig. 11-15. The p^+ channel-stop diffusion in Fig. 11-15a is the same p^+ diffusion shown in Fig. 11-2. The use of an oxide step requires that the threshold voltage of the thick oxide area outside the channel be such that the surface is never depleted. It is compatible with p-channel devices on (111) wafers with a low-resistivity n substrate because the fixed positive oxide charge increases the threshold voltage of the thick oxide area. The diffusion channel-stop method is most widely used and is applicable to both n- and p-channel CCDs.

11-6 BULK- (BURIED-) CHANNEL CHARGE-COUPLED DEVICES [10]

The devices described so far store and transfer the charge in potential wells at the silicon surface under the silicon dioxide and are known as *surface-channel* charge-coupled devices (SCCDs). As discussed earlier, surface states may have a strong influence on the transfer loss and noise, particularly if the signal level is low. These difficulties can be alleviated if the channel is moved away from the Si–SiO$_2$ interface. This results in a *bulk-* or *buried-channel* CCD (BCCD) as shown in Fig. 11-16. The channel is formed by a thin epitaxial or diffused n layer on a p substrate. The energy-band diagram under thermal equilibrium is drawn in Fig. 11-17a. Now if a large positive voltage is applied to the channel via the input and output diodes, the majority carriers in the channel will be completely depleted. A depletion layer is formed on both the reverse-biased p-n junction and under the surface of the MOS capacitors.

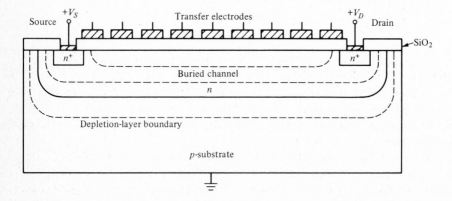

Figure 11-16 The structure and biasing of a buried-channel charge-coupled device (BCCD).

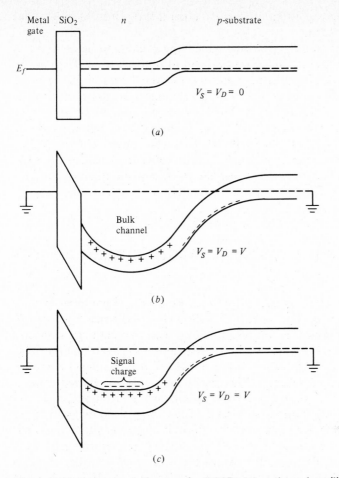

Figure 11-17 Energy-band diagram of a BCCD: (*a*) at thermal equilibrium, (*b*) under complete depletion of all mobile carriers in the channel by reverse bias, and (*c*) with additional signal charge.

The energy-band diagram in Fig. 11-17*b* illustrates the potential well in the *n*-type layer. The corresponding channel is shown as the dashed lines in Fig. 11-16. A mobile signal charge (electron) injected from the source will stay in the potential well to produce a flat portion, as depicted in Fig. 11-17*c*. This well can store and transfer charge by applying appropriate clock pulses to the electrodes, just as in an SCCD.

The advantages of the BCCD include the elimination of the surface-state trapping effect and increased carrier mobility. Therefore, a transfer efficiency of 99.99 percent or higher is realizable without fat zero at room temperature. Furthermore, these devices can be constructed with large fringing fields since the bulk channel is farther away from the electrodes. As a result, a high clock frequency in the range of 100 MHz is obtainable. The only drawback is the added processing complexity and smaller capacitance that reduces signal-handling capability.

11-7 INTEGRATED BUCKET-BRIGADE DEVICES [11]

As pointed out in Sec. 11-1, a bucket-brigade device can be built using either discrete or integrated components with the MOS transistors acting as ON-OFF switches. An integrated BBD is illustrated in Fig. 11-18a. By overlapping the gate structure as shown, the capacitors are combined with the transistors so that a diffused region serves as the drain for one transistor and the source for its neighboring transistor. The structure can be analyzed in terms of the potential wells and barrier diagrams similar to that of a two-phase CCD. In the storage mode, the voltage on ϕ_1 and ϕ_2 electrodes is the same, and all potential wells are about equal, as seen in Fig. 11-18b. When a positive voltage is applied to the ϕ_2 electrodes, i.e., for $V_2 > V_1$, the potential barrier between the first and second wells is cut down and charge will flow from the first to the second n region, as depicted in Fig. 11-18c.

The BBD can also be realized by using JFETs or bipolar transistors as the

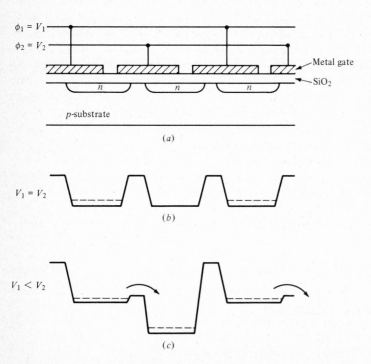

Figure 11-18 An integrated bucket-brigade device: (a) cross section of BBD, (b) surface-potential distribution in storage mode, and (c) surface-potential distribution during charge transfer.

switches. The bipolar transistor has low input impedance and high base current and is not suitable for BBD applications. The JFET BBD corresponds to the BCCD in that the channel is formed in the bulk. Its speed and transfer efficiency are superior to the MOSBBD. The improved performance is achieved at the expense of increased processing complexity. However, the charge-transfer inefficiency, typically 10^{-3}, is substantially worse than that of BCCDs, which can be as small as 10^{-5}.

11-8 CHARGE-COUPLED IMAGERS [12]

An imaging device converts an optical image into electrical signals. In a standard Vidicon TV camera system, the optical image is converted into charge pulses by a photodiode matrix upon exposure [13]. The collected charge in the diode matrix is then discharged by an electron beam which scans the individual diodes and produces discharging current pulses. A *charge-coupled imager* (CCI) is a self-scanning system that does away with the electron beam. It is a new device that will undoubtedly find more applications in the future.

To convert an optical signal into electrical pulses, we can use either the front illumination shown in Fig. 11-19 or back illumination shown in Fig. 11-12b. Electron-hole pairs are generated by the light source. If we introduce storage potential wells at all the CCD stages by applying appropriate clock pulses, the photogenerated minority carriers will be collected in these wells for a time called the *optical integration time*. The collected charge packets are shifted down the CCD register and are converted into current or voltage pulses at the output terminal. Usually, the optical integration time is much longer than the total shifting time of all stored charge packets to avoid smearing the image. Presently, two types of CCD imagers are available commercially, the *line imager* and *area imager*.

A practical CCD line imager is shown in Fig. 11-20. The shaded areas represent photosensing elements with potential wells for optical charge integration. After charge packets are accumulated, they are first transferred into two parallel CCD shift registers; then they are shifted to the output, following the directions of the arrows. This is the basic feature of a 256-stage buried-

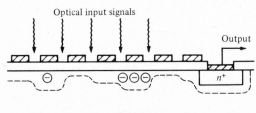

Output

n^+

p substrate

Figure 11-19 Carrier generation in a CCI by front illumination.

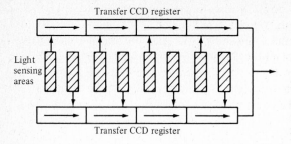

Transfer CCD register

Light sensing areas

Transfer CCD register

Figure 11-20 Charge-coupled line imager with two parallel CCD shift registers.

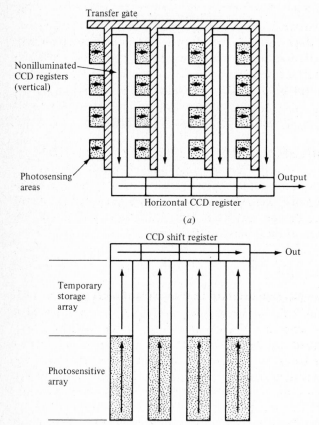

Transfer gate

Nonilluminated CCD registers (vertical)

Photosensing areas

Output

Horizontal CCD register

(a)

CCD shift register

Out

Temporary storage array

Photosensitive array

(b)

Figure 11-21 Charge-coupled area imagers with *(a)* interline transfer and *(b)* frame transfer. *(After Séquin and Tompsett [12].)*

channel line imager produced by Fairchild. Line imagers with 1500 elements or more have been reported in the literature.

The two types of CCD area imaging systems are distinguished by the method of transfer, *interline transfer* and *frame transfer*, shown in Fig. 11-21. In essence, the interline-transfer CCI can be visualized as stacking line imagers in parallel. Charge packets are transferred into parallel shift-register lines and sequentially shifted down to the output register, which is read by transferring the charge horizontally. The frame-transfer system has an opaque temporary storage array with the same number of elements as the photosensing array. The charge packets under the photosensing array are transferred over to the temporary storage array as a frame of a picture. Subsequently, the information in the temporary storage array is shifted down one by one to the output shift register and transferred out horizontally.

The limitations of a CCI result from transfer inefficiency and dark current. These undesirable effects introduce noise and distortion of the input video signal. The dark current is produced by thermally generated minority carriers in the absence of light. This current should be a small fraction of the expected video current, particularly if it is nonuniform over the imaging area. There are three principal dark-current components: one arising from carriers generated in the bulk depletion region, one from the neutral bulk, and one from the semiconductor surface. For quality bulk material with long lifetimes, say $100 \mu s$, the dark current is essentially surface-dominated and is on the order of $30 \, nA/cm^2$.

11-9 APPLICATIONS IN SIGNAL PROCESSING AND MEMORY

In addition to applications in imaging, CCDs can be used as an analog signal processor or a digital memory. Since this is a fast-developing field, it is not possible to describe these devices in detail, and the following discussion is intended as an introduction to these new applications.

Analog Signal Processing [14]

A CCD or BBD can be considered as a controllable delay line. The time delay is controlled by the clock pulse rate applied at the electrodes. An analog signal voltage is usually sampled and converted into charge packets. These charge packets are delayed through the CTD with a delay time readily adjustable by varying the clock frequency. It is necessary, however, that the maximum frequency component of the input signal be less than one-half the sampling frequency, as specified by Shannon's theory. The analog delay line is just a simple linear CTD identical to the shift register discussed earlier.

According to the time-delay principle, a CTD can be used for a time-division multiplexing system, as illustrated in Fig. 11-22. Data from many input channels are fed into the CTD multiplexer in parallel and subsequently read out in serial form. After passing through a transmission medium, the serial data are converted back to parallel form in the demultiplexer, as shown.

If a CTD delay line has a built-in sensing element in each stage, it can be used as a tapped delay line. A tapped delay line can be built into a frequency-sensitive device known as the *transversal filter* [15]. A schematic diagram of

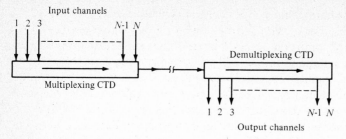

Figure 11-22 A charge-transfer-device time-domain multiplexing system.

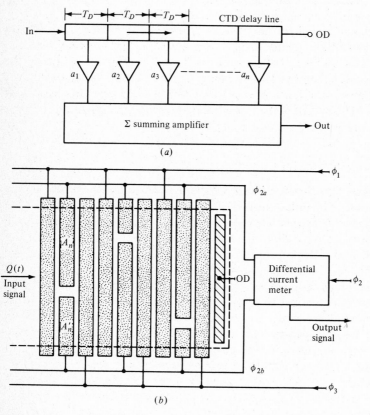

Figure 11-23 (*a*) Schematic diagram and (*b*) implementation of a CCD transversal filter. (*After Buss et al. [15].*)

the transversal filter is shown in Fig. 11-23. Other applications include waveform generation, correlation, and real-time Fourier transformation.

Let us study the operating principle of the transversal filter. The multiply-and-add function is implemented by the split-electrode technique, shown in Fig. 11-23*b*, in which the electrodes of one phase, in this case ϕ_2, are split (only three stages are shown here). One side of each split electrode is connected to the ϕ_{2a} clock line and the other to the ϕ_{2b} clock line. The two ϕ_2 lines are clocked simultaneously with the same voltage clock pulses. The principle of operation of the split-electrode technique is that as charge is transferred within the silicon substrate into the region under an electrode, an opposite charge must flow onto the electrode from the clock line. The charge flow into each section of the split electrode is proportional to the signal charge and the area of the split section of the electrode. By measuring the difference in currents in the two clock lines, we can sample the signal charge with a weighting coefficient for each stage: $a_n = (A_n - A_n')/(A_n + A_n')$, where A_n and A_n' are the area of the upper and lower sections, respectively. For example, let us consider an electrode that is exactly split in the middle, i.e., $A_n = A_n'$ and then $a_n = 0$. Similarly, we find the weighting coefficient $a_n = 1$ for $A_n = 1$ and $A_n' = 0$, and $a_n = -1$ for $A_n = 0$ and $A_n' = 1$. The value of the weighting coefficient of the split electrodes in Fig. 11-23b ranges between $+1$ and -1. The accuracy of the tap weight is limited by the tolerance of photolithography. The summing function takes place automatically because the ϕ_{2a} electrodes are connected together and the ϕ_{2b} electrodes are tied in the same way. The output of the filter is obtained by integrating the difference current, as shown by the differential current integrator in Fig. 11-23b.

Digital Memory Systems [16]

Since a shift register is basically a temporary memory, a CTD can be used for digital data storage. The only limitation is caused by the loss of charge during storage and transfer. For this reason, it is necessary to refresh or regenerate the memory periodically so that the stored bits can be maintained.

Although it is possible to make a CTD RAM, the serial nature of the CTDs makes them suitable in truly serial, multiple-loop or serial-parallel-serial systems in which the data are recirculated and periodically refreshed. The expected access time is between 10 and 100 μs for a 10^5-bit system and 10 ms for a 10^8-bit system. A possible structure is the serpentine organization shown in Fig. 11-24a, which has signal-refreshing stages along the path of travel. Alternatively, parallel operation of smaller storage loops requires shorter access time, but a decoder is needed to address the individual storage loop. The serial-parallel-serial (SPS) organization is shown in Fig. 11-24c. Data are written into the memory via the upper shift register horizontally at the full bit rate f_c, and each line of information is stepped down vertically one by one. Thus, the vertical transfer is at a much slower rate, that is, f_c/M if there are M stages in the upper horizontal register. Readout is accomplished by means of a lower shift register, also operated at the rate of f_c. Therefore, in a memory with $M \times N$ cells in the vertical array, the total number of transfers for any given charge packet is only $m(M + N)$, where m is the number of phases per cell. The SPS system requires the least amount of refreshing and hence has a higher packing density and lower cost per bit. However, since it has only one single access point, the access time is long.

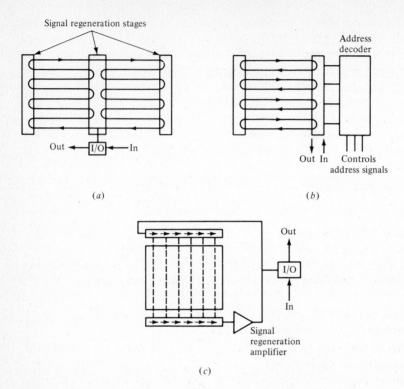

Figure 11-24 CCD memory systems with (*a*) single-loop serpentine organization, (*b*) multiple-loop parallel operation, and (*c*) a single-loop series-parallel-series organization. *(After Kosonocky [16].)*

PROBLEMS

11-1 Sketch the energy-band diagram and charge distribution of an MOS capacitor fabricated on an *n* substrate under the condition of (*a*) deep depletion and (*b*) thermal equilibrium.

11-2 Verify Eq. (11-7).

11-3 An MOS capacitor has the following parameters: substrate with $N_a = 10^{15}$ cm^{-3}, $V_{FB} = 2$ V, $x_o = 1000$ Å, electrode area of 10 by 20 μm. Calculate (*a*) the oxide capacitance, (*b*) the surface potential at $V_G = 10$ V, (*c*) the depletion-layer depth of (*a*), and (*d*) the depletion-layer charge.

11-4 Verify Eq. (11-10).

11-5 (*a*) Calculate the capacitance between the electrode and the substrate of the MOS capacitor of Prob. 11-3 at $V_G = 10$ and $Q_{\text{sig}} = 0$.

    (*b*) A signal-charge packet is injected into the potential well, and the resulting C_{GS} is found to double the value obtained in (*a*). Find the total number of injected electrons and the electron density.

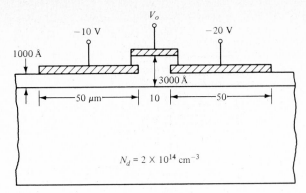

V_o

-10 V $\qquad$ -20 V

1000 Å

3000 Å

$\leftarrow$ 50 μm $\rightarrow$ | 10 | $\leftarrow$ 50 $\rightarrow$

$N_d = 2 \times 10^{14}$ cm^{-3}

Figure P11-10.

11-6 A CCD is fabricated on a p-type substrate with $N_a = 2 \times 10^{14}$ cm^{-3}. The oxide thickness is 1500 Å, and the electrode area is 10 by 20 μm.

(*a*) Calculate the surface potential and depletion-layer depth for two adjacent electrodes biased at $V_G = 10$ and 20 V, respectively. Assume $V_{FB} = 0$ and $Q_{sig} = 0$.

(*b*) Repeat (*a*) after 10^6 electrons are introduced into the cell.

(*c*) Sketch the potential-well diagram for (*b*).

11-7 (*a*) What is the fringing field at the electrode boundary in Prob. 11-6*a* if the interelectrode spacing is 3 μm?

(*b*) Assume that before the charge transfer, 10^6 electrons are uniformly distributed in a potential well with $V_G = 10$ V. Estimate the time required to transfer all the electrons to an adjacent well with $V_G = 20$ V by means of the fringing-field current.

11-8 (*a*) Sketch the cross-sectional diagram of a CCD two-stage shift register together with the necessary charge-generation and sensing devices. Assume an interelectrode spacing of 3 μm and electrode area of 10 by 50 μm.

(*b*) Find the maximum clock rate of the device for both n- and p-channel structures if carriers are assumed to be transferred by diffusion. Use $\mu_p = 200$ cm^2/V-s and $\mu_n = 650$ cm^2/V-s.

11-9 A three-phase CCD analog delay line has 1000 stages and is operated at a clock of 1 MHz. The transfer inefficiency is 10^{-4}. Determine the attenuation and phase-shift degradation for 100-kHz signal and for its third harmonic.

11-10 Assume an interface density (Q_{ss}) of 10^{11} cm^{-2} in Fig. P11-10. (*a*) Sketch the potential diagram for the CCD cell shown with $V_o = 0$. (*b*) What is the required V_o to facilitate charge transfer?

REFERENCES

1. G. F. Amelio, W. J. Bertram, Jr., and M. F. Tompsett, Charge-coupled Imaging Devices: Design Considerations, *IEEE Trans. Electron. Devices*, **ED-18**: 986 (1971).

2. C-K Kim and E. H. Snow, *p*-Channel Charge Coupled Devices with Resistive Gate Structure, *Appl. Phys. Lett.*, **20**: 514 (1972).

3. W. J. Bertram et al., A Three-Level Metallization Three-Phase CCD, *IEEE Trans. Electron. Devices*, **ED-21**: 758 (1974).

4. W. F. Kosonocky and J. E. Carnes, Design and Performance of Two Phase CCDs with Overlapping Polysilicon and Aluminum Gates, *Int. Electron. Device Meet., Washington, 1973, Tech. Dig.*, p. 123.

5. C-K Kim, Two-Phase Charge Coupled Linear Imaging Devices with Self-aligned Barrier, *Int. Electron. Device Meet., Washington, 1974, Tech. Dig.*, p. 55.

6. C-K Kim and M. Lenzlinger, Charge Transfer in Charge Coupled Devices, *J. Appl. Phys.*, **42**: 3586 (1971).

7. J. E. Carnes, W. E. Kosonocky, and E. G. Ramberg, Drift-aiding Fringing Fields in Charge-coupled Devices, *IEEE J. Solid-State Circuits*, **SC-6**: 322 (1971).

8. W. B. Joyce and W. J. Bertram, Linearized Dispersion Relation and Green's Function for Discrete Charge Transfer Devices with Incomplete Transfer, *Bell Syst. Tech. J.*, **50**: 1741 (1971).

9. M. F. Tompsett, A Simple Charge Regenerator for Use with Charge-Transfer Devices and the Design of Functional Logic Arrays, *IEEE J. Solid-State Circuits*, **SC-7**: 237 (1972).

10. C-K Kim, Design and Operation of Buried Channel Charge Coupled Devices, *CCD Appl. Conf. Proc. Nav. Electron. Lab., San Diego, Calif., September 1973.*

11. F. L. J. Sangster, Integrated MOS and Bipolar Analog Delay Line Using Bucket-Brigade Capacitor Storage, *IEEE Solid-State Circuits Conf., 1970, Dig.*, p. 74.

12. C. H. Séquin and M. F. Tompsett, "Charge Transfer Devices," chap. 5, Academic, New York, 1975.

13. M. H. Crowell and E. F. Labuda, The Silicon Diode Array Camera Tube, *Bell Syst. Tech. J.*, **48**: 1481 (1969).

14. D. D. Buss, W. H. Bailey, R. W. Brodersen, and C. R. Reeves, Signal Processing Applications of Charge Coupled Devices, *WESCON Rec.*, pap. 2.2 (1974).

15. D. D. Buss, D. R. Collins, W. H. Bailey, and C. R. Reeves, Transversal Filtering Using Charge-Transfer Devices, *IEEE J. Solid-State Circuits*, **SC-8**: 138 (1973).

16. W. F. Kosonocky, Charge-Coupled Devices: An Overview, *WESCON Rec.*, pap. 2.1 (1974).

ADDITIONAL READINGS

Amelio, G. F.: Charge Coupled Devices, *Sci. Am.*, **230**(8): 22 (1974).

Boyle, W. S., and G. E. Smith: Charge Coupled Devices: A New Approach to MIS Device Structures, *IEEE Spectrum*, **8**: 18 (1971).

Séquin, C. H., and M. F. Tompsett: "Charge Transfer Devices," Academic, New York, 1975.

ATOMS, ELECTRONS, AND ENERGY BANDS

This appendix has been prepared for students not previously exposed to the theories of modern physics. It begins with the atomic model according to Bohr, which is followed by a discussion of the concept of wave-particle duality. Schrödinger's equation is presented, along with quantization of energy levels. Electronic structures of elements and the energy-band theory are then introduced.

A-1 THE BOHR ATOM

In the simplest model, the atom consists of a positively charged nucleus and negatively charged electrons. The total charge of all the electrons is equal to that of the nucleus, so that the atom as a whole is electrically neutral. Because the nucleus contains nearly all the mass of the atom, it is essentially immobile, whereas the electrons circle around in closed orbits.

Let us consider the case of hydrogen, which has only one electron per atom (Fig. A-1). Two forces are established between the nucleus and the electron. The first is the force of attraction described by Coulomb's law

$$F_1 = \frac{q^2}{4\pi\epsilon_o r^2} \tag{A-1}$$

where q = electronic charge
ϵ_o = permittivity of free space
r = separation of charged particles

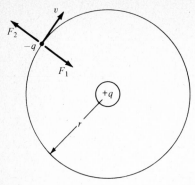

Figure A-1 The hydrogen atom according to the Bohr model.

The second is the centrifugal force resulting from the orbital motion of the electron,

$$F_2 = \frac{mv^2}{r} \tag{A-2}$$

where m is the free-electron mass and v is the speed of the electron in its circular orbit. Since a stationary orbit is reached when the two forces are equal,

$$\frac{q^2}{4\pi\epsilon_o r^2} = \frac{mv^2}{r} \tag{A-3}$$

By choosing the potential energy at infinity as zero reference, we find that the potential energy of the electron is

$$U = \int_{\infty}^{r} F_1 \, dr = -\frac{q^2}{4\pi\epsilon_o r} \tag{A-4}$$

Since the kinetic energy of the electron is $mv^2/2$, the total energy is

$$E = \frac{mv^2}{2} - \frac{q^2}{4\pi\epsilon_o r} = -\frac{q^2}{8\pi\epsilon_o r} \tag{A-5}$$

The last term in this equation, obtained by making use of Eq. (A-3), demonstrates that the energy of the electron becomes smaller, i.e., more negative, as it approaches the nucleus.

According to Bohr, the electron may assume only certain discrete levels of energy in a stationary orbit. As long as the electron maintains a particular orbit, it cannot radiate or absorb energy. However, radiation does take place when the electron makes a transition from an orbit of higher energy E_2 to one of lower energy E_1. The radiation consists of a quantum of light, or *photon*, whose energy is given by

$$E_2 - E_1 = h\nu \tag{A-6}$$

where h = Planck's constant
ν = frequency of radiated energy
$h\nu$ = photon energy

Bohr further postulated that a stationary orbit is determined by the condition in which the angular momentum of the electron is quantized and given by

$$mvr = \frac{nh}{2\pi} \tag{A-7}$$

where n is an integer greater than or equal to 1.

Solving Eqs. (A-3), (A-5), and (A-7) to eliminate r and v, we obtain

$$E_n = -\frac{q^4 m}{8h^2 \epsilon_o^2 n^2} \tag{A-8}$$

For each value of n there is a corresponding energy E_n, which is discrete. Thus, the quantization of the angular momentum leads to discrete allowable energy levels. Substituting Eq. (A-8) into Eq. (A-6), we can find the frequencies of emitted photons corresponding to the transitions between different energy levels. For each frequency, the corresponding wavelength λ is given by c/v, where c is the velocity of light.

The discrete energies and some of the transition photon wavelengths (in angstrom units) are shown in Fig. A-2. The energy level with $n = 1$, called the *ground* or *normal state*, is occupied by the electron when the temperature is absolute zero and there is no incident radiation. The higher energy levels correspond to n greater than 1 and are known as *excited states*. The excitation is accomplished by radiant energy, and the incident photon must have an energy exactly equal to the difference between the two energy levels involved. Thus,

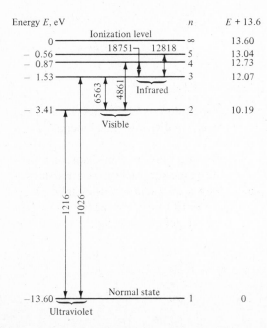

Figure A-2 The lowest five energy levels and the ionization energy of hydrogen. The spectral lines are expressed in angstroms.

$$E_2 = E_1 + h\nu \tag{A-9}$$

If this condition is not satisfied, the photon will not be absorbed.

If sufficient energy is imparted to the electron, it may reach the zero-total-energy level ($n = \infty$ in Fig. A-2) and the electron is said to be free from the influence of the nucleus. The required energy to reach this state, known as the *ionization energy*, is 13.6 eV for hydrogen.

A-2 THE WAVE PROPERTIES OF MATTER; SCHRÖDINGER'S EQUATION

The photon has the dual character of wave and particle. The wave properties are evidenced by the wavelength associated with the photon energy and are based on optical-interference experiments. On the other hand, when a photon is absorbed by an atom, the event takes place in a localized point of space, so that the photon may be visualized as a particle. Since the energy of a photon is $h\nu$ and its velocity is c, the momentum p is $h\nu/c$. Therefore, we have

$$p = \frac{h\nu}{c} = \frac{h}{\lambda} = \frac{hk}{2\pi} \tag{A-10}$$

where the relation $k = 2\pi/\lambda$ has been used to obtain the last expression and k is known as the *wave number*. Equation (A-10) is known as the *de Broglie relation*.

The de Broglie relation is found to be applicable to electrons and atoms, which also exhibit the dual character of wave and particle. A general theory describing the wave properties of matter is given by *Schrödinger's equation* as follows:

$$\nabla^2\psi(x, y, z) + \frac{8\pi^2 m}{h^2}[E - U(x, y, z)]\psi(x, y, z) = 0 \tag{A-11}$$

with

$$\nabla^2 = \frac{\partial^2}{\partial x^2} + \frac{\partial^2}{\partial y^2} + \frac{\partial^2}{\partial z^2} \tag{A-12}$$

where $\psi(x, y, z)$ is known as the *wavefunction*, E is the total energy of the particles, and $U(x, y, z)$ is the potential energy. Physically, the absolute square of ψ defines the probability of finding the electron in a certain space. Thus, $|\psi|^2 \, dx \, dy \, dz$ represents the probability of finding the electron in the volume $dx \, dy \, dz$ around point (x, y, z) in space. Since the probability has a value ranging from 0 to 1, the wavefunction must be normalized so that

$$\int |\psi|^2 \, dx \, dy \, dz = 1 \tag{A-13}$$

The term $E - U$ represents the kinetic energy and is given by

$$E - U = \frac{p^2}{2m} = \frac{h^2 k^2}{8\pi^2 m} \tag{A-14}$$

where Eq. (A-10) has been used. For $U = 0$, we can plot E vs. k and obtain Fig. A-3, known as the E-k diagram. It expresses the energy-momentum relationship for a free particle.

As an example of the solution of Schrödinger's equation, let us consider a potential well in the form of a cube, each side of which has length L. Assume that the inside of the well is at zero potential and the potential is infinite outside of well, so that no electrons can escape. Therefore, the probability of finding an electron outside of the well must be zero; that is, $\psi = 0$ at $x, y, z < 0$ and $x, y, z > L$. For the one-dimensional case (Fig. A-4), we find, using Eqs. (A-11) and (A-14),

$$\frac{d^2\psi(x)}{dx^2} + k^2\psi(x) = 0 \tag{A-15}$$

where

$$k^2 = \frac{8\pi^2 mE}{h^2} \tag{A-16}$$

The general solution of Eq. (A-15) is

$$\psi(x) = A \sin kx + B \cos kx \tag{A-17}$$

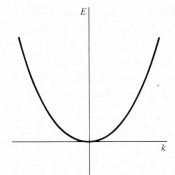

Figure A-3 The E-k diagram for a free particle.

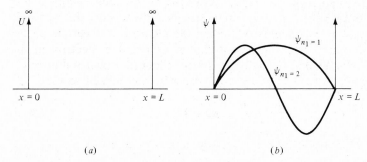

Figure A-4 (a) One-dimensional potential well and (b) the first two eigenfunctions from Eq. (A-22).

Since $\psi(x) = 0$ at $x = 0$, we have $B = 0$. Using the second boundary condition at L, we find

$$kL = n_1\pi \qquad \text{or} \qquad k_n = \frac{n_1\pi}{L} \tag{A-18}$$

where n_1 is a positive integer. Note that either $A = 0$ or $n_1 = 0$ represents the trivial solution in which ψ would vanish everywhere. Using Eq. (A-18), we can rewrite Eq. (A-16) as

$$E = \frac{h^2 k^2}{8\pi^2 m} = \frac{n_1^2 h^2}{8mL^2} \tag{A-19}$$

To complete solution of the Schrödinger equation, normalize ψ by rewriting (A-13) as

$$\int_0^L |\psi(x)|^2 \, dx = 1 \tag{A-20}$$

Thus, we have

$$\int_0^L A^2 \sin^2 \frac{n_1\pi x}{L} \, dx = \frac{A^2 L}{2} = 1 \tag{A-21}$$

or $A = (2/L)^{1/2}$. Consequently, the solution, known as *eigenfunction*, is

$$\psi = \left(\frac{2}{L}\right)^{1/2} \sin \frac{n_1\pi x}{L} \tag{A-22}$$

The eigenfunctions for $n_1 = 1, 2$ are plotted in Fig. A-4b.

For the three-dimensional potential well, solution of the Schrödinger equation is obtained by the technique of separation of variables, and the result is

$$\psi = \left(\frac{8}{L^3}\right)^{1/2} \sin k_1 x \sin k_2 y \sin k_3 z \tag{A-23}$$

where

$$k_1 = \frac{n_1\pi}{L} \qquad k_2 = \frac{n_2\pi}{L} \qquad k_3 = \frac{n_3\pi}{L} \tag{A-24}$$

and

$$k^2 = k_1^2 + k_2^2 + k_3^3 = \frac{8\pi^2 m}{h^2} E \tag{A-25}$$

In Eq. (A-25), each energy state is defined by three wave numbers, k_1, k_2, and k_3. Each set of wave numbers yields a different eigenfunction ψ. From other experimental evidence it is found that an eigenfunction has associated with it two electron states of opposite *spin* sign. Thus, four quantum numbers (n_1, n_2, n_3, s) are needed to define an energy state.

A-3 THE DENSITY OF STATES

In the calculation of electron concentration in a metal or semiconductor, it is necessary to know how many states there are between the energy levels E

and $E + dE$. Let us consider the potential-box problem described in the preceding section and rewrite Eq. (A-25) as

$$E = \frac{h^2}{8mL^2}(n_1^2 + n_2^2 + n_3^2) = \frac{h^2\mathbf{R}^2}{8mL^2} \tag{A-26}$$

where
$$\mathbf{R}^2 = n_1^2 + n_2^2 + n_3^2 \tag{A-27}$$

Each set of integers (n_1, n_2, n_3) determines both ψ and E, and $\mathbf{R}$ represents a vector to a point (n_1, n_2, n_3) in three-dimensional space. In this space, every unit cube ($L = 1$) specifies a state, so that the number of states in any volume is just equal to the numerical value of the volume. Thus, in a sphere of radius $\mathbf{R}$, the number of free electron states N is

$$N = \frac{4\pi|R|^3}{3} \tag{A-28}$$

Since $\mathbf{R}$ and E are related by Eq. (A-26), we can say that the corresponding number of states having a total energy less than E is

$$\frac{4\pi|R|^3}{3} = \frac{4\pi}{3}\left(\frac{8m}{h^2}\right)^{3/2} E^{3/2} \tag{A-29}$$

where we have set $L = 1$. Differentiating Eq. (A-29) with respect to E, we find that the number of states between E and $E + dE$ is

$$2\pi\left(\frac{8m}{h^2}\right)^{3/2} E^{1/2}\, dE \tag{A-30}$$

Since only positive integers are allowed for n_1, n_2, and n_3, we must divide Eq. (A-30) by a factor of 8; that is, each dimension can assume one-half of all integers. Furthermore, we must multiply the result by a factor of 2 to account for the two allowed values for the spin. Thus, we obtain

$$N(E)\, dE = \frac{4\pi(2m)^{3/2}}{h^3} E^{1/2}\, dE \tag{A-31}$$

This quantity is known as the *density-of-states function* per unit volume.

A-4 ELECTRONIC STRUCTURE OF THE ELEMENTS

Using the principle outlined in the previous section, we can solve Schrödinger's equation for hydrogen or any atom with multiple electrons. The solution for the hydrogen atom can be found in most texts in introductory quantum mechanics and is not given here. The potential energy expressed in Eq. (A-4) is first substituted into Eq. (A-11), and an appropriate coordinate system is chosen to simplify the mathematics. In general, it is found that four quantum numbers are required to define an energy state:

1. The principal quantum number $n = 1, 2, 3, \ldots$
2. The orbital angular-momentum quantum number $l = 0, 1, 2, \ldots, n-1$ for each n
3. The orbital magnetic quantum number $m = 0, \pm 1, \pm 2, \pm 3, \ldots, \pm l$ for each l
4. The electron spin $s = +\frac{1}{2}, -\frac{1}{2}$

Each set of quantum numbers specifies an energy state. According to the *Pauli exclusion principle*, no two electrons can have the same set of quantum numbers because each state can accommodate only one electron. However, different quantum states can be *degenerate*, i.e., have the same energy. For any atom with more than one electron, electrons will occupy different quantum states, and the electron configuration can be classified into *shells* and *subshells*.

All the electrons in an atom which have the same principal quantum number are said to belong to the same electron shell. The first four electron shells corresponding to $n = 1, 2, 3, 4$ are identified by the letters K, L, M, N, respectively. A shell is further divided into subshells, corresponding to the values of the orbital angular-momentum quantum number. The first four subshells corresponding to $l = 0, 1, 2, 3$ are identified as s, p, d, f, respectively. The distribution of electrons in an atom among the shells and subshells is tabulated in Table A-1 for the first four shells. To account for all the chemical elements given in the periodic table in Table A-2 (where the atomic number gives the number of electrons per atom) seven shells are required.

In Table A-1, we have two states in the K shell for $n = 1$ since $l = m = 0$ and there are two spins. These are called the $1s$ states. For the L shell with $n = 2$, we have $l = 0, 1$. In the s subshell corresponding to $l = 0$, there are again two states because of the two spins. This is known as the $2s$ subshell, where 2 stands for $n = 2$. In the p subshell, we have $l = 1$ and $m = 0, \pm 1$ which will make six states when the two spins are taken into account. This is known as the $2p$ subshell. The total number of energy states in the L shell is therefore $2 + 6 = 8$, as given in Table A-1.

Table A-1 Electron shells and subshells

Shell n	K 1	L 2		M 3			N 4			
l Subshell	0 s	0 s	1 p	0 s	1 p	2 d	0 s	1 p	2 d	3 f
Number of electrons	2	2	6	2	6	10	2	6	10	14
	2	8		18			32			

Table A-2 Periodic table of the elements†

Period	Group IA	Group IIA	Group IIIB	Group IVB	Group VB	Group VIB	Group VIIB	Group VIII			Group IB	Group IIB	Group IIIA	Group IVA	Group VA	Group VIA	Group VIIA	Inert gases
1	H 1 1.01																	He 2 4.00
2	Li 3 6.94	Be 4 9.01											B 5 10.81	C 6 12.01	N 7 14.01	O 8 16.00	F 9 19.00	Ne 10 20.18
3	Na 11 22.99	Mg 12 24.31											Al 13 26.98	Si 14 28.09	P 15 30.97	S 16 32.06	Cl 17 35.45	Ar 18 39.95
4	K 19 39.10	Ca 20 40.08	Sc 21 44.96	Ti 22 47.90	V 23 50.94	Cr 24 52.00	Mn 25 54.94	Fe 26 55.85	Co 27 58.93	Ni 28 58.71	Cu 29 63.54	Zn 30 65.37	Ga 31 69.72	Ge 32 72.59	As 33 74.92	Se 34 78.96	Br 35 79.91	Kr 36 83.80
5	Rb 37 85.47	Sr 38 87.62	Y 39 88.90	Zr 40 91.22	Nb 41 92.91	Mo 42 95.94	Tc 43 (99)	Ru 44 101.07	Rh 45 102.90	Pd 46 106.4	Ag 47 107.87	Cd 48 112.40	In 49 114.82	Sn 50 118.69	Sb 51 121.75	Te 52 127.60	I 53 126.90	Xe 54 131.30
6	Cs 55 132.90	Ba 56 137.34	La 57 138.91	Hf 72 178.49	Ta 73 180.95	W 74 183.85	Re 75 186.2	Os 76 190.2	Ir 77 192.2	Pt 78 195.09	Au 79 196.97	Hg 80 200.59	Tl 81 204.37	Pb 82 207.19	Bi 83 208.98	Po 84 (210)	At 85 (210)	Rn 86 (222)
7	Fr 87 (223)	Ra 88 (226)	Ac 89 (227)	Th 90 232.04	Pa 91 (231)	U 92 238.04	Np 93 (237)	Pu 94 (242)	Am 95 (243)	Cm 96 (247)	Bk 97 (247)	Cf 98 (251)	Es 99 (254)	Fm 100 (253)	Nd 101 (256)	No 102 (254)	Lw 103 (257)	

The Rare Earths

Ce 58 140.12	Pr 59 140.91	Nd 60 144.24	Pm 61 (147)	Sm 62 150.35	Eu 63 151.96	Gd 64 157.25	Tb 65 158.92	Dy 66 162.50	Ho 67 164.93	Er 68 167.26	Tm 69 168.93	Yb 70 173.04	Lu 71 174.97

† The number to the right of the symbol for the element gives the atomic number. The number below the symbol for the element gives the atomic weight.

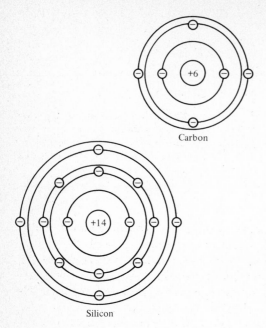

Carbon

Silicon

Figure A-5 Schematic diagrams of carbon and silicon atoms.

The energy levels of an atom are filled from the lowest energy state to the highest energy state. The K shell has the lowest energy and is first occupied. As the atomic number increases, the next subshell will be filled and then the next shell. Thus, for carbon in column IVA in the periodic table with an atomic number of 6, the $1s$ and $2s$ subshells are filled, and only two states in the $2p$ subshell are occupied. The electronic structure of carbon can be designated by $1s^2 2s^2 2p^2$, and its schematic diagram is shown in Fig. A-5, where the shell structure of silicon is also shown. Note that both silicon and carbon have four electrons in the outermost shell, a characteristic common to all elements in the fourth column of the periodic table. These electrons are called the *valence* electrons, and carbon and silicon as well as germanium are known as *tetravalent* elements.

A-5 THE ENERGY-BAND THEORY OF CRYSTALS

In the previous sections we have considered the single, isolated atom, and the theory is applicable to gases in which individual atoms are sufficiently far apart. In a crystalline solid (see Sec. 1-1), the atoms are close enough to interact, and the potential energy $U(x, y, z)$ becomes a periodic function of space, as shown in Fig. A-6. The eigenfunctions of each atom obtained from the solution of Schrödinger's equation overlap those of its neighbors because of the close proximity of atoms. Each energy level in the outermost shell now splits into a number of discrete levels. This phenomenon is similar to that when two *LC*

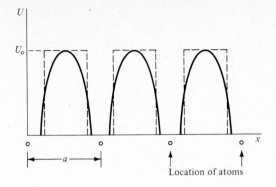

Figure A-6 Variation of the electron's potential energy in a one-dimensional crystal (*solid curves*) along with an idealized model (*dashed lines*) suitable for analytical calculations.

tuning circuits with identical resonant frequencies are brought together, and the resonant curve becomes double-peaked instead of single-peaked.

Let us examine the carbon atom in further detail. The carbon atom has two shells. The first shell has two states which are filled. These are low-energy states, and electrons in these states are tightly bound to the nucleus. The second shell has eight states. Four of them are filled, and the other four are empty. If we were able to vary the atomic spacing between adjacent carbon atoms at will, we would obtain the energy-level distribution shown in Fig. A-7.

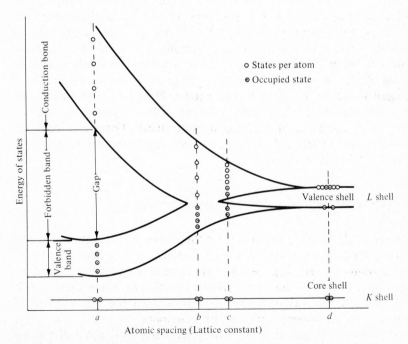

Figure A-7 The energy bands of the carbon lattice as a function of the atomic spacing at $T = 0$ K.

At atomic spacing d, the atoms are sufficiently far apart to behave like isolated atoms. As the atoms are brought closer together at c, the two upper levels split into two bands, having two and six states, respectively. At atomic spacing b, the two bands overlap. As we reduce the atomic spacing further, two of the states from the upper band drop into the lower band, as indicated at atomic spacing a. It turns out that a represents the actual atomic spacing for carbon (as well as for silicon and germanium). At this atomic spacing, the four electrons fill the lower band, which is known as the *valence band*. The upper band is empty, and is known as the *conduction band*. In Fig. A-7, the K shell is shown as a discrete level. Actually, it is also split into a narrow band, but its interaction with neighboring atoms is small and can be neglected.

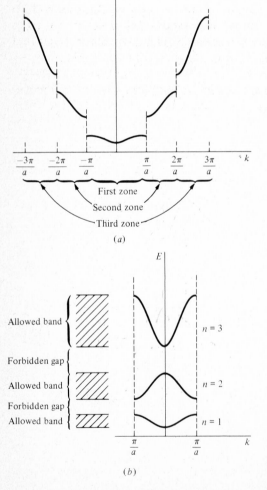

(a)

(b)

Figure A-8 (*a*) The energy as a function of wave number k for the Kronig-Penney model. The discontinuities in energy occur at $k = n\pi/a$, with $n = \pm 1, \pm 2, \pm 3, \ldots$ (*b*) The E-k diagram reduced to the first Brillouin zone. Allowed bands and forbidden gaps are shown.

Mathematically, we can use the function $U(x) = U(x + a)$ to represent the periodic potential energy shown in Fig. A-6 and then solve Schrödinger's equation accordingly. This is known as the *Kronig-Penney model*. The solution, called the *Bloch wave function*, exhibits discrete bands of energy levels as depicted in the *E-k* diagram in Fig. A-8a. The solid curves represent *allowed energy levels* or *bands*. The discontinuities in the diagram indicate that the probability of finding an electron is zero in certain energy regions called *forbidden gaps*. The zones indicated in the diagram are called *Brillouin zones*; they can be simplified to the *reduced* Brillouin zone scheme shown in Fig. A-8b.

The Kronig-Penney model is the first step in understanding the energy-band theory of solids. In crystals such as silicon and gallium arsenide, the periodic potential energy is more complicated, leading to more complex energy-band diagrams. Because these crystals are anisotropic, the *E-k* diagram along the principal axes is not necessarily the same and a three-dimensional picture is required to show all the features of the energy-band model. In practice, two-dimensional band diagrams are used to represent GaAs and Si (Fig. A-9). In these diagrams, a crystalline direction is designated along the *k* axis. Note that the minimum energy of the conduction band is *directly* above the maximum energy of the valence band in GaAs. Therefore, GaAs is known as a *direct-gap* semiconductor. In Si, the conduction-band minimum does not align with $k = 0$ and therefore it is known as an *indirect-gap*

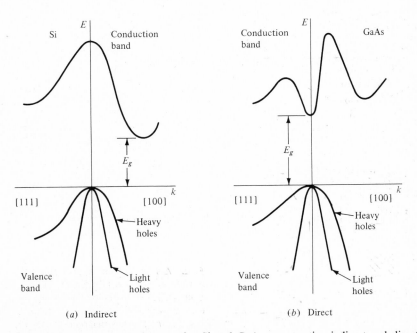

(a) Indirect   (b) Direct

Figure A-9 Energy-band diagrams for Si and GaAs representing indirect and direct band-gap materials, respectively.

semiconductor. The energy-momentum relation shown in Fig. 1-5 is based on these diagrams.

One important piece of information we can obtain from the E-k diagrams is the *effective mass* (see Chap. 1) in a solid. Because of the influence of the periodic potential, the effective mass differs from the electronic mass in free space. By taking the partial derivative of E with respect to k twice in Eq. (A-14) we have

$$\frac{\partial^2 E}{\partial k^2} = \frac{h^2}{4\pi^2 m_e} \tag{A-32}$$

which can be rewritten as

$$m_e = \frac{h^2/4\pi^2}{\partial^2 E/\partial k^2} \tag{A-33}$$

By applying Eq. (A-33) to the conduction band in Fig. A-9 we obtain the effective mass of electrons in the conduction band. Similarly, the effective mass of *holes* (see text) in the valence band can be obtained. Note that the valence bands of Si and GaAs are both degenerate. The heavy-hole band has a smaller $\partial^2 E/\partial k^2$, which yields a larger effective mass from Eq. (A-33). The light-hole band has a larger $\partial^2 E/\partial k^2$ and a smaller effective mass. These finer features in the E-k diagram, though important in specifying the effective masses, are ignored in most device analyses.

ADDITIONAL READINGS

Eisberg, R. M.: "Fundamentals of Modern Physics," Wiley, New York, 1961.
Sproull, R. L.: "Modern Physics," 2d ed., Wiley, New York, 1963.

$$\frac{\partial}{\partial t} \int p \, d\,volume = -\frac{1}{q} \oint J_p \cdot dS + \int (G_L - u) \, d\,volume$$

$$= -\frac{1}{q} \int \nabla \cdot J_p \, d\,volume + \int (G_L - u) \, d\,volume$$

$$\frac{1}{q} \nabla \cdot J_p + u - G_L = \frac{-\partial p}{\partial t}$$

For $\frac{\partial p}{\partial t} = 0$ (1 dimensional problem)

$$\frac{1}{q} \frac{d J_p}{dx} + u = \frac{1}{q} \frac{d J_p}{dx} + \frac{p_n - p_{no}}{\tau_p} = G_L$$

for $E = 0$: $J_p = -q D_p \frac{dp}{dx}$ $D_p \frac{d^2 p}{dx^2} + \frac{p_n - p_{no}}{\tau_p} = G_L$

$$\frac{d^2 p}{dx^2} + \frac{p_n - p_{no}}{L_p^2} = \frac{G_L}{D_p}$$

$$p_n(x) - p_{no} = k_1 e^{-x/L_p} + k_2 e^{x/L_p} + G_L \tau_p$$

$$= G_L \tau_p + k_1 e^{-x/L_p} \quad (k_2 = 0 \text{ for finite } p(x))$$

$$0 < x < +\infty$$

DERIVATION OF THE
THERMIONIC-EMISSION CURRENT

We can rewrite Eq. (1-1) as

$$E = \frac{P^2}{2m} = \frac{1}{2m}(p_x^2 + p_y^2 + p_z^2) \tag{B-1}$$

where m is the free-electron mass. Differentiating (B-1) leads to

$$dE = \frac{P \, dP}{m} \tag{B-2}$$

The conversion to rectangular coordinates is obtained by considering the incremental volume in momentum space between a sphere of radius P and $P + dP$, and the relationship is

$$4\pi P^2 \, dP = dp_x \, dp_y \, dp_z \tag{B-3}$$

Substituting Eqs. (B-1) to (B-3) into Eq. (5-8) yields

$$
\begin{aligned}
I &= \frac{2qA}{mh^3} \int_{P_{xo}}^{\infty} \int_{p_y=-\infty}^{\infty} \int_{p_z=-\infty}^{\infty} p_x e^{-(p_x^2+p_y^2+p_z^2-2mE_f)/2mkT} \, dp_x \, dp_y \, dp_z \\
&= \frac{2qA}{mh^3} \int_{P_{xo}}^{\infty} e^{-(p_x^2-2mE_f)/2mkT} p_x \, dp_x \int_{-\infty}^{\infty} e^{-p_y^2/2mkT} \, dp_y \int_{-\infty}^{\infty} e^{-p_z^2/2mkT} \, dp_z
\end{aligned} \tag{B-4}
$$

where $p_{xo}^2 = 2m(E_f + q\phi_m)$. The last two integrals can be evaluated by using the definite integral

$$\int_{-\infty}^{\infty} e^{-ax^2} \, dx = \left(\frac{\pi}{a}\right)^{1/2} \tag{B-5}$$

and each integral yields $(2\pi mkT)^{1/2}$. The first integral is evaluated by setting

$$\frac{p_x^2 - 2mE_f}{2mkT} = u \tag{B-6}$$

Therefore we have

$$du = \frac{p_x \, dp_x}{mkT} \tag{B-7}$$

The lower limit of the first integral can be written as

$$\frac{2m(E_f + q\phi_m) - 2mE_f}{2mkT} = \frac{q\phi_m}{kT} \tag{B-8}$$

so that the first integral becomes

$$mkT \int_{q\phi_m/kT}^{\infty} e^{-u} \, du = mkTe^{-q\phi_m/kT} \tag{B-9}$$

Substituting the results of (B-5) and (B-9) into Eq. (B-4), we obtain

$$I = A\frac{4\pi mqk^2}{h^3} T^2 e^{-q\phi_m/kT} = ART^2 e^{-q\phi_m/kT} \tag{B-10}$$

where $R = \dfrac{4\pi mqk^2}{h^3}$ and is known as Richardson's constant.

$$\sinh x \equiv \frac{1}{2}(e^x - e^{-x})$$

$$\cosh x \equiv \frac{1}{2}(e^x + e^{-x})$$

SOME PROPERTIES OF THE ERROR FUNCTION

$$\operatorname{erf} z \equiv \frac{2}{\sqrt{\pi}} \int_0^z e^{-a^2}\, da \qquad \operatorname{erfc} z \equiv 1 - \operatorname{erf} z$$

$$\operatorname{erf} 0 = 0 \qquad \operatorname{erf} \infty = 1$$

$$\operatorname{erf} z \approx \frac{2}{\sqrt{\pi}}\, z \qquad \text{for } z \ll 1$$

$$\operatorname{erfc} z \approx \frac{1}{\sqrt{\pi}} \frac{e^{-z^2}}{z} \qquad \text{for } z \gg 1$$

$$\frac{d \operatorname{erf} z}{dz} = \frac{2}{\sqrt{\pi}} e^{-z^2}$$

$$\int_0^z \operatorname{erfc}(z')\, dz' = z \operatorname{erfc} z + \frac{1}{\sqrt{\pi}}(1 - e^{-z^2})$$

$$\int_0^\infty \operatorname{erfc}(z)\, dz = \frac{1}{\sqrt{\pi}}$$

erfc z

z	erfc z	z	erfc z	z	erfc z	z	erfc z
0	1.000 00	1.00	0.157 30	2.00	0.004 68	3.00	0.000 022 09
0.10	0.887 54	1.10	0.119 80	2.10	0.002 98	3.10	0.000 011 65
0.20	0.777 30	1.20	0.089 69	2.20	0.001 86	3.20	0.000 006 03
0.30	0.671 37	1.30	0.065 99	2.30	0.001 14	3.30	0.000 003 06
0.40	0.571 61	1.40	0.047 72	2.40	0.000 689	3.40	0.000 001 52
0.50	0.479 50	1.50	0.033 90	2.50	0.000 407	3.50	0.000 000 743
0.60	0.396 14	1.60	0.023 65	2.60	0.000 236	3.60	0.000 000 356
0.70	0.322 20	1.70	0.016 21	2.70	0.000 134	3.70	0.000 000 167
0.80	0.257 90	1.80	0.010 91	2.80	0.000 075	3.80	0.000 000 77
0.90	0.203 09	1.90	0.007 21	2.90	0.000 041	3.90	0.000 000 35

$$n = n_i \, e^{E_f - E_i' / kT} = n_i \, e^{\psi - \phi / V_T}$$

$$p = n_i \, e^{-(E_f - E_i')/kT} = n_i \, e^{-(\psi - \phi)/V_T}$$

$$n_{no} = n_i \, e^{\psi_n - \phi / V_T} \qquad p_{no} = n_i \, e^{-\frac{(\psi_n - \phi)}{V_T}}$$

$$n_{po} = n_i \, e^{\psi_p - \phi / V_T} \qquad p_{po} = n_i \, e^{-(\psi_p - \phi)/V_T}$$

$$n = n_{po} \, e^{\psi - \psi_p / V_T} = n_{no} \, e^{-(\psi_n - \psi)/V_T}$$

$$p = p_{no} \, e^{\psi_n - \psi / V_T} = p_{po} \, e^{-(\psi - \psi_p)/V_T}$$

$$\left. \begin{array}{c} \psi_n - \psi \\ \psi - \psi_p \end{array} \right\} \text{always positive}$$

$$n_{po} \, p_{po} = n_{no} \, p_{no} = n_i^2$$

$$\psi_o = \psi_n - \psi_p = V_T \, \ln \frac{n_{no}}{n_{po}}$$

$$n(x_2) = n(x_1) \, e^{\frac{\psi(x_2) - \psi(x_1)}{V_T}} \qquad p(x_2) = p(x_1) \, e^{\frac{\psi(x_1) - \psi(x_2)}{V_T}}$$

ANSWERS

ANSWERS

Chapter 1

3. (632)

6. (*a*) 4; (*b*) $\dfrac{a}{\sqrt{2}}$; (*c*) 74%

7. (*a*) 2.35 Å; (*b*) $2.0 \times 10^{23}/cm^3$

8. (*b*) Si $5.0 \times 10^{22}/cm^3$; Ge $4.41 \times 10^{22}/cm^3$

9. (*b*) $m_e = 1.12m$

10. At 77 K: $p = 8.75 \times 10^{14}/cm^3$; $n \simeq 0$

                $E_g = 1.19\,eV$; $E_f - E_v = 48.7\,meV$

   At 600 K: $p = 8.52 \times 10^{15}/cm^3$; $n = 7.52 \times 10^{15}/cm^3$

                $E_g = 1.04\,eV$; $E_f - E_v = 0.42\,eV$

12. Si: $N_d > 1.4 \times 10^{18}/cm^3$; $N_a > 5 \times 10^{17}/cm^3$

   GaAs: $N_d > 2.3 \times 10^{16}/cm^3$; $N_a > 3.5 \times 10^{17}/cm^3$

14. Ge 0.78 eV; Si 1.12 eV; GaAs 1.47 eV

15. Ge 6.6 meV; Si 10.2 meV

16. (*a*) $E_c - E_f = 0.264$; 0.086; -0.033 eV

    (*b*) $N_d^+ = N_d$; $0.68N_d$; $0.02N_d$

17. (*a*) Ge 43 Ω-cm; Si 2.28×10^5 Ω-cm

    (*b*) Ge 3.63 Ω-cm; Si 9.26 Ω-cm

18. (*a*) at 200 K: $n = 10^{13}/cm^3$; $p \simeq 0$

                     $\mu_n = 4.0 \times 10^3\ cm^2/V\text{-}s$

                     $\sigma = 6.4 \times 10^{-3}/\Omega\text{-}cm$

       at 300 K: $n = 10^{13}/cm^3$; $p = 2.25 \times 10^7/cm^3$

                     $\mu_n = 1.3 \times 10^3\ cm^2/V\text{-}s$

                     $\sigma = 2.1 \times 10^{-3}/\Omega\text{-}cm$

    (b) at 200 K: $n = 10^{18}/\text{cm}^3$; $p \approx 0$
                       $\mu_n = 3.8 \times 10^2 \text{ cm}^2/\text{V-s}$
                       $\sigma = 61/\Omega\text{-cm}$
     at 300 K: $n = 10^{18}/\text{cm}^3$; $p = 2.25 \times 10^2/\text{cm}^3$
                       $\mu_n = 3.15 \times 10^2 \text{ cm}^2/\text{V-s}$
                       $\sigma = 50/\Omega\text{-cm}$

19. $\mu_n = 11.5 \text{ cm}^2/\text{V-s}$
20. (a) $p = 10^{16}/\text{cm}^3$; $n = 2.25 \times 10^4/\text{cm}^3$
     $E_f - E_v = 0.18 \text{ eV}$; $\rho = 1.3 \text{ }\Omega\text{-cm}$
   (b) $n = 2 \times 10^{15}/\text{cm}^3$; $p = 1.13 \times 10^5/\text{cm}^3$
     $E_c - E_f = 0.246 \text{ eV}$; $\rho = 2.3 \text{ }\Omega\text{-cm}$
21. (a) $1.35 \times 10^5 \text{ cm/s}$

ANSWERS

Chapter 2

4. $n = 10^{15}/\text{cm}^3$; $p = 10^{11}/\text{cm}^3$
5. (b) $E_c - E_t = 0.50 \text{ eV}$
10. (a) $\mathscr{E} = -V_T/x$; (b) $\mathscr{E} = V_T a$
12. (a) $\sigma = 0.23/\Omega\text{-cm}$; $E_{fn} - E_i = 0.288 \text{ eV}$
                           $E_i - E_{fp} = 0.209 \text{ eV}$
   (b) $G_L = 10^{21}/\text{cm}^3\text{-s}$; $E_{fn} - E_i = 0.304 \text{ eV}$
         $\sigma = 0.509/\Omega\text{-cm}$; $E_i - E_{fp} = 0.287 \text{ eV}$

ANSWERS

Chapter 3

1. $T = 1126°C$
2. $x_j = 5.84 \text{ }\mu\text{m}$
4. (c) 1.18 h; (d) $x_j = 3.41 \text{ }\mu\text{m}$
5. (a) $27 \text{ }\mu\text{m}$; (b) 7.7 min
6. 48.6 min
7. 2300 Å; 3120 Å
8. 3200 Å; 4360 Å
9. 37.7 min
10. (a) 220 keV; 333 μA-s; (b) 33.3 s
11. (a) $5 \times 10^{17} \exp(-x^2/14)/\text{cm}^3$; x in μm; (b) $1.66 \times 10^{14}/\text{cm}^2$
12. (a) $2.67 \text{ }\Omega/\square$; ($b$) 187 mil

ANSWERS

Chapter 4

1. (a) 0.906 V
   (b) $W = 1.09 \text{ }\mu\text{m}$; $|\mathscr{E}_m| = 1.67 \times 10^4 \text{ V/cm}$

14. 345
16. $\psi_o = 1.57$ V; $N_d = 5.03 \times 10^{16}/\text{cm}^3$
17. (a) 4.38 pF; 2.45 pF
    (b) $f_r = 1.7$ MHz; 2.27 MHz
18. (a) 1.36 h; (b) 1.5×10^{-4} pF/μm^2
19. (a) 30 ns
20. $1.1 \times 10^{18}/\text{cm}^3$

ANSWERS

Chapter 5

1. (a) 1.2 μm; (b) 87 pF; (c) 1.84×10^5 V/cm
2. (a) $N_d = 5.63 \times 10^{15}/\text{cm}^3$; $\psi_o = 0.68$ V; $q\phi_b = 0.79$ eV (b) 0.80 eV
4. $\psi_s = 0.48$ V
5. (a) $q\phi_b = 0.95$ eV; $\psi_o = 0.71$ V; $W = 0.96$ μm; (b) 0.12 mA/cm^2
6. (a) $V_n = 0.18$ V; $\psi_o = 0.62$ V; (b) 4.2×10^4
7. 5.04×10^5
9. (a) $\Delta\phi = 1.9 \times 10^{-2}$ V; $x_o = 95$ Å; (b) $\Delta\phi = 1.1 \times 10^{-2}$ V; $x_o = 55$ Å
10. (b) -1.67 mV/°C
11. $f_c = 503$ MHz

ANSWERS

Chapter 6

1. (a) Ge 1.82 μm; Si 1.11 μm; GaAs 0.87 μm; (b) 2.25 eV; 1.82 eV
2. (a) 9×10^{-3} J/s; (b) 2.57×10^{-3} J/s; (c) 2.81×10^{16}/s
5. (b) $R = 4.35$ Ω; 0.8
7. $V = 0.0258 \ln[1 + 7.4 \times 10^{-3} N_a]$
8. (b) 84 μA/cm^2
11. (a) 1.34%; (b) 5.56%; 4.15
12. (a) 55%; (b) at -23°C: 7.78%; at 77°C: 3.14%
13. (a) 2460 fL; (b) 3810 fL; (c) 3720 fL
14. 50 Å

ANSWERS

Chapter 7

1. (a) 0.76 V; (b) 2.31 V; (c) 6.4 m℧; (d) 3.2 m℧

2. $I_D = \dfrac{4q\mu_n N_d a^2}{L} \left\{ V_D + \dfrac{K\epsilon_o}{qN_d a^2} [(V_D + \psi_o - V_G)^2 - (\psi_o - V_G)^2] \right.$

$\left. - \dfrac{4}{3} \left(\dfrac{2K\epsilon_o}{qN_d} \right)^{1/2} [(V_D + \psi_o - V_a)^{3/2} - (\psi_o - V_G)^{3/2}] \right\}$

3. $I_D = \dfrac{2q\mu_n N_d Z}{L}\left\{ V_D - \dfrac{1}{3a}\left(\dfrac{2K\epsilon_o}{qN_d}\right)^{1/2}[(V_D + \psi_o - V_{G1})^{3/2} - (\psi_o - V_{G1})^{3/2}\right.$

$\left. + (V_D + \psi_o - V_{G2})^{3/2} - (\psi_o - V_{G2})^{3/2}]\right\}$

7. (a) 61 MHz; (b) 6.1 MHz; (c) 38.3 GHz
8. 137 kΩ

ANSWERS

Chapter 8

2. (a) $Q_B = -[4qK_s\epsilon_o N_a V_T \ln(N_a/n_i)]^{1/2}$; (b) $\psi_{si} = 2V_T \ln(Na/n_i)$;

(c) $\mathscr{E}|_{x=0}\left[\dfrac{4qN_a V_T \ln(N_a/n_i)}{K_s\epsilon_o}\right]^{1/2}$

3. (a) 17.7 nF/cm²; (b) 35.4 nF/cm²; (c) 17.6 nF/cm²
5. −0.726 V
6. (a) $\Delta V_{FB} = -5.08$ V; (b) $\Delta V_{FB} = -10.16$ V; (c) $\Delta V_{FB} = -3.69$ V
7. −1.48 V
8. 1.4×10^{11} cm⁻²
11. 2.17 mA
12. (a) 36 MHz; (b) 1.4 MHz

13. (a) $I_{DS} = \dfrac{\mu_n C_o}{L}(V_G - V_{TH})^2$; (c) 2.8 kΩ; (d) 28 kΩ

14. (a) $V_{TH} = -4.0$ V; (b) 5.5×10^{11} borons/cm²

ANSWERS

Chapter 9

1. (c) $I_E = I_{pE} + I_{nE} + I_{rg}$
$-I_B = I_{nE} + I_{rg} + (I_{pE} - I_{pC}) - I_{CO}$
$-I_C = I_{pC} + I_{CO}$
3. (a) 3.8×10^{12} cm⁻³; (b) 0.53 V; (c) 1
4. $\alpha = 0.94$; $h_{FE} = 17$
5. I_E: 10^{-3}; 10^{-2}; 10^{-1}; 1; 10
β: 41.0; 39.5; 28.9; 7.9; 0.95
9. 1.43×10^{19} cm⁻²

10. $1 - \dfrac{x_B{}^2}{L_n a}\left[1 - \dfrac{1}{a}(1 - e^{-a})\right]$

11. $n(x) = \dfrac{1}{a}\left\{1 - \exp\left[\dfrac{a(x - x_B)}{x_B}\right]\right\}$

12. 1.6 Ω

13. $r_{B'} = \dfrac{\rho_B w}{12x_{B1}l} + \dfrac{\rho_B d}{2x_{B2}l}$

14. (a) $\tau_B = x_B^2/2D_n$;　(b) $\tau_B = \dfrac{x_B^2}{D_n}\dfrac{1}{a}\left[1 - \dfrac{1}{a}(1 - e^{-a})\right]$

15. **300 MHz**

17. (a) $I_D/I_i\big|_{V_o=0} = \dfrac{g_{b'e} - g_m}{g_{b'e} + j\omega(C_{TC} + C_D + C_{TE})}$;　(b) $\omega_\beta = \dfrac{g_m}{(C_{TC} + C_O + C_{TE})h_{FE}}$

18. $g_m = 77$ m℧; $g_{b'e} = 1.4$ m℧; $C_D = 26$ pF; $C_{TC} = 0.2$ pF
19. 25.7 ms
20. $BV_{CBO} = 30$ V; $BV_{CEO} = 12.5$ V; $BV_{PT} = 3.2 \times 10^4$ V

22. $I_A = \dfrac{\alpha_2 I_G + I_{CO1} + I_{CO2}}{1 - \alpha_1 - \alpha_2}$

23. (a) $I_A/I_G = -\dfrac{\alpha_2}{1 - (\alpha_1 + \alpha_2)}$;　(b) $\alpha_2 \to 1$; $\alpha_1 + \alpha_2 \to 1$

ANSWERS

Chapter 10

2.

	Masks	Oxidation	Diffusion	Epitaxy
EDP	6	4	4	1
CDI	5	3	3	1
BDI	5	3	3	1

3. 28.3 μm
5. $C_{TE} = 5$ pF; $C_{TC} = 0.74$ pF; $C_{TS} = 3.18$ pF; $r_{sc} = 75$ Ω; $BV_{CBO} = 55$ V; $BV_E = 7$ V; $BV_S = 55$ V
6. 214 MHz
9. (a) 0.79;　(b) 0.24
14. 2.4 V

ANSWERS

Chapter 11

3. (a) 34 nF/cm^2;　(b) 6.62 V;　(c) 2.93 μm;　(d) 9.38 $\times 10^{-14}$ C
5. (a) 3.21 nF/cm^2;　(b) 4.12 $\times 10^{-13}$ coul
6. (a) at $V_G = 10$ V: $\phi_s = 8.93$ V; $x_d = 7.62$ μm; at $V_G = 20$ V: $\phi_s = 18.46$ V; $x_d = 10.95$ μm;　(b) at $V_G = 20$ V: $\phi_s = 15.08$ V; $x_d = 9.90$ μm
7. (a) 3.18 $\times 10^4$ V/cm;　(b) 63 ps
8. (b) $\tau_n = 64$ ns; $\tau_p = 207$ ns
9. $A_n/A_o = 0.944$; $\Delta\phi = 0.176$; $A_n 3_{rd}/A_o = 0.675$; $\Delta\phi_3 = 0.285$
10. (b) -5.22 V

Page references in **boldface** refer to illustrations or tables; page references in *italic* refer to problems.

I

flow

$$I = I_0 (e^{V/V_T} - 1) + I_{rec}$$

$V > 0, \quad pn > n_i^2 \quad u > 0 \quad \text{forward bias}$

$V < 0, \quad pn < n_i^2 \quad u < 0 \quad \text{reverse bias}$